T0182152

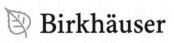

Methods of Solving Number Theory Problems

Ellina Grigorieva

Methods of Solving Number Theory Problems

 Birkhäuser

Ellina Grigorieva
Department of Mathematics
   and Computer Science
Texas Woman's University
Denton, TX
USA

ISBN 978-3-030-08130-0   ISBN 978-3-319-90915-8   (eBook)
https://doi.org/10.1007/978-3-319-90915-8

Mathematics Subject Classification (2010): 00A07, 97U40, 11D04, 11D41, 11D72, 03F03, 11A41, 11-0

This book is published under the imprint Birkhäuser, www.birkhauser-science.com by the registered
company Springer Nature Switzerland AG part of Springer Nature
The registered company address is: Gewerbestrasse 11, 6330 Cham, Switzerland

*To the Memory of My Father, **Valery Grigoriev**, A warm-hearted Humanist,*

*to my wonderful mother, **Natali Grigorieva**,*

*and to my beautiful daughter, **Sasha**. Your encouragement made this book possible.*

*And to my university mentor and scientific advisor academician, **Stepanov Nikolay Fedorovich**. Without your help and brilliant mind my career as a scientist would not be successful!*

Preface

I recall that some basis of elementary number theory was introduced to me as a child in public school through challenging problems posted in class or through my own math Olympiad experience. One of the interesting problems was this one: Take a number between 1 and 100, divide it by 2 until it is possible to get a natural number; if it is impossible, then multiply the number by 3 and add to it 1, then divide by 2, and continue the process. For example, let us take 17. Right away we need to multiply it by 3 and add 1, which results in 52, dividing by 2 we have 26, then 13, then by multiplying it by 3 and adding 1 we will get 40, then 20, then 10, then 5, then $5 \cdot 3 + 1 = 16$, then 8, then $4 \rightarrow 2 \rightarrow 1$. If we continue to multiply 1 by 3 and add 1, then we will again get 4, and eventually would end up in the same cycle. A similar chain can be obtained for the even number 20:

$$20 \rightarrow 10 \rightarrow 5 \rightarrow 16 \rightarrow 8 \rightarrow 4 \rightarrow 2 \rightarrow 1 \rightarrow 4 \rightarrow 2 \rightarrow 1 \ldots$$

Surprisingly, whatever natural number n was originally taken by me or by any of my classmates, we would end up in the cycle,

$$4 \rightarrow 2 \rightarrow 1.$$

That cycle could not be escaped by any additional trial. This problem was invented by Ancient Greeks and this strange behavior of all natural numbers (not only between 1 and 100) resulting in the same ending scenario remains unexplained. Using powerful computers it is possible to take very big natural numbers and run the algorithm. All natural numbers up to 2^{60} were checked for convergence to 1 and it looks like no other ending for any natural number can be found.

Then, in high school, we often solved linear, quadratic or exponential equations in two or more variables in integers using simple rules of divisibility or algebraic identities, without any knowledge of modular arithmetic or even Fermat's Little Theorem. However, we did talk about Fermat's Last Theorem and believed that the equation,

$$x^n + y^n = z^n$$

does not have a solution in integers for all natural powers, $n > 2$.

We believed that it would be proven during our lifetime. My mother was a physicist and knew much more about modern science at the time than I did. I remember her excitement about what Pierre Fermat wrote on the margins of his copy of Diophantus' *Arithmetica*, "This is how easily I have proven my Theorem; however, there is not enough space on the book margins to show all its proof."

Many mathematicians have dedicated their life to finding the proof of Fermat's Last Theorem. Euler, who really admired the genius of Fermat and proved most of his theorems and conjectures left without proof, attempted his own proof for Fermat's Last Theorem with the ideas and some methods that were found in Fermat's notes. The fact that the great Euler was unable to find the proof for Fermat's Last Theorem (he proved a particular case for $n = 3$) may have been the excuse other mathematicians needed to give up. The thing is that most of them, including myself as a teenager, believed that Fermat did have a proof of his theorem. The genius of British mathematician Andrew Wiles was in ignoring Fermat's note and assuming that Fermat could not prove his Last Theorem using the knowledge and apparatus of the time. Could he? Maybe…Mathematicians should not give us too quickly. And even an easy problem in number theory allows several methods of solving it and this is what is so fascinating about number theory.

The great British mathematician G. H. Hardy stated that elementary number theory should be considered one of the best subjects for the initial mathematical education. It requires very little prior knowledge and its subject is understandable. The methods of reasoning adopted by it are simple, common, and few. Among the mathematical sciences there is no equal in its treatment of natural human curiosity. Indeed, many questions are put so specifically that they usually permit experimental numeric validation. Many of the rather deep problems allow visual interpretation, for example, finding Pythagorean triples. In addition, elementary number theory best combines deductive and intuitive thinking which is very important in the teaching of mathematics. Number theory gives clear and precise proofs and theorems of an irreproachable rigor, shapes mathematical thinking, and facilitates the acquisition of skills useful in any branch of mathematics. Often the solution of its problems requires overcoming significant difficulties, mathematical ingenuity, finding new methods, and ideas that are being continued in modern mathematics. In favor of the study of the theory of numbers, it is fair to say that for every kind of deep mathematical investigation in different fields, we often encounter relatively simple number-theoretic facts.

What is the Subject of Number Theory?

Number theory is the study of numerical systems with their relations and laws. First of all, it is focused on natural numbers that are the basis for constructing other numerical systems: integers, rational and irrational, real and complex. Number theory is a branch of mathematics which deals with the properties of numbers. One of the main problems of number theory is the study of the properties of integers. The main object of number theory is natural numbers. Their main property is divisibility.

Number theory studies numbers from the point of view of their structure and internal connections and considers the possibility of representing certain numbers through other numbers, simpler in their properties. The questions of the rigorous logical justification of the concept of a natural number and its generalizations, as well as the theory of operations associated with them, are considered separately.

The Main Topics of Number Theory

The problems and challenges that have arisen in number theory can be categorized as follows:

1. The solution of Diophantine equations, i.e. the solution in integers of algebraic equations with integer coefficients or systems of such equations for which the number of unknowns is greater than the number of equations;
2. Diophantine approximations, i.e. the approximation of real numbers by rational numbers, the solution in integers of all kinds of inequalities, the theory of transcendental numbers or the study of the arithmetic nature of different classes of irrational numbers with respect to transcendental numbers;
3. Questions of distributing prime numbers in a series of natural numbers and other numerical sequences;
4. Additive problems, i.e. the decomposition of integer (usually large) numbers into summands of a certain type;
5. Algorithmic problems of number theory, e.g. cryptography.

Famous Unsolved Problems, Hypotheses in Number Theory

There are still unsolved problems in number theory besides the one previously mentioned that is known as the **Collatz conjecture**. Maybe one of them could be solved by you. Some of the famous unsolved problems are:

1. The Odd Perfect Number Conjecture
2. Are there Fermat prime numbers for $n > 4$?
3. The Twin Prime Conjecture
4. Goldbach's Conjecture
5. The Riemann Hypothesis

The Odd Perfect Number Conjecture. Are there odd "perfect" numbers? The Odd Perfect Number Conjecture has escaped proof for centuries.

Perfect numbers are positive integers that are the sum of their proper divisors. For instance, 6 is a perfect number because the sum of its proper divisors, 1, 2, and 3 equals 6 $(1 + 2 + 3 = 6)$. Euclid first found a way to construct a set of even perfect numbers in Book IX of The Elements. In his book, Euclid showed that if $2^p - 1$ is prime where p is prime, then $2^{p-1} \cdot (2^p - 1)$ is a perfect number.

Prime Fermat Numbers. There are prime Fermat numbers given by $F_n = 2^{2^n} + 1$ for $n = 0, 1, 2, 3, 4$. Leonard Euler showed that $F_5 = 4294967297 = 641 \cdot 6700413$ is not prime. Are there still such numbers for other n values? Many mathematicians believe that there are no new Fermat prime numbers.

The Twin Prime Conjecture. Are there infinitely many twin prime numbers? Twin primes are pairs of primes of the form $(p, p+2)$. Examples are (3,5), (5,7), (11,13), (17,19), and (197,199). The largest known twin prime pair at the time of writing is $(2996863034895 \pm 1) \cdot 2^{1290000}$ was discovered in September 2016 by PrimeGrid. A straightforward question now arises that since there are an infinite number of primes, are there also an infinite number of twin primes? One of the reasons that this question is interesting is that we know that the gap between the primes increases for larger numbers (from the prime number theorem, we learn that the "average" gap between primes smaller than p is $\ln(p)$). Nevertheless, is there an infinite number of twin primes anyway? As of January 2016, the largest known prime is $2^{74207281} - 1$ (more than 22 million digits, also discovered by GIMPS).

The Binary Goldbach's Conjecture. Every even integer greater than 2 can be written as the sum of two primes. You can check that

$$4 = 2+2, \quad 6 = 3+3, \quad 8 = 5+3, \quad 10 = 5+5, \quad 12 = 7+5, \ldots$$

Note that for many even integers this representation is not unique. For example,

$$52 = 5+47 = 11+41 = 23+29,$$
$$100 = 3+97 = 11+89 = 17+83 = 29+71 = 41+59 = 47+53.$$

Euler first stated this conjecture but failed to prove it. There is no doubt that this conjecture is true, and its validity was checked up to every even number less than or equal to $4 \cdot 10^{18}$. However, so far nobody has given the complete proof of this conjecture.

The Riemann Hypothesis. Riemann expanded Euler's zeta function to the field of complex numbers and stated that the real part of any nontrivial zero of the zeta function is 1/2. Though this topic is out of the scope of this book, some information about the Riemann zeta function can be found in Section 1.4.

The twin prime and Goldbach's Conjecture together with the Riemann Hypothesis are included as Hilbert Problem 8 and remain the most famous open mathematical problems.

What is this Book About?

It is known that students have a hard time in trying to solve math problems involving integers, perhaps due to the fact that they study numbers in elementary school and basically never touch the topic again throughout the entire math curriculum. Many don't find arithmetic problems interesting or of much use in our everyday life since many believe that we don't need to know number theory for such fields as engineering or programming.

History develops in spirals and goes through cycles, and similar thoughts came to people's minds hundreds or even thousands years apart. The number theory that was an excitement for Archimedes, Pythagoras and Diophantus, but was abandoned for thousands of years, was refreshed by the work of Descartes and Fermat, was not of interest for two hundred years after them as most mathematicians wanted to develop mathematical analysis and solve engineering problems instead of solving Diophantine equations or other problems involving integers that did not seem to be of importance. The genius of Euler refreshed great interest in number theory again. His work showed that number theory is not just a mental exercise for people who like a challenge. Number theory problems, while simple in their formulation, often require development of new field of mathematics. Thus, the simple difference of squares method used by Pierre Fermat for factoring big numbers is fundamental to the data encryption we use today. Search for the proof of the Last Fermat Theorem gave life to algebraic geometry. The Euler zeta function was expanded to the field of complex numbers by Riemann over 150 years ago and is now a popular problem once again.

However, these days, in my opinion, the general public has again lost interest in number theory and many believe that this field of mathematics is only of interest to specialists and that knowledge of the methods of solving Diophantine equations, Pell's equations or the proof of Riemann's hypothesis are not topics of general interest. Number theory courses became elective subjects in many math under-graduate programs. Moreover, most mathematics teachers who specialize in teaching mathematics in middle or high schools never had number theory in their life and most of them cannot help their students to solve any problem that appear in math contests, such as AIME or the USAMO. Those who did take a number theory course as graduate students are usually surprised about its importance. Those who had number theory earlier in their life know that number theory is important, gives us a power over the physical world we live in, and for that reason has been studied since the beginning of recorded history.

Solving math contest problems involving numbers and studying number theory helps us to think as mathematicians. We no longer view math as a disjointed collection of formulas and facts to memorize; instead, we learn to approach abstract problems in a systematic and creative manner. Additionally, we lose any inhibitions and fear of failure from our faulty memories. Solving problems empowers us. Moreover, participating in different math Olympiads can lead a student to admis-sion at the most prestigious universities and basically change his or her life.

This book has been designed as a self-study guide or for a one semester course in introductory number theory problem solving. The reader is only required to know algebra and arithmetic, and some calculus. The idea of this book is to develop the mathematical skills of the readers and enable them to start solving unusual math problems. We will demonstrate how problems involving numbers can develop your creative thinking and how experience in solving some challenging problems will give you confidence in the subject matter.

For over 25 years, whenever I spotted an especially interesting or tricky prob-lem, I added it to my notebook along with an original solution. I have accumulated thousands of these problems. I use them every day in my teaching and include many of them in this book. Most math books start from theoretical facts, give one or two examples, and then a set of problems sometimes with no answers. In this book every statement is followed by problems. You are not just memorizing a theorem, but applying it immediately. Upon seeing a similar problem in the homework section you will be able to recognize and solve it. The book consists of five chapters covering important but often overlooked topics.

The book is written so that it can be understood by middle and high school students who are good in algebra and arithmetic, but not necessarily familiar with some notation and symbols of modern number theory. The apparatus of number theory is introduced when it is needed and often different methods are proposed for a solution to a problem with discussion of their advantage and disadvantage.

Chapter 1 is entitled "Numbers: Problems Involving Integers" and will enable you to solve such unusual problems as

- For what values of n is the number $N = \underbrace{99\ldots9}_{n\,\text{digits}}7$ divisible by 7?
- Prove that $n^3 + 6n^2 - 4n + 3$ is divisible by 3 for any natural number n.
- Find the sum of the digits of all natural numbers from one to 10^5
- The number obtained by striking the last four digits is an integer times less than the original number. Find all such numbers ending in 2019
- How many zeros in $1 \cdot 2 \cdot 3 \cdot 4 \cdot \ldots \cdot 2016 \cdot 2017 = 2017!$?

These problems cannot be solved with a calculator but by using approaches developed in this book you will be able to easily handle them. This chapter covers even and odd numbers, divisibility, division with a remainder, the greatest common divisor, least common multiple, and primes. You will learn about proper divisors, perfect, amicable, triangular, and square numbers known to Archimedes. At the end of Chapter 1 congruence notation and its application are introduced.

The Euclidean algorithm, representation of a number by finite or infinite continued fractions, approximation of an irrational number, finding the ending digits of a number raised to a large power or to several consecutive powers, the Fermat's Little Theorem, and Euler's formula are covered in Chapter 2. The reader will be challenged to find a solution to the following problems:

- What is the last digit of $2^{3^{4^5}}$?
- What are the last three digits of 2^{2018}?
- What is the remainder of $222^{555} - 555^{222}$ divided by 7?
- Reduce the fraction $\dfrac{16n + 60}{11n + 41}$.
- Prove that there is no infinite arithmetic progression of only prime numbers.
- Evaluate $1 + \cfrac{1}{1 + \cfrac{1}{1 + \cfrac{1}{1 + \cfrac{1}{1 + \ldots}}}}$.

In Chapter 2, you will continue to learn about integer numbers and their representation in different bases. Most of us know that every even number in base ten ends in even digit. You will prove that a number written in an odd base, for example, base 7, can be also an even number if and only the sum of its digits is even. Chapter 2 goes into the theory of continued fractions, representation of irrational numbers by nonperiodic continued fractions, quadratic irrationalities, and the Lagrange Theorem. Mathematical induction and other methods of proof including the method of generating functions and a derivation of the formula for Catalan numbers conclude Chapter 2.

Chapter 3 is mainly devoted to linear and nonlinear equations in integers, Diophantine equations in several variables, extracting the integer portion of a quotient, applications of factoring, Fermat's methods of factoring numbers based on difference of squares formula, homogeneous polynomials, and challenging problems with factorials and exponents. For example, you will learn how to find all natural solutions to the equations like these:

- $3^{2x} - 2^y = 1$
- $x^2 + 5y^2 + 34z^2 + 2xy - 10xz - 22yz = 0$
- $5x^4 - 40x^2 + 2y^6 - 32y^3 + 208 = 0$
- $x! + 4x - 9 = y^2$

Chapter 3 introduces methods of solving Fermat-Pell's equations and demonstrates how this equation connects algebra and geometry. By solving Pell's equations you will see how linear algebra and quadratic forms can be used to solve nonlinear equations in integers. Additionally, working with factorials, you will be able to find the last two digits, for example, of the following huge number:

$$N = 1! + 2! + 3! + \cdots + 2016! + 2017!$$

and apply Wilson's Theorem to solve some challenging problems with factorials.

In Chapter 4, the connection between algebraic and geometric methods will be continued when finding Pythagorean triples and quadruples, i.e. $x^2 + y^2 + z^2 = w^2$. Some introduction to algebraic geometry and stereographic projection will be given here as well.

Waring's Problem i.e. the representation of a number by the sum of other numbers raised to the same natural power, will be introduced here. You will learn why, for example, the prime number 19 cannot be written by the sum of two squares but another prime number, 13, can be so written, i.e. $13 = 2^2 + 3^2$. You will also see the advantage of using complex numbers in solving the problems as follows.

- Given n natural numbers $x_1, x_2, x_3, \ldots, x_n$. Prove that the number

$$N = (1 + x_1^2) \cdot (1 + x_2^2) \cdot \ldots \cdot (1 + x_n^2)$$

can be represented by the sum of two squares.

Here the reader will solve challenging problems like these:

- Find 11 consecutive natural numbers, the sum of the squares of which is a perfect square.
- Solve the following equation over the set of integers:

$$x_1^4 + x_2^4 + x_3^4 + \cdots + x_{14}^4 = 100000000002015.$$

Chapter 4 of the book is also devoted to some basic concepts, such as quadratic residues, and Legendre and Jacobi symbols. It is recommended to all college students taking an introductory course in number theory. This chapter will help you grasp different concepts of number theory by solving the same problem in several ways. The last section of Chapter 4 is a collection of word problems involving integers.

Many problems in this book can be understood and performed by a high school or a middle school student who wishes to become confident in problem solving. Knowledge of caculus is needed but not prerequisite for better understanding of certain topics. There are some overlaps in knowledge and concepts between chapters. These overlaps are unavoidable since the threads of deduction we follow from central ideas of the chapters are intertwined well within our scope of interest. For example, we will on occasion use the results of a particular lemma or theorem in a solution but wait to prove the statement until it becomes essential to the thread at hand. If you know the property you can follow along right away and, if not, then you may find it in the following sections or in the suggested references.

Throughout the book, the fascinating historical background will be given that shows a connection between modern methods of number theory and the methods known by Ancient Greeks, Babylonians, or Egyptians. This book contains 265 challenging problems and 110 homework problems with hints and detailed solutions. A historical overview of number theory is given in Appendix 1. From the point of view of methods and applications, seven main directions in number theory can be distinguished as summarized in Appendix 2.

If students learn an elegant solution and do not apply the approach to other problems, they forget it as quickly as an unused phone number. However, if a teacher uses and reuses the same approach throughout the entire curriculum, students will remember and begin to see and appreciate the beauty of mathematics. Although each chapter of the book can be studied independently, I mention the same patterns throughout the book. This helps you see how math topics are connected. This book can be helpful in two ways:

1. For self-education, by people who want to do well in math classes, and to be prepared for any exam or competition.
2. For math teachers and college professors who would like to use it as an extra source in their classroom.

How to Use this Book?

Here are my suggestions about how to use this book. Read the corresponding section and try to solve the problem without looking at my solution. If you find any question or section too difficult, skip it and go to another one. Later you may come back and try to understand it. Different people react differently to the same question and the reaction sometimes is not related to intelligence or education. Return to difficult sections later and then solve all the problems. Read my solution when you have completed your own solution or when you feel like you are just absolutely stuck. Think about similar problems that you would solve using the same or similar approach. Create your own problem and write it down along with your original solution. Now, you have made this powerful method your own and will be able to use it when needed.

I promise that this book will make you successful in problem-solving. If you do not understand how a problem was solved or if you feel that you do not understand my approach, please remember that there are always other ways to do the same problem. Maybe your method is better than one proposed in this book. I hope that upon finishing this book you will love math and its language as I do. Good luck and my best wishes to you!

Denton, TX, USA

Ellina Grigorieva, Ph.D.
Professor of Mathematics
Department of Mathematics
and Computer Science

Acknowledgements

During the years working on this book, I received feedback from my friends and colleagues at Lomonosov Moscow State University, for which I will always be indebted.

I want to thank Mr. Takaya Iwamoto for beautiful graphics.

I am especially grateful for the patient and conscientious work of Ms. Jennifer Anderson and the contributions of Dr. Paul Deignan in the final formatting and preparation, and for the multitude of useful and insightful comments on its style and substance.

I appreciate the extremely helpful suggestions from the reviewers: your feedback made this book better!

I would also like to acknowledge the editors Benjamin Levitt and Christopher Tominich at Birkhäuser who has always been encouraging, helpful, and positive.

Denton, TX, USA   Ellina Grigorieva, Ph.D.

Contents

1 Numbers: Problems Involving Integers 1
  1.1 Classification of Numbers: Even and Odd Integers 1
    1.1.1 Other Forms of Even and Odd Numbers 5
  1.2 Multiples and Divisors: Divisibility 6
  1.3 The Decimal Representation of a Natural Number: Divisibility
    by 2, 3, 4, 5^n, 8, 9, 10, and 11 13
  1.4 Primes: Problems Involving Primes 23
  1.5 Greatest Common Divisor and Least Common Multiple 36
  1.6 Special Numbers: Further Classification of Numbers........... 43
    1.6.1 Divisors. Proper Divisors. Perfect and Amicable
      Numbers 43
    1.6.2 Triangular and Square Numbers 50
  1.7 Congruence and Divisibility 53

2 Further Study of Integers 63
  2.1 Euclidean Algorithm................................ 64
    2.1.1 Finding the gcd, lcm and Reducing Fractions 64
    2.1.2 Other Bases................................. 70
    2.1.3 Continued Fractions and Euclidean Algorithm 78
  2.2 Representations of Irrational Numbers by Infinite Fractions...... 85
    2.2.1 Quadratic Irrationalities 87
  2.3 Division with a Remainder 95
  2.4 Finding the Last Digits of a Number Raised to a Power 101
  2.5 Fermat's Little Theorem 108
    2.5.1 Application of Fermat's Little Theorem to Finding
      the Last Digit of a^b 115
  2.6 Euler's Formula 117
    2.6.1 Application of the Euler's Formula to Finding
      the Last Digits of a^b 120

2.7 Methods of Proof 123
  2.7.1 Geometric Proof by Ancient Babylonians and Greeks 124
  2.7.2 Direct Proof 127
  2.7.3 Proofs by Contradiction 129
  2.7.4 Mathematical Induction 132
  2.7.5 Using Analysis and Generating Functions 135

3 Diophantine Equations and More 141
  3.1 Linear Equations in Two and More Variables 142
    3.1.1 Homogeneous Linear Equations 143
  3.2 Nonhomogeneous Linear Equations in Integers 144
    3.2.1 Using Euclidean Algorithm to Solve Linear Equations.... 146
    3.2.2 Extracting an Integer Portion of a Quotient 149
  3.3 Solving Linear Equations and Systems Using Congruence 156
    3.3.1 Solving Linear Congruence Using Continued Fractions ... 159
    3.3.2 Using Euler's Formula to Solve Linear Congruence 162
  3.4 Nonlinear Equations: Applications of Factoring 165
    3.4.1 Newton's Binomial Theorem 166
    3.4.2 Difference of Squares and Vieta's Theorem 171
    3.4.3 Homogeneous Polynomials 179
  3.5 Second-Order Diophantine Equations in Two or Three
      Variables 196
    3.5.1 Methods and Equations 196
    3.5.2 Problems Leading to Pell's Equation 201
    3.5.3 Fermat-Pell's Equation 204
    3.5.4 Pell's Type Equations and Applications 224
  3.6 Wilson's Theorem and Equations with Factorials 237

4 Pythagorean Triples, Additive Problems, and More 245
  4.1 Finding Pythagorean Triples. Problem Solved by Ancient
      Babylonians 247
    4.1.1 Method 1. An Arithmetic Approach................. 248
    4.1.2 Method 2. Using Algebraic Geometry 251
    4.1.3 Method 3. Using Trigonometry 255
    4.1.4 Integer Solutions of $a^2 + b^2 + c^2 = d^2$. 257
  4.2 Waring's and other Related Problems 265
    4.2.1 Representing a Number by Sum of Squares 265
    4.2.2 Sum of Cubes and More 289
  4.3 Quadratic Congruence and Applications.................. 293
    4.3.1 Euler's Criterion on Solution of Quadratic
          Congruence 303
    4.3.2 Legendre and Jacobi Symbols 306
  4.4 Word Problems Involving Integers..................... 316

5   Homework   335
   5.1   Problems to Solve   335
   5.2   Answers and Solutions to the Homework.   340

Erratum to: Pythagorean Triples, Additive Problems, and More E1

References   377

Appendix 1. Historic Overview of Number Theory   381

Appendix 2. Main Directions in Modern Number Theory   385

Index   389

Chapter 1
Numbers: Problems Involving Integers

Many important properties of integers were established in ancient times. In Greece, the Pythagorean school (6th century BC) studied the divisibility of numbers and considered various categories of numbers such as the primes, composite, perfect, and amicable. In his *Elements*, Euclid (3rd century BC) gives an algorithm for determining the greatest common divisor of two numbers, outlines the main properties of divisibility of integers, and proves the theorem that primes form an infinite set. Also in the 3rd century BC, Eratosthenes discovered an algorithm to extract prime numbers from a series of natural numbers now called, "The Sieve of Eratosthenes". There are many great names associated with the development of number theory such as Diophantus, Fermat, Descartes, Euler, Legendre, Gauss, and Lagrange, to list just a few.

Over the centuries, an interest in numbers has not been entirely lost. Problems involving numbers are often included in math contests, some of which look simple but require a tremendous amounts of nonstandard thinking. Other problems look scary at first glance but allow beautiful and often short solutions.

The idea of this chapter is to develop the mathematical skills of the readers and enable them to solve unusual math problems involving integers and their properties. We will demonstrate how problems involving numbers can develop your creative thinking and how experience in solving some challenging problems will give you confidence in the subject matter.

1.1 Classification of Numbers: Even and Odd Integers

The positive whole numbers: $1, 2, 3, \ldots, N, \ldots$ are called the **natural numbers** and are used for counting, e.g., "there are five apples in the basket" and ordering, e.g., "China is the largest country in the world by population". This is the oldest defined category of numbers and the simplest in membership. In mathematical notation, we may describe the sequence of natural numbers as $1, 2, 3, 4, \ldots \in \mathbb{N}$.

© Springer International Publishing AG, part of Springer Nature 2018
E. Grigorieva, *Methods of Solving Number Theory Problems*,
https://doi.org/10.1007/978-3-319-90915-8_1

The second set of numbers in order of composition is the **set of integers** that include natural numbers and their negatives, such as $-1, -2, -3, \ldots -N, \ldots$ and zero 0 in order to make the operations of addition and subtraction closed on the set. The notation for this set is $\mathbb{Z}$ which comes from the German word, *zahlen*, for "numbers".

The next set is the **set of rational numbers**, $\mathbb{Q}$ for "quotient". We define a number to be rational if it can be written as a fraction $\frac{m}{n}$ where $m \in \mathbb{Z}$ and $n \in \mathbb{N}$. This way we do not have to deal with divisibility by zero. Obviously, natural numbers and integers are also rational but not vice versa.

The numbers that cannot be represented by a fraction are the **irrational numbers**, such as $\sqrt{2}, e, \pi$. This set does not have a special notation. Rational numbers and irrational numbers together form the **set of real numbers**, $\mathbb{R}$. Descartes and Fermat (17th century) were the first to use the coordinate method to represent real numbers on the number line. Integers were discrete points on the line with whole coordinates while real numbers span a continuum, i.e., they fill out the entire number line without gaps. The introduction of real numbers allows us to perform the operations of addition, subtraction, multiplication, division (except by zero), and to raise real numbers to a power with a resultant that is also a member of the real numbers.

Finally, there is the largest set of numbers, the set of complex numbers $\mathbb{C}$. This set contains numbers of the type $a + ib$, where $a, b \in \mathbb{R}$ and i is an imaginary unit, such that $i^2 = -1$. Complex numbers geometrically are points in the plane formed by the Real and Imaginary axis. If $a, b \in \mathbb{Z}$, then the complex number $a + ib$ is a lattice point. We say that $\mathbb{N} \subset \mathbb{Z} \subset \mathbb{Q} \subset \mathbb{R} \subset \mathbb{C}$. The relationship between sets (excluding $\mathbb{C}$) is shown in Figure 1.1.

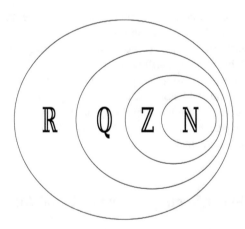

Fig. 1.1 Sets of Numbers

All natural numbers can be divided into two groups: **even numbers** and **odd numbers**. Most of us know that $3, 11, 27$ are odd numbers and $20, 36$, or 100 are even numbers. Let us find the general form for odd and even numbers. What is common to all even numbers such as $2, 4, 20, 2000$?

Correct. They are multiples of the number 2. Therefore, we can say that every even number can be written as $2n$, where $n = 1, 2, 3, \ldots$. Different values of n create different even numbers, for example, $340 = 2 \cdot 170$, $6 = 2 \cdot 3$, etc.

Now let us find the general rule for an odd number. If you write all even and odd numbers together in ascending order from 1 to 100 as $1, 2, 3, 4, \ldots, 98, 99, 100$, you notice that every even number is surrounded on the left and on the right by two odd numbers that are one less or one more than that particular even number. So if $2k$ is an even number, its neighbors are $(2k - 1)$ and $(2k + 1)$.

Conclusion. Every odd number $1, 3, 5, \ldots, 11007 \ldots$ can be written as $2n + 1$, where $n = 0, 1, 2, 3, \ldots$ or in the form $2m - 1$, where $m = 1, 2, 3, \ldots$. So,

$$13 = 2 \cdot 6 + 1, \ n = 6$$
$$17 = 2 \cdot 8 + 1, \ n = 8.$$

We notice that different values of n produce different odd numbers. We will call $2n$ a generator of even numbers and $(2n - 1)$ a generator of odd numbers.

Problem 1

For a natural number, k, any even number may be represented by $2k$ and any odd number by $2k - 1$. Construct an argument to support the conjecture that the sum of any two even numbers and the sum of any two odd numbers are both even numbers.

Solution. Let us take a few numbers. $16 + 6 = 22$ is an even number, $19 + 7 = 26$ is an even number. Sounds good! Now let us prove that the conjecture works in any case.

Part 1. The sum of any even numbers is an even number. Let $2n$ be one even number and $2m$ the other, then their sum is $2n + 2m = 2(n + m)$ a multiple of 2, an even number.

Part 2. The sum of any two odd numbers is an even number. Adding two different odd numbers $(2n - 1)$ and $(2m - 1)$ yields $2n - 1 + 2m - 1 = 2n + 2m - 2 = 2(n + m - 1)$ which is a multiple of 2, an even number again.

Remark. Please show for yourself that the difference of two even numbers or two odd numbers is always an even number.

Two consecutive numbers are usually denoted by either n and $n + 1$ or $n - 1$ and n. Because such two numbers differ by one, then depending on the value or the expression for the first one, we can always create a corresponding consecutive number that is one more or one less of the given number. For example, if $N = 3k + 5$

the following number will be $N + 1 = 3k + 5 + 1 = 3k + 6$. Obviously, $n^2 - 1$ and $n^2 + 1$ differ by 2 so they are not consecutive but they are either consecutive even or consecutive odd numbers. Therefore any two numbers that differ by 2 are either both even or both odd. This little, obvious fact allows us to solve many problems right away. Let me demonstrate it by solving Problem 2.

Problem 2

Find integer solutions to the equation $k^2 - 4m - 3 = 0$.

Solution.
Method 1. Let us rewrite this equation in a different form and try to factor it using the difference of squares formula,

$$k^2 = 4m + 3$$
$$k^2 - 1 = 4m + 2$$
$$(k + 1)(k - 1) = 2(2m + 1).$$

The two numbers on the left are either both even or both odd.

   Case 1. Assume that the numbers $k - 1$ and $k + 1$ are both even. Then the left side of the equation is divisible by 4, but the right side is not divisible by 4, just by 2 (because $2m + 1$ is odd). There is no solution.

   Case 2. Assume that both factors on the left are odd. Then the equation will have no solutions because the right side is even.
Method 2. Let us rewrite the given equation again as $k^2 = 4m + 3$, but instead of factoring it, investigate whether or not a square of an integer number divided by 4 can leave a remainder of 3. Since all numbers are either even or odd, we will replace k by $2n$ and $2n + 1$, respectively, square, and look at their remainders under division by 4,

$$k^2 = 4m + 3$$
$$(2n)^2 = 4n^2 = 4t \neq 4m + 3,$$
$$(2n + 1)^2 = 4n^2 + 4n + 1 = 4s + 1 \neq 4m + 3, \quad t, s \in \mathbb{Z}.$$

Therefore no perfect square divided by 4 leaves a remainder of 3.
Answer. There are no solutions.

Remark. In solving the previous problem by the first method, we factored $k^2 - 1 = (k - 1) \cdot (k + 1)$, and obtained two factors that differ by 2 and obviously are either both even or both odd. The same conclusion can made if we factor any difference of two integer squares,

$$n^2 - m^2 = (n - m) \cdot (n + m).$$

The factors differ by an even number $2m$ and so again are either both even or both odd. This fact can be very useful in solving problems in integers.

1.1.1 Other Forms of Even and Odd Numbers

Since $2^n = 2 \cdot 2 \cdot 2 \cdots 2$ is an even number, then 2^n is surrounded by $2^n - 1$ and $2^n + 1$ on the left and on the right, respectively, which are both odd. Thus, $2^{10} = 1024$ is an even number while $2^{10} - 1 = 1023$ and $2^{10} + 1 = 1025$ are odd numbers. Moreover, if we consider only odd powers of 2, then $2^n + 1$ will be always a multiple of 3. For example, $2^3 + 1 = 9$, $2^5 + 1 = 33$, and $2^9 + 1 = 513$. Let us prove this statement by solving Problem 3.

> **Problem 3**
>
> Prove that a number that is one more than 2 raised to any odd power is always a multiple of 3.

Proof. Consider $k = 2^{2n+1} + 1$ and let us show that it is always divisible by 3. Using the properties of exponentiation, the number k can be written as

$$
\begin{aligned}
k &= 2^{2n+1} + 1 = 2 \cdot 2^{2n} + 1 = 2 \cdot 4^n + 1 = 2 \cdot (3+1)^n + 1 \\
&= 2 \cdot (\{3^n + n \cdot 3^{n-1} + \tfrac{n(n-1)}{2} \cdot 3^{n-2} + \cdots + n \cdot 3\} + 1) + 1 \\
&= 2 \cdot 3m + (2+1) = 3s + 3 = 3l, \ l \in \mathbb{N}.
\end{aligned}
$$

We split 4 as $(3+1)$ and then applied the Newton Binomial Theorem (see Section 3.4.1) to its nth power. Clearly the expression inside the braces is a multiple of 3. Multiplying 2 by this quantity and by 1 (using the distributive law), we then add 2 to the last term, 1, which gives 3.

Finally, $k = 3l, l \in \mathbb{N}$. The proof is complete.

You can also prove this fascinating fact after reading about Fermat's Little Theorem in Chapter 2 of the book. On the other hand, because 3^n is an odd number for any natural n, its preceding number $3^n - 1$ will be even as well as the proceeding term, $3^n + 1$. Let us try to solve Problem 4 which was offered at the Russian Math Olympiad in 1979.

> **Problem 4**
>
> Can the number $2^n + n^2$ have a 5 as a last digit for some natural number n?

Solution.

Part 1. Let us consider the even number, $n = 2k$. Because 2^n is always even and n^2 is even for any even n, the number $2^n + n^2$ cannot have a 5 as its unit's (the last) digit. As we proved before, a sum of two even numbers is an even number and the last digit of any even number is either 0 or even. For example, if $n = 6$, then

$2^n + n^2 = 2^6 + 6^2 = 100$. The last digit is 0. Or if $n = 10$ (another even number), then $2^n + n^2 = 1124$. The number 4 is even.

Part 2. Assume that n is an odd number, written as $n = 2k - 1$. Its last digit can be 1, 3, 5, 7, or 9. The last digit of n^2 can be 1, 5, or 9 (check it!). The unit digit of 2^n for odd n can be either 2 or 8. Noticing that the last digit of $2^n + n^2$ is the sum of last digits of both terms, let us consider,

Row 1: 1,   5, 9 (last digit of n^2)

Row 2:   2,   8 (last digit of 2^n)

We see that there is no combination of elements of Row 1 and Row 2 that yields 5. **Answer.** The number cannot end in 5.

After you read and practice more, you will be able to prove this more rigorously.

1.2 Multiples and Divisors: Divisibility

If a number n is divisible by m (n is a multiple of m), then it can be written as $n = m \cdot k$, where n, k, and m are integers. If m is an arbitrary integer, the multiples of m are all numbers 0, $\pm m$, $\pm 2m$, $\pm 3m$, $\pm 4m$,..., $\pm km$, where m is an integer. If a natural number n is not divisible by a number m, we say that n divided by m leaves a remainder r and can be written in the form $n = m \cdot k + r$, where r is some integer such that $1 \leq r \leq (m-1)$.

Example. You have 20 apricots and have to divide them equally between three children. Everyone can get 6 apricots and two apricots will stay in the basket. If originally we had 19 apricots, then again every child would get 6 with one remaining in the basket. Three children with 0 remaining would equally divide 18 apricots.

Let us write down all multiples of 3 between 1 and 20: 3, 6, 9, 12, 15, and 18. If a natural number n is a multiple of 3 it can be represented in the form $n = 3k$ where $k = 1, 2, 3, \ldots$. Thus, $18 = 3 * 6$, $9 = 3 * 3$, $300 = 3 * 100$ etc. If a natural number is not a multiple of 3, this number divided by 3 gives a remainder of 1 or 2. As a check, $20 = 3 \cdot 6 + 2$ (2 is a remainder), $301 = 3 \cdot 100 + 1$ (1 is a remainder).

Conclusion. All natural numbers that are not multiples of 3 can be written in the form $n = 3k + 1$ or $n = 3k + 2$. For example, 554 is not a multiple of 3 and $554 = 3 \cdot 184 + 2$.

We can visualize all natural numbers as those that are multiples of 3, $(3k)$, those that divided by 3 leave a remainder of 1, $(3k + 1)$, and those that when divided by 3 leave a remainder of 2, $(3k + 2)$. It is like dividing a big pie of all natural numbers into three pieces. (See Fig. 1.2)

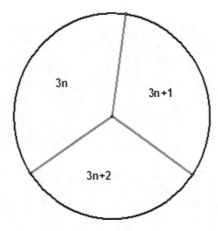

Fig. 1.2 Divisibility by 3

In general, we can expand this idea into division by any number k. Thus, the pie of all natural numbers will be divided into k pieces, marked as $km, km + 1, km + 2, \ldots, km + (k-1)$.

It is useful to know the properties of divisibility.

Property 1

A product of any two consecutive integers is a multiple of 2. Therefore, $n(n+1)$ and $(n-1)n$ are divisible by 2.

Proof. All integers are either even or odd. Among two consecutive numbers either the first is an odd and the second is an even or vice versa, hence their product is always divisible by 2.

1. Assume that $n = 2k - 1$ is odd. Then is

$$n(n+1) = (2k-1) \cdot (2k-1+1) = 2k(2k-1)$$

   a multiple of 2? Yes! For example, $15 \cdot 16 = 240$, $6 \cdot 7 = 42$.
2. Assume that $n = 2k$ is even, then is $n(n+1) = 2k(2k+1)$ is a multiple of 2? Yes!

> **Property 2**
>
> A product of any three consecutive integers is a multiple of 3 or
>
> $$n(n+1)(n+2)$$
>
> is divisible by 3.

> **Property 3**
>
> A product of any three consecutive integers is divisible by 6.

Proof. Consider any possible product of three consecutive integers, i.e.,

$$3k(3k+1)(3k+2), \ (3k-1)(3k)(3k+1), \ (3k-2)(3k-1)(3k).$$

All products are divisible by 3. Moreover, among all factors there are at least one that is even. For products in which the middle factor is an odd, there are two even numbers surrounding it. Therefore, a product of three consecutive integers is always divisible by $3! = 6$. For example, $2 \cdot 3 \cdot 4 = 24$, $4 \cdot 5 \cdot 6 = 24k$, etc.

> **Property 4**
>
> A product of any k consecutive integers or
>
> $$n(n+1)(n+2)\ldots(n+k-1)$$
>
> is a multiple of k and of $k!$.

The last property is very important and we can demonstrate that it works. Assuming that $n = 1$, we can represent a product of k consecutive integers as

$$1 \cdot 2 \cdot 3 \cdots (k-1) \cdot k = k!$$

This is k factorial. Because one factor equals k, then the entire expression is divisible by k.

Let us prove the first part of Property (4).

Proof. Consider the product $p = n(n+1)(n+2)\ldots(n+k-3)(n+k-2)(n+k-1)$, where $n \in \mathbb{N}$. Then n can be written as one of the following:

$$n = k \cdot m,$$
$$n = k \cdot m + 1,$$
$$n = \dots$$
$$n = k \cdot m + k - 2,$$
$$n = k \cdot m + k - 1.$$

Substituting any of the numbers from the list into formula for p, we always obtain at least one factor that is a multiple of k. For example, when $n = km + k - 1$, then $p = (km + k - 1)(km + k)\dots(km + 2k - 3)(km + 2k - 2)$. If $n = km + 1$, then $p = (km + 1)(km + 2)\dots(km + k - 1)(km + k)$, etc.

Next, let us show that a product of k consecutive numbers is divisible by $k!$. From Newton's Binomial Theorem it follows that because

$$C_n^k = \frac{n(n-1)(n-2)\dots(n-k+1)}{k!} = \frac{n!}{k!(n-k)!}$$

is a natural number, then $n(n-1)(n-2)\dots(n-k+1)$ is always divisible by $k!$

Problem 5

Prove that $n^3 + 6n^2 - 4n + 3$ is divisible by 3 for any natural number n.

Solution. Rewriting the given number as $(n^3 - 4n) + 6n^2 + 3$ we can see that it is enough to prove that the expression within parentheses is divisible by 3.

$$n^3 - 4n = n^3 - n - 3n = n(n^2 - 1) - 3n = (n-1)n(n+1) - 3n \qquad (1.1)$$

Expression (1.1) is divisible by 3 because $3n$ is divisible by 3 and $(n-1)n(n+1)$ is a multiple of 3 as a product of three consecutive numbers. Then

$$n^3 + 6n^2 - 4n + 3 = (n-1)n(n+1) - 3n + 6n^2 + 3 = (n-1)n(n+1) + 3(2n^2 - n + 1)$$

Because our number can be written as a sum of two multiple of 3, the number is divisible by 3.

Problem 6

Prove that $n^2 + 1$ is not divisible by 3 for any natural number n.

Proof. If some number $m = n^2 + 1$ is not divisible by 3 it can be written in one of the following forms $m = 3k + 1$ (remainder 1) or $m = 3k + 2$ (remainder 2). This is our intention: We want to show that for any natural n, a number $m = n^2 + 1$ divided by 3 will give a remainder of 1 or 2. All natural numbers are either divisible by 3 or not divisible by 3 (remainder of 1 or 2) so we will consider the value of m for all possible natural numbers, n.

Part 1. Assume that n is a multiple of 3 or $n = 3k$. Then

$$m = (3k)^2 + 1 = 9k^2 + 1 = 3 \cdot 3k^2 + 1.$$

We can see that m is a sum of a multiple of 3 and the number 1. So for $n = 3k$, a number m will give a remainder of 1 and is not divisible by 3.

Part 2. Assume that $n = 3k + 1$, then

$$m = (3k+1)^2 + 1 = 9k^2 + 6k + 2.$$

In this case m is a sum of three terms: the first two are multiples of 3 and the last one (2) is not. The number m also can be written as $3 \cdot (3k^2 + 2k) + 2$ which is a form of a number that divided by 3 leaves a remainder of 2. Again m is not divisible by 3.

Part 3. Let us consider the last possible situation, $n = 3k + 2$. Then

$$m = (3k+2)^2 + 1 = 9k^2 + 12k + 4 + 1 = (9k^2 + 12k + 3) + 2$$
$$= 3(3k^2 + 4k + 1) + 2 = 3n + 2.$$

We notice that this m divided by 3 also leaves a remainder of 2. This proves that the number $m = n^2 + 1$ is not divisible by 3 for any natural n so we can conclude that 3 doesn't divide $(m = n^2 + 1)$ for all $n \in \mathbb{N}$. Alternatively, all numbers can be written as $3m$ or $3m \pm 1$, then

$$n^2 + 1 = (3m)^2 + 1 = 3l + 1$$

or

$$(3m \pm 1)^2 + 1 = 3l + 2$$

and neither form is divisible by 3.

> **Problem 7**
>
> Prove that number $m = n^3 + 20n$ is divisible by 48 for any even natural n.

Proof. The idea: We don't know any special properties of divisibility by 48. However, $48 = 6 \cdot 8$, and if we can prove that $m = n^3 + 20n$ is divisible by the product of 6 and 8, then we'll be able to prove that $m = n^3 + 20n$ is divisible by 48.

Given that n is even, then $n = 2k$. Now we can rewrite m as

$$m = (2k)^3 + 40k = 8k^3 + 40k = 8k(k^2 + 5).$$

This number is divisible by 8. Now a number m is a product of 8 and $k(k^2 + 5)$. Following the idea it is enough to show that $k(k^2 + 5)$ is a multiple of 6 and we will prove that m is divisible by 48.

Let us rewrite the second factor of m as

$$k(k^2+5) = k(k^2-1+6) = k(k^2-1)+6k = (k-1)k(k+1)+6k.$$

$$m = 8 \cdot 6l = 48 \cdot l.$$

Here we applied a difference of squares formula to the first term. If we look at the expression on the right closely, we can notice that the first term is a product of three consecutive numbers. By Property 3 this term is divisible by 6 and the second term $(6k)$ is a multiple of 6, so the entire expression is divisible by 6.

We have proved that $m = n^3 + 20n$ is divisible by $6 \cdot 8 = 48$.

Problem 8

Part 1. Prove that the square of any even number greater than 2 is either a multiple of 8 or leaves a remainder of 4 when divided by 8.
Part 2. Prove that the square of any odd number divided by 8 leaves a remainder of 1.

Proof.

Part 1. The square of any even number $n = 2k$, $k > 1$ (remember that a number must be greater than 2) is

$$n^2 = 4k^2, \quad k > 1, \quad k \in \mathbb{N}. \tag{1.2}$$

It is obvious from (1.2) that n is divisible by 4, but we intend to prove that n is divisible by 8. Let us look at (1.2) for even ($k = 2l$) and odd ($k = 2l+1$) values of the natural number k:

$$n^2 = 4k^2 = \begin{cases} 4(2l)^2 & = 16l^2, \\ 4(2l+1)^2 = 4(4l^2+4l+1) = 16(l^2+l)+4. \end{cases} \tag{1.3}$$

From (1.3) we notice that yes, the square of any even number is either a multiple of 8 or leaves a remainder of 4 when divided by 8. Check. Let us take some even numbers for n randomly: 4, 6, 24, and 66.

- $n^2 = 4^2 = 16$ is a multiple of 8,
- $n^2 = 6^2 = 36 = 8 \cdot 4 + 4$ divided by 8 gives a remainder of 4,
- $n^2 = 24^2 = 576 = 8 \cdot 72$ is divisible by 8,
- $n^2 = 66^2 = 4356 = 8 \cdot 272 + 4$ gives a remainder of 4.

Part 2. As we know, any odd number can be written in the form $m = 2n+1$. Any number that divided by 8 gives a remainder of 1 can be represented as $(8k+1)$. It means that we have to prove that

$$(2n+1)^2 = 8k+1. \tag{1.4}$$

Expanding the left side of (1.4) we obtain

$$(2n+1)^2 = 4n^2 + 4n + 1 = 4(n^2+n)+1 = 4n(n+1)+1 = 8k+1. \quad (1.5)$$

It is easy to see that $4n(n+1) = 8k$ where k is some natural number because $n(n+1)$ is a multiple of 2 as a product of two consecutive terms and can be written as $2m$ so $4 \cdot 2m = 8k$. To check, let our odd number be 15. We have $15^2 = 225 = 8 \cdot 28 + 1$. It works! Now you can try other odd numbers. The following problem will give you more experience in the properties of natural numbers and their divisibility.

Problem 9

Some natural number divided by 6 gives a remainder of 4 and when divided by 15 gives a remainder of 7. Find the remainder of this number when divided by 30.

Solution. Let n be the unknown number. Then

$$n = 6k + 4, \quad (1.6)$$

$$n = 15m + 7. \quad (1.7)$$

Multiplying (1.6) by 5 and (1.7) by 2 we obtain

$$5n = 30k + 20, \quad (1.8)$$

$$2n = 30m + 14. \quad (1.9)$$

Because $n = 5n - 2 \cdot 2n$, combining (1.8) and (1.9) we have

$$\begin{aligned}
n &= 30k + 20 - 2(30m + 14) \\
&= 30(k - 2m) - 8 \\
&= 30(k - 2m) - 30 + 22 \\
&= 30(k - 2m - 1) + 22.
\end{aligned} \quad (1.10)$$

Expression (1.10) is the other form of the number n and 22 is the remainder.

Problem 9 is interesting because we solved it and found the exact remainder without finding the number n itself. Knowledge of mathematics is a great tool that gives us an opportunity to solve problems in general. Thus, form (1.10) for n gives us an algorithm for finding n. The smallest n appears when $(k - 2m - 1) = 0$ and $n = 22$. It is true that

$$22 = 6 \cdot 3 + 4 \text{ and}$$
$$22 = 1 \cdot 15 + 7.$$

The next n can be found when $(k - 2m - 1) = 1$ and $n = 30 \cdot 1 + 22 = 52$. You can check that

$$52 = 6 \cdot 18 + 4 \text{ and}$$
$$52 = 3 \cdot 15 + 7.$$

For $(k - 2m - 1) = 100$ we obtain $n = 3022$, etc.
Answer. 22.

Problem 10

Prove that among n consecutive numbers, only one number is divisible by n.

Solution. Consider n consecutive integers such as $a, a+1, a+2, ..., a+n-1$. Each of these numbers can be written as

$$a_k = a + r, \ 0 \leq r \leq n-1.$$

Assume that two numbers of the sequence a_l and a_s are divisible by n. Hence, their difference $a_l - a_s = nl - ns = n(l - s)$ also must be divisible by n. However, the maximum difference between two numbers of the sequence is $n - 1$ which is not divisible by n. We have reached a contradiction and completed the proof.

Remark. See how you can use this result to prove Property 4.

1.3 The Decimal Representation of a Natural Number: Divisibility by $2, 3, 4, 5^n, 8, 9, 10,$ and 11

Any natural $(n+1)$ digits number can be written in the form $N = a_n a_{n-1} \ldots a_2 a_1 a_0$, where

$$a_0 \ -\text{the digit of ones}$$
$$a_1 \ -\text{the digit of tens}$$
$$a_2 \ -\text{the digit of hundreds, etc.}$$

Thus $395 = 3 \cdot 100 + 9 \cdot 10 + 5$ is a three digit number and by analogy

$$\begin{aligned} N &= a_n a_{n-1} \ldots a_2 a_1 a_0 \\ &= a_n \cdot 10^n + a_{n-1} \cdot 10^{n-1} + \cdots + a_2 \cdot 10^2 + a_1 \cdot 10^1 + a_0. \end{aligned} \tag{1.11}$$

Form (1.11) is the base ten representation of a natural number. With it we can prove the divisibility properties of natural numbers.

Divisibility by 2.

> **Property 5** (Divisibility by 2)
>
> An integer is divisible by 2 if its last digit is divisible by 2.

Proof. Let $N = a_n a_{n-1} \ldots a_2 a_1 a_0$ be some arbitrary natural number. Obviously a_0 is the last digit of the number. Rewriting N in the form (1.11) we see that N is the sum of $(n+1)$ terms. The first n terms are multiples of 2 (10^n, 10^{n-1}, ..., 100, and 10 are even numbers). In order that the number N is a multiple of 2, a_0 must be either 0 or a multiple of 2. So we have proved the property. For example, 32, $15,754$ and $1,998,978$ are divisible by 2.

Divisibility by 4.

> **Property 6** (Divisibility by 4)
>
> A natural number N is divisible by 4 if its last two digits form a multiple of 4.

Proof. The proof of this is similar to divisibility by 2.

$$N = a_n a_{n-1} \ldots a_2 a_1 a_0$$
$$= \{a_n \cdot 10^n + a_{n-1} \cdot 10^{n-1} + \cdots + a_2 \cdot 10^2\} + a_1 \cdot 10^1 + a_0$$
$$= 4k + a_1 a_0.$$

If the expression inside braces is divisible by 4, then only the number written by the last two digits determines whether or not the entire number is a multiple of 4.

Divisibility by 8.

> **Property 7** (Divisibility by 8)
>
> An integer is divisible by 8 if the number written by the last three digits is divisible by 8.

The proof of this statement is similar to the preceding proofs. Please try it by yourself.

Divisibility by 3.

> **Property 8** (Divisibility by 3)
>
> An integer N is divisible by 3 if the sum of its digits is divisible by 3.

Proof. This proof demands a fresh idea. Let

$$N = a_n a_{n-1} \ldots a_2 a_1 a_0$$
$$= a_n \cdot 10^n + a_{n-1} \cdot 10^{n-1} + \cdots + a_2 \cdot 10^2 + a_1 \cdot 10^1 + a_0.$$

Notice that

$$
\begin{aligned}
10^1 &= & 10 = 9 + 1 \\
10^2 &= & 100 = 99 + 1 \\
10^3 &= & 1000 = 999 + 1 \\
&\cdots& \\
10^k &= & 999\ldots99 + 1.
\end{aligned}
$$

Replacing each 10^k as $999\ldots99 + 1$ allows us to rewrite the number N in the form,

$$N = a_n \cdot \underbrace{999...99}_{n \text{ digits}} + a_{n-1} \cdot \underbrace{999...99}_{(n-1) \text{ digits}} + \ldots$$
$$+ a_2 \cdot 99 + a_1 \cdot 9 + (a_n + a_{n-1} + \cdots + a_2 + a_1 + a_0).$$

Each term containing $a_k \cdot 999\ldots99$ is a multiple of 3. This means that the sum of the first N terms is a multiple of 3 as well. In order for N to be a multiple of 3, the sum of all its digits (expression within parentheses) must be a multiple of 3. Therefore,

$$(a_n + a_{n-1} + \cdots + a_2 + a_1 + a_0)$$

is divisible by 3.

Example. Is 9748 a multiple of 3?
Answer. No. $9 + 7 + 4 + 8 = 28$, 28 is not divisible by 3, therefore 9748 is not a multiple of 3.

Divisibility by 9.

> **Property 9** (Divisibility by 9)
>
> A number is divisible by 9 if the sum of its digits is divisible by 9.

The proof of this statement is similar to the proof of divisibility by 3.

Example. We can show that $\underbrace{2323...23}_{2016 \text{ times}}$ is not divisible by 9: Adding $3 + 2 = 5$ we multiply it by 2016 to obtain the sum of all digits $5 \cdot 2016$. Since 5 is not divisible by 9, we need to check the divisibility of 2016. The sum of its digits is 9, then the given number with 4032 digits is a multiple of 9.

Divisibility by 5.

Property 10 (Divisibility by 5)

An integer is divisible by 5 if its last digit is either zero or 5.

The proof of this statement will be given below as Property 11.

Problem 11

Find the largest integer that divides any product of five consecutive odd numbers.

Solution. A product of five consecutive odd integers can be written as

$$N = (2n+1)(2n+3)(2n+5)(2n+7)(2n+9).$$

Let us prove that N is always divisible by 3 and 5.

The fact that this number is always divisible by 3 and by 5 can be done by Induction. Obviously, if $n = 1$, and $N = 3 \cdot 5 \cdot 7 \cdot 9 \cdot 11$ is divisible by 3 and 5. Assume that $(2k+1)(2k+3)(2k+5)(2k+7)(2k+9) = 15m$ for $n = k$, $k \in \mathbb{N}$.

Let us demonstrate that for any consecutive number $n = k+1$, the new product will be always divisible by 5 and 3. First, we will show that it will be divisible by 5, i.e.,

$$(2k+3)(2k+5)(2k+7)(2k+9)(2k+11) = 5t.$$

Because $11 = 1 + 10$, the last factor can be written as $2k + 1 + 10$ and the number as the sum of two multiples of 5,

$$
\begin{aligned}
&(2k+3)(2k+5)(2k+7)(2k+9)(2k+11) \\
&= (2k+3)(2k+5)(2k+7)(2k+9)(2k+1+10) \\
&= (2k+1)(2k+3)(2k+5)(2k+7)(2k+9) \\
&+10 \cdot (2k+3)(2k+5)(2k+7)(2k+9) = 5t.
\end{aligned}
$$

Second, let us demonstrate that the same expression will be divisible by 3. For this we will decompose 11 as $11 = 3 + 8$,

$$
\begin{aligned}
&(2k+3)(2k+5)(2k+7)(2k+9)(2k+11) \\
&= (2k+3)(2k+5)(2k+7)(2k+9)((2k+8)+3) \\
&= (2k+3)(2k+5)[(2k+7)(2k+8)(2k+9)] \\
&+3 \cdot (2k+3)(2k+5)(2k+7)(2k+9) = 3s.
\end{aligned}
$$

In the expression above, the number is divisible by 3 because it consists of two terms divisible by 3. Note that the first term has a factor of three consecutive integers so it is divisible by 6, while the second factor is divisible by 3. Hence, the number is always divisible by 3. Therefore the product of five consecutive odd numbers is always divisible by 15.

Hence the largest divisor is 15.

Answer. $N = 15K$, and 15 is the largest divisor.

Divisibility by $5, 25, 125, 625,..., 5^k$.

> **Property 11** (Divisibility by $5, 25, 125, 625,..., 5^k$)
>
> A number N is divisible by 5^k if the number written by its last k digits is divisible by 5^k.

Proof. Consider the n-digits number N

$$N = a_n a_{n-1} \ldots a_2 a_1 a_0$$
$$= \{a_n \cdot 10^n + a_{n-1} \cdot 10^{n-1} + \cdots + a_k \cdot 10^k\}$$
$$+ a_{k-1} \cdot 10^{k-1} + \cdots + a_2 \cdot 10^2 + a_1 \cdot 10^1 + a_0.$$

The expression inside braces is a multiple of 5^k so the divisibility of N by 5^k is determined only by

$$a_{k-1} \ldots a_2 a_1 a_0.$$

Divisibility by 11.

> **Property 12** (Divisibility by 11)
>
> A number $N = a_n a_{n-1} \ldots a_2 a_1 a_0$ is divisible by 11 if and only if a number $|a_0 - a_1 + a_2 - a_3 + \cdots + (-1)^n a_n|$ is divisible by 11. In other words, 11 divides N if and only if 11 divides the alternating sum of the digits of N.

Proof. Let us prove divisibility by 11. We will do some analysis first. A number is divisible by 11 if it is written by even number of 9 digits, 99, 9999, 999999, etc. Also a number is divisible by 11 if it is 11, 1001, 100001, 10000001, etc. (Check!) Also, numbers with odd numbers of 9 digits are not divisible by 11. (See 9, 999, 99999, etc.) Let us see if a number 65321 is divisible by 11.

$$65321 = 6 \cdot 10000 + 5 \cdot 1000 + 3 \cdot 100 + 2 \cdot 10 + 1,$$
$$65321 = 6(9999 + 1) + 5(1001 - 1) + 3(99 + 1) + 2(11 - 1) + 1$$
$$= 6 \cdot 9999 + 5 \cdot 1001 + 3 \cdot 99 + 2 \cdot 11 + 6 - 5 + 3 - 2 + 1.$$

The numbers in the odd positions are parts of the multiple $(999\ldots 99 + 1)$ and so are added to the total. The numbers in the even positions are part of the multiple $(100\ldots 01 - 1)$ and so get subtracted. The sum of odd numbered digits minus the sum of even numbered digits is $6+3+1-(5+2) = 3$ which is not divisible by 11. On the other hand, 8030209 is a multiple of 11, since $8+3+2+9-0 = 22$. Another proof of this property (using congruence) will be given later in this chapter.

Remark. Obviously, a number $\underbrace{aaa...aa}_{2n \text{ digits}}$ is always divisible by 11, i.e. $22 = 2 \cdot 11$ or $3333 = 303 \cdot 11$. On the order hand, $\underbrace{aaa...aa}_{2n+1 \text{ digits}}$ divided by 11 will leave a remainder $1 \le a \le 9$. For example, $55555 = 5050 \cdot 11 + 5$

Let us consider the following problems.

Problem 12

Prove that $\underbrace{22...2}_{2017 \text{ digits}} 1$ is not a prime number.

Proof. Adding all the digits, $2 \cdot 2017 + 1 = 4035 = 3 \cdot n$, we obtain a multiple of 3, hence, the given number is divisible by 3 and cannot be prime.

Remark. Read more about prime numbers in the following section.

Problem 13

Prove that the number $n = 2^{30}$ has at least two repeated digits.

Solution. We have only 10 digits $(0,1,2,3,4,5,6,7,8,9)$ and obviously any number with more than ten digits will have repeated digits. Let us find out how many digits are in the given number. We know the tenth power of 2 is 1024. So,

$$1000 = 10^3 < 1024 = 2^{10} < 2000 = 2 \cdot 10^3.$$

Raising both sides of of the inequality to the third power, we obtain the inequality,

$$10^9 < 2^{30} < 8 \cdot 10^9.$$

Thus, $n < 8,000,000,000$, and has ten digits! Assume that all the digits (the order does not matter, but the first digit obviously must be less than 8) are different. Then their sum must be $0+1+2+3+4+...+9 = 45$ which is a multiple of 3. However, $2^{30} \ne 3n$, because no power of 2 is divisible by 3. We obtained a contradiction and have proved that the base ten representation of $n = 2^{30}$ must have at least two repeated digits.

Remark. Above we used the so-called **Pigeonhole Principle** that is also known as the **Dirichlet Box Principle**. I learned it as follows: Assume that n rabbits are placed into m cages. If $n > m$ then at least one cage will have more than one rabbit.

Problem 14

Can a number written by 100 zeros, 100 ones, and 100 digits of 2 be a perfect square?

Solution. Such a number is huge and has 300 digits, the sum of which is $1 \cdot 100 + 2 \cdot 100 = 300 = 3n$. Hence this number must be divisible by 3. If a number is a perfect square and divisible by 3, then it must be divisible by 9. Obviously, 300 is not a multiple of 9. Therefore no number containing 100 zeros, 100 ones, and 100 twos can be a perfect square.

Problem 15

Find the sum of the digits of all numbers from 1 to 10^5.

Solution. There are $100,000$ numbers from 1 to 10^5, which are all different: some with two digits, some with three digits, etc. Finding how many of them are with two digits or three digits is not that hard, but it would not help in finding the answer to the problem.

Let us try to organize all the numbers in a such way that we would be able to find the answer to the problem. All the numbers except the last one, $100,000$, can be written as $\overline{abcde}$. For example, $1 = 00001$, $34 = 00034$, 87246, or $23458 = 023458$. Each such number will be paired with the number $\overline{(9-a)(9-b)(9-c)(9-d)(9-e)}$. Then the sum of the digits of each such a pair would be easily calculated as

$$a+b+c+d+e+9-a+9-b+9-c+9-d+9-e = 5 \cdot 9 = 45.$$

For example, the number $80,001$ will be paired with $19,998$,

$$
\begin{array}{r}
80,001 \\
+\ \ \ 19,998 \\
\hline
99,999
\end{array}
$$

and the number $273 (00273)$ with $99,726$, etc.

$$00,273$$
$$+$$
$$99,726$$
$$- -$$
$$99,999$$

Among all the numbers between 1 and 100,000, there are precisely 50,000 such interesting pairs, each pair with the sum of digits of 45. The total sum of the digits of all numbers from 1 to 99,999 is $45 \cdot 50,000$, then we need to add 1 for a separate number of $100,000$. Finally, the answer is $45 \cdot 50,000 + 1 = 2,250,001$.
Answer. The sum is $2,250,001$.

Problem 16

Can $N = n^2 + 17n - 2$ be divisible by a) 11? and b) 121?

Solution.
Part 1. Let us rewrite the given number in the form,

$$N = n^2 + 6n + 9 - 11 + 11n = (n+3)^2 + 11 \cdot (n-1).$$

Since the number N contains two terms, one of which is multiple of 11, it will be divisible by 11 if $n + 3 = 11k$, then $n = 11k - 3$ will make N divisible by 11.
Part 2. Because N must be divisible by 121 and $121 = 11^2$, then we need to consider only such n for which N is already divisible by 11,

$$N = (11k - 3 + 3)^2 + 11(11k - 3 - 1) = 121k^2 + 11(11k - 4)$$
$$= 121k^2 + 121k - 44 \neq 11^2 \cdot m.$$

N is never divisible by 121.
Answer. Part 1. If $n = 11k - 3$, then $N = n^2 + 17n - 2$ is divisible by 11. **Part 2.** N is not divisible by 121.

Problem 17

Is a number written by eighty-one digits of one, such as

$$\underbrace{111\ldots11}_{81 \text{ digits}},$$

divisible by 81?

Solution.

$$\underbrace{111\ldots11}_{81\ \text{digits}} = 10^{80} + 10^{79} + 10^{78} + \cdots + 10^2 + 10 + 1$$

$$= \underbrace{99\ldots9}_{80\ \text{digits}} + 1 + \underbrace{99\ldots9}_{79\ \text{digits}} + 1 + \cdots + 99 + 1 + 9 + 1 + 1$$

$$= 81 \cdot 1 + 9\{\underbrace{11\ldots1}_{80\ \text{digits}} + \underbrace{11\ldots1}_{79\ \text{digits}} + \cdots + 11 + 1\}.$$

If we prove that 9 divides the expression within braces, then 81 divides $\underbrace{111\ldots11}_{81\ \text{digits}}$.

Let us rewrite this sum as

$$11\ldots1 + \cdots + 111 + 11 + 1 = 1 \cdot 10^{79} + 2 \cdot 10^{78} + 3 \cdot 10^{77} + \cdots$$
$$+ 78 \cdot 10^2 + 79 \cdot 10 + 1 \cdot 80.$$

We can do this because adding eighty numbers on the left is equivalent to adding 80 ones, 79 tens, 78 hundreds, and so on. Then rewriting each power of ten as 1 plus a number written by the corresponding number of nines we obtain,

$$1(\underbrace{99\ldots9}_{79\ \text{digits}} + 1) + 2(\underbrace{99\ldots9}_{78\ \text{digits}} + 1) + \cdots + 78(99 + 1) + 79(9 + 1) + 80 =$$

$$(\underbrace{99\ldots9}_{79\ \text{digits}} + 2 \cdot \underbrace{99\ldots9}_{78\ \text{digits}} + \cdots + 78 \cdot 99 + 79 \cdot 9) + [1 + 2 + \cdots + 78 + 79 + 80].$$

The first term within parentheses is obviously divisible by nine. The second, within the brackets, is the sum of the first eighty natural numbers and is $\frac{1+80}{2} \cdot 80 = 81 \cdot 40$ and is divisible by 81. Therefore, 81 divides $\underbrace{111\ldots11}_{81\ \text{digits}}$.

Let us solve Problem 18.

Problem 18

The number obtained by striking the last four digits is an integer number of times less than the original number. Find all such numbers ending in 2019.

Solution. Suppose that the original number is $abc2019$, then after deleting the last four digits, we must get abc so

$$a \cdot 10^6 + b \cdot 10^5 + c \cdot 10^4 + 2019 = k \cdot (a \cdot 10^2 + b \cdot 10^1 + c),$$

which can be rewritten as

$$10^4(a \cdot 10^2 + b \cdot 10^1 + c) + 2019 = k \cdot (a \cdot 10^2 + b \cdot 10^1 + c).$$

Without loss of generality, assuming the a number written by deleting four last digits is x, the condition of the problem can be written as

$$10^4 \cdot x + 2019 = k \cdot x.$$

or in a factorized form,

$$2019 = x \cdot (k - 10000). \tag{1.12}$$

Since $2 + 0 + 1 + 9 = 12$, we know that 2019 is a multiple of 3. Because $2019 = 3 \cdot 673$, where 673 is prime, then the solution to the equation (1.12) can be obtained by solving the four systems,

1.

$$\begin{cases} x = 3 \\ k - 10000 = 673 \end{cases} \Leftrightarrow k = 10673, \ x = 3, \ N = 3 \cdot 10673 = 32019.$$

2.

$$\begin{cases} x = 673 \\ k - 10000 = 3 \end{cases} \Leftrightarrow k = 10,003, \ x = 673, \ N = 673 \cdot 10003 = 6,732,019.$$

3.

$$\begin{cases} x = 1 \\ k - 10000 = 2019 \end{cases} \Leftrightarrow k = 12,019, \ x = 1, \ N = 12,019.$$

4.

$$\begin{cases} x = 2019 \\ k - 10000 = 1 \end{cases} \Leftrightarrow k = 10,001, \ x = 2019, \ N = 10,001 \cdot 2019 = 20,192,019.$$

Answer. $N = 32,019$; $N = 6,732,019$; $N = 12,019$; $N = 20,192,019$.

Problem 19

A mother was born in 1961 and her daughter in 1986. They created two numbers related to their birthdays. Which of the two numbers $\underbrace{999\ldots9}7$ or
 1986 digits

$\underbrace{999\ldots9}7$ is divisible by 7?
1961 digits

Solution. Consider the number created by the daughter and rewrite it using base ten representation,

$$\underbrace{999\ldots9}_{1986 \text{ digits}}7$$

$$= 7 + 9 \cdot 10 + 9 \cdot 10^2 + \ldots + 9 \cdot 10^{1986}$$
$$= 7 + 9 \cdot 10(1 + 10 + 100 + \ldots + 10^{1985})$$
$$= 7 + 9 \cdot 10 \cdot \underbrace{111\ldots1}_{1986 \text{ times}}$$

Since 7 is a multiple of 7, we need to find out if $\underbrace{111\ldots1}_{1986\ times}$ is divisible by 7. We do not
know anything special about divisibility by 7. However, any number written by an
even number of ones will be divisible by 11. It is interesting that $111,111 = 11 \cdot k$ and
on the other hand, it is obviously divisible by 111 ($111,111 = 111 \cdot 1001$). However,
111 is not divisible by 11, then the second factor, 1001 must be divisible by 11, and
as we can see from its prime factorization $1001 = 11 \cdot 7 \cdot 13 = 7 \cdot m$, then $111,111 = $
is also divisible by 7. Next, we can expand the number $N = \underbrace{111\ldots1}_{1986\ times}$ by blocks by 6

digits of one,

$$M = \underbrace{111\ldots1}_{6\ times} = 1 \cdot 10^5 + 1 \cdot 10^4 + 1 \cdot 10^3 + 1 \cdot 10^2 + 1 \cdot 10 + 1$$

that are all multiples of 7. Because 1986 is evenly divisible by 6, the entire number
N is divisible by 7.

$$N = \underbrace{111\ldots1}_{1986\ times} = 10^{1985} + 10^{1984} + 10^{1983} + 10^{1982} + 10^{1981} + \ldots$$
$$+10^6 + 10^5 + 10^4 + \ldots + 10 + 1$$
$$= 10^{1980} \cdot (10^5 + 10^4 + \ldots + 10 + 1) + 10^{1974} \cdot (10^5 + 10^4 + \ldots + 10 + 1) + \ldots$$
$$+(10^5 + 10^4 + \ldots + 10 + 1)$$
$$= 10^{1980} \cdot M + 10^{1974} \cdot M + \ldots + M$$
$$= M \cdot (10^{1980} + \ldots + 1)$$
$$= 7 \cdot l.$$

Using the same arguments and because $1961 = 6 \cdot 178 + 3$ is not divisible by 6, then
$\underbrace{999\ldots9}7$ is not divisible by 7. Moreover, $\underbrace{99\ldots9}7$ is divisible by 7 if $n = 6k$.
$\quad$1961 digits $\qquad\qquad\qquad\qquad\qquad$ n digits
Answer. Only the number $\underbrace{999\ldots9}7$ is divisible by 7.
$\qquad\qquad\qquad\qquad$ 1986 digits

1.4 Primes: Problems Involving Primes

We call an integer p a prime number when its only divisors are the trivial ones, ± 1
and $\pm p$. Let's list some positive primes:

$$2,\ 3,\ 5,\ 7,\ 11,\ 13,\ 17,\ 19,\ 23,\ 29,\ 31,\ 37,\ldots$$

Prime numbers are fundamental to our number system. I like how Dr. J. Keating
from the University of Bristol compared prime numbers with pieces of Lego: "You
have individual blocks of Lego which you cannot break down any further. The smallest blocks of Lego come in different sizes but you cannot break them in half. There

are the primes. Out of those blocks you can begin building, you can build Lego objects." Similar to Lego objects, any arbitrary natural number N can be uniquely represented as a product of primes raised to some powers,

$$N = p_1^{n_1} \cdot p_2^{n_2} \cdots \cdots p_n^{n_l}, \tag{1.13}$$

where n_i are natural numbers and p_i are primes. This is called a prime factorization. To get this prime factorization of a positive integer, keep breaking it down into factors until all the factors are primes.

Since ancient times, people wanted to know the answers to the following questions:

1. How to find out whether or not a given number is prime?
2. Are there infinite or finite number of primes?
3. How are the prime numbers distributed among all natural numbers?

Question 1. Mathematicians wanted to establish a rule to determine if N is prime or not. Some rules are well known. For example, in order to find out whether or not a number is prime, we can check all divisibility properties. If it is divisible by any prime number, then it is not prime. We do not have to check divisibility by all prime numbers. Assume that c is the number we need to investigate. If it is not prime, then it can be represented as $a \cdot b$, where one of the factors is less than or equal to $\sqrt{c}$. (Both factors equal $\sqrt{c}$, if c is perfect square) This can be easily proven by contradiction. Assume that both factors are greater than $\sqrt{c}$, i.e., $a > \sqrt{c}$, $b > \sqrt{c}$, then their product will be greater than c, $ab > c$, which contradicts the fact that it must be equal to c. Therefore, we can check only divisibility of c by any prime number $p < \sqrt{c}$. If it is not divisible by any such prime, then the number is itself prime.

Example. Assume that we do not know that 101 is prime number. Because the number is odd, and since $[\sqrt{101}] = 10$ (here denote by $[a]$ the greatest integer less than or equal to a), we need to check divisibility of 101 only by 3, 5, 7. Obviously, it is not divisible by any of the listed numbers, hence 101 is prime.

Example. Is 239 prime or composite number? Because $[\sqrt{239}] = 15$, we need to check divisibility of 239 only by $3, 5, 7, 11, 13$. Since the answer is "no" in each case, the number is prime.

Similarly, consider the number of the current year, 2017. Since $[\sqrt{2017}] = 44$, then by checking divisibility of 2017 by all the primes before 44, i.e.,

$$2, 3, 5, 7, 11, 13, 17, 19, 23, 29, 31, 37, 41, 43$$

so we can state that 2017 is prime.

Of course, when the given number is much greater, this method would be difficult to implement. The French mathematician Pierre Fermat succeeded in the factorization of relatively large natural numbers. His method is based on the difference of squares formula and is discussed in Chapter 3 of this book. Let us see how the difference of squares method works for the factorization of 899. We notice that it is one less than 900 and hence, it can be factored as

$$899 = 900 - 1 = 30^2 - 1^2 = (30 - 1) \cdot (30 + 1) = 29 \cdot 31.$$

Fermat's factorization method is essential to modern data encryption.

If a number N is selected and we want to know if it is prime or a composite number, the Fermat's Little Theorem can be used. This theorem will be studied later in the book. However, the necessary condition for a number to be prime will be formulated here.

Lemma 1 (Necessary Condition for N to be Prime)

For a natural number N to be prime, it is necessary that

$$a^{N-1} \equiv 1 \pmod{N}, \qquad (1.14)$$

is true, where a is any prime number less than N. If this condition does not hold, then N is not prime.

If the condition of this Lemma 1 does not hold, then number N is not prime and can be factored. However, Lemma 1 is not a sufficient condition for a number N to be a prime number. In some cases such as if $a^{N-1} \equiv 1 \pmod{N}$, the number N can still be a composite number. For example, $N = 11 \cdot 31$ is not prime but $2^{340} \equiv 1 \pmod{341}$. There are other numbers that pass the necessary condition but are not primes. For example, $N = 645$ is obviously not prime because it is divisible by 5 and 3 and $645 = 3 \cdot 5 \cdot 43$. However, $2^{644} \equiv 1 \pmod{645}$. In order to disprove that a number is not prime it is enough to demonstrate that the condition (1.14) does not hold for one prime, for example, $a = 2$. Computer algorithms usually check condition (1.14) for all appropriate prime numbers a.

Another method is the application of the Wilson's theorem and can be also formulated as a lemma.

Lemma 2 (Necessary and Sufficient Condition for N to be Prime)

A natural number N is prime if and only if

$$(N-1)! + 1 \qquad (1.15)$$

is divisible by N.

This lemma would not be suitable for finding prime numbers by hand. Even with the invention of powerful computers and creation of efficient computational algorithms, the application of Wilson's Theorem is not the fastest way to find previously unknown large prime numbers. It takes more than 24 hours to evaluate $(N-1)!$ and only a couple of hours to check the condition given by the Fermat's Little Theorem for huge numbers written by tens of thousands of digits. However, this does not diminish the importance of Wilson's Theorem at all, and in fact, this is one of the very popular theorems that are often used in different math contests. For example, using the obvious fact that 101 is prime, we can state that

$$100! \equiv -1 \pmod{101} \Rightarrow 100! = 101n - 1.$$

This observation and other applications of Wilson's Theorem to problem-solving are given in Chapter 3.

Question 2. There are infinitely many primes. The first known proof of the fact was given by Euclid in the third century BC. Let us prove this by contradiction. Assume that there are only a finite set of primes and p_n is the very last prime number. Consider the product of all prime numbers plus one,

$$P = 2 \cdot 3 \cdot 5 \cdot \ldots \cdot p_n + 1 \tag{1.16}$$

Because P is not divisible by any prime number 2, 3, 5,... , p_n, then it is itself a new prime number or a composite number that is not divisible by any primes, $p_1, p_2, ..., p_n$, but is divisible by prime $p_k > p_n$ which is a contradiction to the initial assumption. Therefore p_n is not the last prime number.

Example. Consider, for example, $2 \cdot 3 \cdot 5 \cdot 7 \cdot 11 \cdot 13 + 1 = 30,031 = 59 \cdot 509 = 59K$ is not a prime.

Assuming again that N is the last known prime number, consider the product of all natural numbers from 1 to N, called $N!$ (factorial) and add to it one,

$$1 \cdot 2 \cdot 3 \cdot 4 \cdot 5 \cdot \ldots \cdot (N-2) \cdot (N-1) \cdot N + 1 = N! + 1.$$

This number will be much greater than N, but also when it divided by any existing prime less than or equal to N it will leave a remainder of 1. Therefore, $(N! + 1)$ is either a new prime or a composite number divisible by a prime bigger than N. Therefore, N is not the largest prime.

We can prove that for any given number N there are precisely N consecutive natural numbers, with only one prime number among them. Here is such a sequence.

$$N! + 1, \; N! + 2, \; N! + 3, \; ..., \; N! + N - 1, \; N! + N,$$

where only the first number could be a prime but not necessarily. For example, $6! + 1 = 720 + 1 = 721$ is not a prime number but obviously divisible by 7, i.e., $721 = 7 \cdot 103$.

Another ingenious proof of the existence of infinitely many primes was given by Leonard Euler. His proof can be summarized as follows: Assume that $p_1, p_2, p_3, ... p_k$ are the finite set of all existing prime numbers. Consider the infinite geometric series,

$$1 + \frac{1}{p_1} + \frac{1}{p_1^2} + \ldots + \frac{1}{p_1^n} + \ldots = \frac{1}{1 - \frac{1}{p_1}}$$
$$1 + \frac{1}{p_2} + \frac{1}{p_2^2} + \ldots + \frac{1}{p_2^n} + \ldots = \frac{1}{1 - \frac{1}{p_2}}$$

$$\ldots$$

$$1 + \frac{1}{p_k} + \frac{1}{p_k^2} + \ldots + \frac{1}{p_k^n} + \ldots = \frac{1}{1 - \frac{1}{p_k}}$$

(1.17)

Multiplying these convergent geometric series, the left and the right sides, we will again obtain convergent series,

$$\frac{1}{1 - \frac{1}{p_1}} \cdot \frac{1}{1 - \frac{1}{p_2}} \ldots \frac{1}{1 - \frac{1}{p_k}} = \sum \frac{1}{p_1^{\alpha_1} p_2^{\alpha_2} \ldots p_k^{\alpha_k}}$$

(1.18)

Consider the denominator of the fractions on the right. From arithmetics, we know that any number can be represented by factors $m = p_1^{\alpha_1} p_2^{\alpha_2} \ldots p_k^{\alpha_k}$ uniquely. Hence the sum on the right contains all reciprocals $\frac{1}{m}$ of all possible natural numbers m. Moreover, we can put these fractions, for example, in the order by increasing denominators and obtain infinite harmonic series,

$$1 + \frac{1}{2} + \frac{1}{3} + \ldots + \frac{1}{n} + \ldots$$

(1.19)

The left side of (1.18) is a finite number and the right side is a divergent series. We obtain a contradiction! This proves that there are infinitely many prime numbers.

Question 3. Mathematicians of all times were very fascinated with primes and how primes are spaced out into a number line. The very first primes 2, 3, 5, 7 are very close together and there are bigger gaps between say 31 and 37 or 47 and 53. However, there is no regular pattern in the distribution of the primes although there are certain regularities. Looking at the line of natural numbers, the primes get farther apart as we move along toward infinity. It is not so difficult to find some long gap in the primes. In fact, we can find the existence of $(N - 1)$ non-prime numbers in a row (composite numbers) if instead of adding one to $N!$, we would add 2, 3, 4, ..., N to it. Thus,

$$N! + 2 \text{ is a multiple of } 2$$
$$N! + 3 \text{ is a multiple of } 3$$
$$N! + N \text{ is a multiple of } N.$$

Problem 20

Prove that there exist 500 consecutive composite numbers.

Proof. Let us take a sequence of 500 numbers, such that $501! + 2, 501! + 3, 501! + 4$, ..., $501! + 500, 501! + 501$. There are consecutive natural numbers and there is no prime among them!

On the other hand, the size of the gaps in the primes seems to jump around in a very strange manner. For example, there are many so-called **twin primes**, separated by just one even number, such as 3 and 5, 11 and 13, 59 and 61, etc.

Problem 21

Can all twin primes be described by the formula $(6n-1, 6n+1)$?

Solution. Let us reformulate this problem and prove or disprove two statements:

Statement 1. A pair $(6n-1, 6n+1)$ is always a twin prime pair.
This statement is wrong. It is enough to substitute for $n = 6k^2$ and the first number of the pair will be a composite number, not prime,

$$6n - 1 = 36k^2 - 1 = (6k - 1)(6k + 1).$$

Statement 2. If there is a pair of twin prime numbers, then this pair can be described by $(6n-1, 6n+1)$. This statement is also false because the very first twin prime pair $(3, 5)$ cannot be represented by the formula. However, if we start thinking about all other possible cases, then they will fit the formula. Any number greater than 3, in the form

$$p_1 \cdot p_2 \cdot p_3 \cdot \ldots \cdot p_n + 1 = 2 \cdot 3 \cdot \ldots + 1 = 6n + 1$$

is a candidate to be twin prime number if

$$n \neq 6k^2; \quad n \neq 36m^3.$$

The distribution of prime numbers among all natural numbers does not follow any regular pattern. The greatest mathematician of all times, Leonard Euler said "Mathematicians have tried in vain to this day to discover order in the sequence of prime numbers, and we have reason to believe that it is a mystery onto which the human mind will never penetrate". Euler spent some time working on the distribution of primes and introduced the **Euler zeta function**. Thus, formula (1.18) for infinitely many primes can be rewritten as

$$\prod_p \frac{1}{1 - \frac{1}{p^s}} = \sum_{n=1}^{\infty} \frac{1}{n^s}, \quad s > 1$$

where p represents all primes. This is the famous Euler equality. If $s > 1$ the series on the right is convergent and for $s = 1$ we have divergent harmonic series. The function

$$\zeta(s) = \sum_{n=1}^{\infty} \frac{1}{n^s}, \tag{1.20}$$

these days is called the Riemann ζ function because the function was further investigated in the work of the great German mathematician Bernhard Riemann

(1826–1866). Riemann used it with the complex variable s in order to prove an asymptotic law of the prime numbers distribution. He observed that the frequency of prime numbers is very closely related to the behavior of $\zeta(s)$. The Riemann hypothesis asserts that all nontrivial solutions of the equation $\zeta(s) = 0$ lie on a straight line with $Re(s) = 1/2$. A proof of this hypothesis if it is true would give the answer to the mysterious distribution of prime numbers. It would also pay a million dollars to one who first proves or disproves the Riemann hypothesis. This was announced on May 24, 2000 by the Clay Institute of Cambridge Massachusetts in order to stimulate work on the Riemann hypothesis. Many outstanding mathematicians, such as Louis de Branges de Bourcia, a Distinguished Professor of Purdue University (in 1989 he proved the Bieberbach conjecture), would probably continue working on the Riemann hypothesis if there was no monetary award. Who knows, maybe you will prove the Riemann hypothesis. Sometimes a simple idea can be missed by all of the experts. I want to encourage you to practice your problem-solving skills and wish you good luck.

Here are some interesting problems involving primes. There are more problems involving primes throughout the entire book.

Problem 22

Some prime numbers can be written by the formula $N = 2^n - 1$. Prove that the power n must also be a prime number.

Proof. Assume a contradiction, i.e., n is not prime and can be factored as $n = k \cdot l$. Then N can be factored as follows

$$N = 2^{k \cdot l} - 1 = (2^k)^l - 1^l = ((2^k)^{l-1} + (2^k)^{l-2} + \ldots + 2^k + 1) \cdot (2^k - 1).$$

Since N is a product of at least two factors, then it cannot be prime. Therefore our assumption was wrong and n must be prime itself in order to generate some prime numbers N.

Remark. Any prime number of the form $N = 2^p - 1$ is called a **Mersenne prime**.

Example. $2^{17} - 1$, $2^{19} - 1$, $2^{31} - 1$ are prime numbers, while $2^{23} - 1 = 8,388,607 = 47 \cdot 178,481$ is not prime.

Problem 23

Prove that there are infinitely many prime numbers in the form of

$$p = 4k + 3.$$

Proof. Assume that all prime numbers $p_1 = 3, p_2 = 5, p_3 = 7, p_4 = 11, \ldots p_n$ represent a finite set of only n numbers. Consider the number $N = 4 \cdot (p_2 \cdot p_3 \ldots \cdot p_n) + 3$. Such a number is not divisible by any of the primes $p_2, p_3, \ldots, p_n$ but is prime of the form $4k + 3$. This is a contradiction and therefore there are infinitely many prime numbers of type $4k + 3$.

Problem 24

When choosing from a list of four different whole numbers, I can select three whose product is 74, and you can select three whose product is 54. What is the product of all four numbers?

Solution. Let us do a prime factorization of 74, i.e.,

$$74 = 1 \cdot 2 \cdot 37.$$

Because 37 is prime, we have already found three of four numbers: 1, 2, and 37. Now 54 can be factored only as $54 = 1 \cdot 2 \cdot 27$. The fourth unknown number is 27 and the product of all four numbers is $1 \cdot 2 \cdot 37 \cdot 27 = 1998$.
Answer. 1998.

Problem 25

Masha states that the number in the form $N = n^2 + n + 41$ is always prime. Prove or disprove her statement.

Solution. Let us rewrite the number as

$$N = n^2 + n + 41 = n \cdot (n+1) + 41.$$

It means that N can be factored if either $n = 40$, $n + 1 = 41$, $N = 40 \cdot 41 + 41 = 41^2$ or $n = 41$, $n + 1 = 42$, $N = 41 \cdot 42 + 41 = 41 \cdot 43$. Obviously, for any other values of $n = 0, 1, 2, \ldots, 39$, the number N will be prime. For example, if $n = 2$, $N = 2 \cdot 3 + 41 = 47$. Therefore, Masha did not create a formula for prime numbers. Note that this formula was first found by Euler.

Let us solve Problem 26.

Problem 26

Consider $2 \cdot 3 \cdot 5 + 1 = 31$ or $2 \cdot 3 \cdot 5 \cdot 7 + 1 = 211$ are prime numbers. Is it true that the product of n consecutive numbers $p_1 \cdot p_2 \cdot p_3 \cdot \ldots \cdot p_n + 1$ is always prime?

Solution. The statement is false because $2 \cdot 3 \cdot 5 \cdot 7 \cdot 11 \cdot 13 + 1 = 30{,}031 = 59 \cdot 509$. In general, $p_1 \cdot p_2 \cdot p_3 \cdot \ldots \cdot p_n + 1$ is either a new prime or a composite number that is not divisible by any of the prime factors, but by a prime $p_k > p_n$, $k > n$.

Problem 27

Find such pairs of primes (x, y) that satisfy the equation $x^2 = 2y^2 + 1$.

Solution. Applying the difference of squares formula, we can factor this equation as

$$2 \cdot y^2 = (x - 1)(x + 1)$$

It follows from the condition of the problem that if x is an odd prime, then $x - 1 = 2k$ and $x + 1 = 2k + 2$ are both even factors (because they differ by 2). If the left side is even, then this equation can have solutions if and only if $x - 1 = 2$.

$$2 \cdot y^2 = 2k \cdot (2k + 2)$$
$$y^2 = 2 \cdot k(k + 1)$$

Therefore, $y = 2$, $x = 3$.
Answer. $x = 3$, $y = 2$.

Remark. If we rewrite the given equation as $x^2 - 2y^2 = 1$, some can recognize in it the so-called Pell's equation. The methods of solving such equations are explained in Chapter 3. You will learn that Pell's equations have infinitely many integer solutions and that it is very important to find the very first, nontrivial solution. Having solved such equation with the restriction on its variables allowed us to find this nontrivial solution. Later you will see that all other possible solutions will be obtained from $x = 3$, $y = 2$.

Problem 28

How many natural numbers less than 1000 are not divisible by 5 or 7?

Solution. There are $\left[\frac{1000}{5} \right] = 200$ multiples of 5 and $\left[\frac{1000}{7} \right] = 142$ multiples of 7. Among these $200 + 142$ natural numbers there are $\left[\frac{1000}{35} \right] = 28$ multiples of 35. There are

$$1000 - 200 - 142 + 28 = 686$$

numbers less than 1000 that are not divisible either by 5 or by 7.
Answer. 686.

Remark. Here we denoted $[a]$ as a greatest integer less than or equal to a. For example, since $\frac{1000}{35} \approx 28.57$, then $\left[\frac{1000}{35}\right] = 28$.

Problem 29 (1996 Moscow State University, oral exam)

Find all ending zeros in the digital form of the number

$$100! = 1 \cdot 2 \cdot 3 \cdot \cdots \cdot 98 \cdot 99 \cdot 100.$$

Solution. Among all factors of the given number $N = 1 \cdot 2 \cdot 3 \cdot \cdots \cdot 99 \cdot 100$, there are 20 factors of 5 ($100/5 = 20$), four of which are divisible by 25 ($20/5 = 4$). We can conclude that N is divisible by $5^{20} \cdot 5^4 = 5^{24}$. Among all factors of N there are 50 even, so the number is divisible by 2^{24}. So, if N is divisible by $5^{24} \cdot 2^{24} = 10^{24}$, then it ends in twenty four zeros!

Answer. 24.

Problem 30

What is the largest power of 2 that divides 800!?

Solution. 800! can be written as $N = 1 \cdot 2 \cdot 3 \cdot \cdots \cdot 799 \cdot 800 = 2^n \cdot m$. Let us first consider a simple number, 20!,

$$20! = 1 \cdot 2 \cdot 3 \cdot 4 \cdot 5 \cdot 6 \cdot 7 \cdot 8 \cdot 9 \cdot 10 \cdot 11 \cdot 12 \cdot 13 \cdot 14 \cdot 15 \cdot 16 \cdot 17 \cdot 18 \cdot 19 \cdot 20$$

and extract all the numbers divisible by 2 (2, 4, 6, 8, 10,..., 20). There are $\left[\frac{20}{2}\right] = 10$ such numbers. But is that all? No. Because $4, 8, 12, 16$, and 20 are divisible by 4 and will contain an additional 2, then among all these ten even numbers there are $\left[\frac{20}{4}\right] = 5$ numbers divisible by 4. Again we did not find all the powers of 2 because among all multiples of 4 there are numbers 8 and 16 (multiples of 8) that will give two additional power of 2, i.e., $\left[\frac{20}{8}\right] = 2$ and finally, among these two multiples of 8 there is a number 16 that is a multiple of 16, i.e., $\left[\frac{20}{16}\right] = 1$. Here again we use the notation $\left[\frac{k}{m}\right]$ as the greatest integer less than or equal to $\frac{k}{m}$.

Adding together all the results, we obtain that $20! = 2^n \cdot m$, where

$$n = \left[\frac{20}{2}\right] + \left[\frac{20}{4}\right] + \left[\frac{20}{8}\right] + \left[\frac{20}{16}\right] = 10 + 5 + 2 + 1 = 18$$

This yields that $20 = 2^{18} \cdot m$. Similar ideas can be used in finding the largest power of 2 in 800!,

$$n = \left[\frac{800}{2}\right] + \left[\frac{800}{4}\right] + \left[\frac{800}{8}\right] + \left[\frac{800}{16}\right] + \left[\frac{800}{32}\right]$$
$$+ \left[\frac{800}{64}\right] + \left[\frac{800}{128}\right] + \left[\frac{800}{256}\right] + \left[\frac{800}{512}\right]$$
$$= 400 + 200 + 100 + 50 + 25 + 12 + 6 + 3 + 1 = 797.$$

$$800! = 2^{797} \cdot k.$$

Answer. The largest power of 2 in 800! is 797.

Problem 31 (October 26, 1999, Texas State Math League)

If x and y are integers, what is the least x for which $\frac{1}{640} = \frac{x}{10^y}$?

Solution. Rewriting the equation in the equivalent form of $10^y = 640x$ and dividing both sides by 10 we obtain

$$10^{y-1} = 64x. \tag{1.21}$$

Finding the prime factorization of 10 and 64 and plugging it into (1.21) we have

$$(2 \cdot 5)^{y-1} = (2)^6 \cdot x \text{ or } 2^{y-1} \cdot 5^{y-1} = 2^6 \cdot x. \tag{1.22}$$

In order for equation (1.22) to have any integer solutions, x has to be at least 5^{y-1} when $y - 1 = 6$ or $x = 5^6 = 15,625$.

Answer. $x = 15,625$.

Problem 32 (Lomonosov Moscow State University Entrance Exam)

Find all integers x and y such that

$$\begin{cases} 7875x^3 = 1701y^3 \\ |x| \leq 5. \end{cases}$$

Solution. Let us do a prime factorization of $7875 = 5 \cdot 5 \cdot 5 \cdot 7 \cdot 9$ and $1701 = 3 \cdot 3 \cdot 3 \cdot 7 \cdot 9$. This can simplify the given system as

$$\begin{cases} (5x)^3 = (3y)^3 \\ |x| \leq 5. \end{cases} \tag{1.23}$$

Relation $5x = 3y$ yields that x is a multiple of 3, i.e., $x = 3k$. Using the inequality of system (1.23) we notice that if $|3k| \leq 5$, then $k = 0$, $k = 1$ or $k = -1$. Three different values of k give us three ordered pair solutions:

$$\begin{array}{lll} k = 0, & x = 0, & y = 0 \\ k = 1, & x = 3, & y = 5 \\ k = -1, & x = -3, & y = -5 \end{array}$$

Answer. $(0,0)$, $(3,5)$, and $(-3,-5)$.

> **Problem 33** (AIME 2000)
>
> Suppose that x, y, z are three positive numbers that satisfy the equations:
>
> $$\begin{aligned} xyz &= 1 \\ x + \frac{1}{z} &= 5 \\ y + \frac{1}{x} &= 29 \\ z + \frac{1}{y} &= \frac{m}{n}, \end{aligned}$$
>
> where m and n are relatively prime positive integers. Find $m+n$.

Solution. Multiplying the 2nd, 3rd, and 4th equations we obtain

$$\left(x + \frac{1}{z}\right) \cdot \left(y + \frac{1}{x}\right) \cdot \left(z + \frac{1}{y}\right) = 5 \cdot 29 \cdot \frac{m}{n}$$

Multiplying the terms in parentheses and replacing xyz by 1 and

$$\left(x + y + z + \frac{1}{y} + y + \frac{1}{z} + \frac{1}{x}\right)$$

by

$$5 + 29 + m/n,$$

we obtain,

$$2 + 5 + 29 + \frac{m}{n} = 5 \cdot 29 \cdot \frac{m}{n}$$

or

$$\frac{m}{n} = \frac{36}{(5 \cdot 29 - 1)} = \frac{1}{4}$$

Finally, the answer is

$$m + n = 1 + 4 = 5.$$

Answer. $m + n = 5$.

Let us solve Problem 34.

> **Problem 34**
>
> How many zeros are there in the digital form of the number 2010!?

Solution. We have

$$2010! = 1 \cdot 2 \cdot 3 \cdot 4 \cdot 5 \cdots \cdot 2009 \cdot 2010.$$

Consider its prime factorization,

$$2010! = 2^{\alpha_1} \cdot 3^{\alpha_2} \cdot 5^{\alpha_3} \dots p^{\alpha_q}.$$

We understand that each pair of 2 and 5 will make one zero in the number N. So if we can find all these pairs, we will know the number of zeros. Once again, a bracket with a number inside represents the greatest integer that is less than or equal to that number, i.e., $[a]$ the greatest integer $\leq a$.

First, let us find all numbers divisible by 5,

$$\frac{2010}{5} = 402.$$

Divisible by 25,

$$\left[\frac{2010}{25}\right] = 80.$$

Divisible by 125,

$$\left[\frac{2010}{125}\right] = 16.$$

Divisible by 625,

$$\left[\frac{2010}{625}\right] = 3.$$

Therefore 2010! contains exactly $402 + 80 + 16 + 3 = 501$ numbers of 5. Now let us find out how many factors of 2 we have in 2010! We have to add all multiples of 2, 4, 8, 16, 32, ..., and 1024.

$$1005 + 502 + 251 + 125 + 60 + 30 + 12 + 7 + 3 + 1 = 1996.$$

However, only 501 of them will be paired with a corresponding 5.

Answer. 2010! ends in 501 zeros.

Remark. By solving the problems 29, 30, and 34, we derived the **Legendre Formula** (1.24):

$$n! = p_1^{\alpha_1} \cdot p_2^{\alpha_2} \cdot \dots \cdot p_s^{\alpha_s}.$$

where each power can be found as

$$\alpha_k = \left[\frac{n}{p_k}\right] + \left[\frac{n}{(p_k)^2}\right] + \left[\frac{n}{(p_k)^3}\right] + \dots.$$

Moreover, we notice that

$$\left[\frac{n}{p^m}\right] = \left[\frac{n}{p^{m-1} \cdot p}\right] = \frac{\left[\frac{n}{p^{m-1}}\right]}{p} \tag{1.24}$$

Using this, let us solve Problem 35.

Problem 35

Find the prime factorization of 20!

Solution. First, the primes less than 20 are: $2, 3, 5, 7, 11, 13, 17, 19$. Second, instead of applying the Legendre formula, we will use formula (1.24). Thus, all powers of 2 can be found as follows.

Evaluate $\left[\frac{20}{2}\right] = 10$, by doing this we find how many multiples of 2 are in 20!, then we divide the result, 10, by 2 again and find how many among these ten even numbers are also of multiples of 4 i.e. $\left[\frac{10}{2}\right] = 5$. Then because some of them are also multiples of $8 = 2^3$ we must evaluate $\left[\frac{5}{2}\right] = 2$, etc. All can be written as

$$\left[\frac{20}{2}\right] + \left[\frac{10}{2}\right] + \left[\frac{5}{2}\right] + \left[\frac{2}{2}\right] = 10 + 5 + 2 + 1 = 18.$$

Powers of 3,

$$\left[\frac{20}{3}\right] + \left[\frac{6}{3}\right] = 6 + 2 = 8$$

Powers of 5,

$$\left[\frac{20}{5}\right] = 4$$

Powers of 7,

$$\left[\frac{20}{7}\right] = 2$$

Obviously, the powers of all other primes less than 20 will be 1. Finally, we have

$$20! = 2^{18} \cdot 3^8 \cdot 5^4 \cdot 7^2 \cdot 11 \cdot 13 \cdot 17 \cdot 19.$$

Answer. $20! = 2^{18} \cdot 3^8 \cdot 5^4 \cdot 7^2 \cdot 11 \cdot 13 \cdot 17 \cdot 19.$

1.5 Greatest Common Divisor and Least Common Multiple

If a number c divides a and b simultaneously, we call it a common divisor of a and b. For example, 7 divides 63 $(63 = 7 \cdot 9)$ and 126 $(126 = 7 \cdot 18)$, then 7 is a common divisor of 63 and 126. Can we find another common divisor of 63 and 126 that is greater than 7?

We are looking for the greatest common divisor (gcd) that is the greatest among the common divisors of two numbers a and b, and has notation $d = (a, b)$. It is obvious that

> **Property 13**
>
> Any common divisor of two numbers divides their greatest common divisor.

Notice that 63 is the greatest common divisor of 63 and 126. Let us find the greatest common divisor of 24 and 56. We can see that 8 is this number, because $24 = 8 \cdot 3$ and $56 = 8 \cdot 7$, and $8 = (24, 56)$.

Studying numbers, we should remember a fact stated here as Property 14.

> **Property 14**
>
> If two numbers a_1 and b_1 are **relatively prime**, then their greatest common divisor is one, i.e. $(a_1, b_1) = 1$.

For example, numbers 15 and 22 are relatively prime. They don't have any common factors and $(15, 22) = 1$. Additionally, two consecutive numbers are always relatively prime, i.e., $(n, n+1) = 1$. Three consecutive numbers are also relatively prime $(n, n+1, n+2) = 1$.

Suppose we have two numbers, a and b, and we want to find their least common multiple. We say that a number m is a common multiple of a and b when it is divisible by both of them. For example, a product of a and b is a common multiple because $m = ab$ is divisible by a and b simultaneously.

> **Property 15** (The Least Common Multiple)
>
> Among all common multiples of a and b there is the smallest one that we call the least common multiple (lcm) of a and b and denote it as $[a, b]$.

> **Property 16**
>
> Every common multiple of two numbers is divisible by their least common multiple. The least common multiple of a and b is equal to their product if, and only if, they are relatively prime.

Let us find the least common multiple of 24 and 36. Yes, it is 72 because $72 = 24 \cdot 3$ and $72 = 36 \cdot 2$. Usually when numbers a and b are bigger than 100, a problem of finding a least common multiple becomes more difficult.

Question. Is there any relation between the gcd and lcm?

To answer this question let us consider two numbers a and b. Suppose that we know that their *gcd* is $d = (a, b)$. Now numbers a and b can be written in terms of d as

$$a = a_1 \cdot d$$
$$b = b_1 \cdot d, \qquad (1.25)$$

where a_1 and b_1 are relatively prime.

Property 17

If a number m is the least common multiple of a and b, then it must be divisible by a and by b.

We can write that $m = ha$, where h is some integer. Using (1.25) m can be written as

$$m = ha = ha_1 d. \qquad (1.26)$$

On the other hand, a number m is divisible by b as well. From (1.26) we see that this can happen only if

$$h = kb_1, \qquad (1.27)$$

where k is some integer.

Combining (1.25), (1.26), and (1.27) yields,

$$m = kb_1 a_1 d = k \cdot \frac{b}{d} \cdot \frac{a}{d} \cdot d = k \cdot \frac{ab}{d}. \qquad (1.28)$$

It is obvious that when $k = 1$ a number m becomes the least common multiple of two numbers a and b. This we can write as

$$m = \frac{ab}{(a,b)} = [a,b] \text{ or } [a,b] \cdot (a,b) = a \cdot b. \qquad (1.29)$$

Using (1.29) let us solve Problem 36.

Problem 36

The greatest common divisor of $54,990$ and $40,300$ is 130. Find the least common multiple of these numbers.

Solution. We have

$$M = \frac{54990 \cdot 40300}{130} = 17046900.$$

Answer. $17,046,900$.

Note that it is easier to find the least common multiple or greatest common divisor of two (or more) numbers using prime factorization. Please examine the following problems.

Problem 37

Find the least common multiple and the greatest common divisor of 2880 and 6048 using prime factorization of the numbers.

Solution. Let us factor 2880. This even number ends in 0 so it is divisible by 2 and 5. Also it is divisible by 9 because the sum of its digits equals 18 which is divisible by 9. Thus we can write $2880 = 2 \cdot 5 \cdot 3 \cdot 3 \cdot \underline{32}$.

We should continue looking for prime factors now for the underlined number 32. It is even, so it is divisible by 2. Keep breaking each underlined number down into factors until you have all the prime factors.

$$2880 = 2 \cdot 5 \cdot 3 \cdot 3 \cdot 32 = 2 \cdot 5 \cdot 3 \cdot 3 \cdot 2 \cdot \underline{16} = 2 \cdot 5 \cdot 3 \cdot 3 \cdot 2 \cdot 2 \cdot 2 \cdot 2 \cdot 2 = 2^6 \cdot 3^2 \cdot 5, \quad (1.30)$$

$$6048 = 2 \cdot 3 \cdot 3 \cdot 336 = 2 \cdot 3 \cdot 3 \cdot 2 \cdot \underline{168} = 2 \cdot 3 \cdot 3 \cdot 2 \cdot 2 \cdot 2 \cdot 2 \cdot 21 = 2^5 \cdot 3^3 \cdot 7. \quad (1.31)$$

From (1.30) and (1.31) we can find the least common multiple of 2880 and 6048

$$[6048, 2880] = 2^6 \cdot 3^3 \cdot 5 \cdot 7 = 60480,$$

and the greatest common divisor

$$(6048, 2880) = 2^5 \cdot 3^2 = 288.$$

Now let us check that the product of *gcd* and *lcm* of 6048 and 2880 is equal to $6048 \cdot 2880$.

$$288 \cdot 60480 = 17418240 = 6048 \cdot 2880.$$

Yes, the formula (1.29) works!

The Theorem 1 will summarize our knowledge.

Theorem 1

Let a and b be two given integers, and let

$$a = p_1^{\alpha_1} p_2^{\alpha_2} \dots p_r^{\alpha_r},$$
$$b = p_1^{\beta_1} p_2^{\beta_2} \dots p_r^{\beta_r}$$

be their prime factorization. The greatest common divisor and the least common multiple of the two numbers a and b, are respectively,

$$(a,b) = \prod p_i^{\min(\alpha_i,\beta_i)},$$
$$[a,b] = \prod p_i^{\max(\alpha_i,\beta_i)},$$

where $\prod$ is the product of all factors p_i raised to their corresponding powers.

The notation $\min(\alpha_i,\beta_i)$ means that between two powers of the same factor p_i we choose the least. Thus, in our problem above, the prime factorization of 2880 contains 2^6, and $6048 - 2^5$, then for the *gcd* we choose 2^5 ($5 = \min(5,6)$). The notation $\max(\alpha_i,\beta_i)$ means that between two powers of the same factor we choose the greatest. Thus, in the problem above, looking for a power of 2 of the *lcm* of 2880 and 6048 we take 6 because $6 = \max(5,6)$, and so on.

Problem 38

Find the greatest common divisor and the least common multiple of two numbers such that $a = 2^5 \cdot 25^2$ and $b = 4^3 \cdot 35^2$.

Solution. Of course, you have already noticed that we cannot apply the previous theorem to these numbers right away because we have to find a prime factorization of each number (factors 4, 25 and 35 are not primes). Recalling the properties of exponents we can rewrite a and b as

$$a = 2^5 \cdot 5^2 = 800 \text{ and } b = 2^6 \cdot 5^2 \cdot 7^2 = 78,400. \tag{1.32}$$

From (1.32) we notice that the *lcm* will contain powers of 2, 5, and 7, but the *gcd* contains only powers of 2 and 5.

$$\begin{aligned}[a,b] &= 2^6 \cdot 5^2 \cdot 7^2 = 126,400 \\ (a,b) &= 2^5 \cdot 5^2 \quad\;\; = 800. \end{aligned} \tag{1.33}$$

Expressions (1.33) tell us that a number a itself is the greatest common divisor (a,b) of a and b.

Answer. The *gcd* = 800 and the *lcm* = 126,400.

Problem 39

Find integers a and b such that the roots of

$$x^2 + (2a+9)x + 3b + 5 = 0$$

are distinct integer numbers and the coefficients $(2a+9)$, $(3b+5)$ are positive primes.

Solution. Using Vieta's theorem (See Section 3.4.2) we have the system,

$$\begin{cases} x_1 \cdot x_2 = 3b+5 \\ x_1 + x_2 = -(2a+9) \end{cases}$$

There are two cases:

Case 1. If $x_1 = 1 \Rightarrow x_2 = 3b+5$. Substituting this into the second equation of the system we obtain

$$3b+5+1 = -2a-9$$
$$3(b+2) = -(2a+9)$$
$$\begin{cases} b+2 = 1 \\ 2a+9 = -3 \end{cases} \Leftrightarrow \begin{cases} b = -1 \\ a = -6 \end{cases}$$

This will not give us a solution.

Case 2. If $x_1 = -1 \Rightarrow x_2 = -3b-5$. Substituting this into the second equation of the system we obtain

$$-3b-5-1 = -2a-9$$
$$3(b+2) = -(2a+9)$$
$$\begin{cases} b+2 = 1 \\ 2a+9 = 3 \end{cases} \Leftrightarrow \begin{cases} b = -1 \\ a = -3 \end{cases}$$

Answer. $a = -3$, $b = -1$.

Problem 40 (Olympiad Lomonosov, 2009)

What is the greatest common divisor of two natural numbers m and n if after increasing the number m by six the greatest common divisor of the two numbers is increased nine times.

Solution. Denote the original greatest common divisor by $(m,n) = d$, and the new one by $(m+6,n) = 9d$. From these equations it follows that the numbers m, n, and

$(m+6)$ are divisible by d. Hence, 6 also divisible by d. Since $6 = 2 \cdot 3$, then 2, 3, and 6 satisfy the condition to the problem.

Answer. $d = 2$, 3 or 6.

Problem 41

Find the least natural number, different from 1, such that when it is divided by 2, 3, 5, and 9 leaves a remainder of 1.

Solution. Denote the number by N, then the following is true:

$$N = 2n + 1 = 3m + 1 = 5k + 1 = 9s + 1,$$

It follows that $N - 1$ is simultaneously divisible by 2, 5, and 9. Finding Least Common Multiple of these three numbers $[2, 5, 9] = 90$, we obtain

$$N - 1 = 90s, \ N = 90s + 1$$

If the number is different from one, then this number is 91.

Answer. 91.

Now, let us consider some simple word problems involving the least common multiple.

Problem 42

One garland of light bulbs on a Christmas tree lights up every 15 seconds and another garland every 12 seconds. At a certain time both garlands are lit simultaneously. How soon thereafter will the light bulbs on the garlands be lit again at the same time?

Solution. This will happen after 60 seconds, that is the least common multiple of two given times, $lcm = 60$.

Answer. 60 seconds.

Problem 43

Each morning three busses line up at the bus station. The busses have different routes such that the first bus returns to the starting point after 90 minutes, the second bus after 50 minutes, and the third one after 70 minutes. How soon will all three busses again gather in the bus station?

Solution. If all three busses go by schedule without an emergency or traffic delay, then the next meeting at the starting point will be after $lcm = (90, 70, 50) = 3150$ minutes, that is equal to 52.5 hours, i.e., after more than two days!
Answer. 52.5 hours.

1.6 Special Numbers: Further Classification of Numbers

All natural numbers can be divided into two groups: prime numbers and composite numbers. A special interest in prime numbers is obvious because they are used to build all composite numbers. There are some particularly interesting prime numbers such as the prime twin numbers discussed earlier in this text. Then there are the Mersenne prime numbers and the Fermat prime numbers. There are some composite numbers that have been of historical note such as the amicable and perfect numbers. Finally, there are the figurate numbers (such as triangular, square, etc.) discovered in 2000 BC in the work of Babylonians which also amused the ancient Greek, Chinese, and Indian mathematicians. This section will reveal some of the properties of these fascinating numbers.

1.6.1 Divisors. Proper Divisors. Perfect and Amicable Numbers

Numbers and their divisibility fascinated people throughout time. The ancient mathematician Pythagoras believed in the magic power of numbers, studied them, and made great contributions to number theory. He, for example, introduced the so-called **proper divisors** of a number that are all divisors excluding the number itself. He factored $6 = 1 \cdot 2 \cdot 3$ and added the proper divisors of 6 (1, 2, and 3) to obtain $1 + 2 + 3 = 6$, the same number. Such a number he called a **perfect number**. Then he took another number, $12 = 1 \cdot 2 \cdot 2 \cdot 3$ and added together all its proper divisors as $1 + 2 + 3 + 4 + 6 = 16$, since $16 > 12$, he called such a number **abundant**. On the other hand, a number $10 = 1 \cdot 2 \cdot 5$ and the sum of its proper divisors is $1 + 2 + 5 = 8$ is less than 10, so this number was called a **deficient** number. Pythagoras was very interested in finding all possible perfect numbers and he found another one as $28 = 1 + 2 + 4 + 7 + 14$. Later, Euclid, in his 13th volume of *Elements* gave the formula to calculate all possible perfect numbers! Moreover, Euclid stated that for some values of the natural number, p under the condition that $2^p - 1$ is prime, there is a perfect number,

$$a_n = 2^{p-1} \cdot (2^p - 1). \tag{1.34}$$

The formula gives a perfect number only for special values of p. Thus we know that before the 15th century for almost 1500 years, there were only four perfect numbers known: $6, 28, 496$, and 8128. Only in the 15th century was the fifth perfect number $a_5 = 3,355,0336$, $p = 13$ found. It took almost two hundred years in order to find new perfect numbers when the French mathematician Marin Mersenne (1588–1648) proposed six new perfect numbers for $p = 17, 19, 31, 67, 127, 257$. Although

Euler (1707–1783) had proven that the formula was correct for the description of an even perfect number, only two perfect numbers of the Mersenne sequence were confirmed,

$$a_6 = 8,589,869,056 = 2^{16} \cdot (2^{17} - 1), \ p = 17$$

and

$$a_7 = 2^{18}(2^{19} - 1) = 137,438,691,328, \ p = 19.$$

The 8th Mersenne's number for $p = 31$ was evaluated as

$$a_8 = 2,305,843,008,139,952,128$$

however, without computers it was almost impossible to check other proposed Mersenne numbers and to establish their perfect properties. The number

$$M_p = (2^p - 1) \tag{1.35}$$

is called a Mersenne prime.

The French mathematician Edouard Lucas (1842–1891) found a criterion by which the number $(2^p - 1)$ is prime and also found that Mersenne correctly predicted a perfect number for $p = 127$, but made a mistake for $p = 67$ and $p = 257$. He also stated that there must be other perfect numbers for $p = 61, p = 89, p = 107$, which were missed by Mersenne. With modern computers, the search for prime Mersenne numbers is much easier now as well as the search for new perfect numbers. Nonetheless, there are only 49 perfect numbers found to the date. One of them is $2^{11,212} \cdot (2^{11,213} - 1)$ and has 3376 digits. The 49th perfect number is

$$2^{74,207,280} \cdot (2^{74,207,280} - 1)$$

and has $44,677,235$ digits!

Perfect numbers have very interesting properties. Let me offer you some problems related to proper divisors and perfect numbers.

Problem 44

Find the sum of all proper divisors for 10125.

Solution. Using prime factorization, our number can be written as $N = 10125 = 3^4 \cdot 5^3$. Instead of finding all its factors, let us consider all such numbers in the form $N = p^4 \cdot q^3$ and organize its factors as

$$\begin{array}{ccccc}
1, & p, & p^2, & p^3, & p^4 \\
q, & pq, & p^2q, & p^3q, & p^4q \\
q^2, & pq^2, & p^2q^2, & p^3q^2, & p^4q^2 \\
q^3, & pq^3, & p^2q^3, & p^3q^3, & \underbrace{p^4q^3}
\end{array}$$

where the last factor is the number N itself. Denote this sum by $\sigma(N)$. Let us add all these factors by adding the five columns and using the formula for the sum of a geometric progression,

$$
\begin{aligned}
\sigma(N) &= (1+q+q^2+q^3) + p(1+q+q^2+q^3) + ... + p^4(1+q+q^2+q^3) \\
&= (1+q+q^2+q^3) \cdot (1+p+p^2+p^3+p^4) \\
&= \frac{q^4-1}{q-1} \cdot \frac{p^5-1}{p-1}.
\end{aligned}
$$

The sum of all proper divisors, T, will be less than $\sigma(N)$ by the value of the given number,

$$
T = \frac{q^4-1}{q-1} \cdot \frac{p^5-1}{p-1} - p^4 q^3.
$$

Therefore, the sum of all proper divisor of 10125, $(q=5,\ p=3)$ is

$$
T = \frac{5^4-1}{5-1} \cdot \frac{3^5-1}{3-1} - 10125 = 18,876 - 10,125 = 8751.
$$

Answer. $T = 8751$.

Using the method of solution for Problem 44, we can evaluate the sum of the proper divisors for all natural numbers if we know their prime factorization. The following properties are valid.

Property 18

Let a natural number N be factored as follows:

$$
N = p_1^{\alpha_1} \cdot p_2^{\alpha_2} \cdot ... \cdot p_n^{\alpha_n},
$$

then the total number of its divisors and the sum of its divisors are

$$
\tau(N) = (\alpha_1+1) \cdot (\alpha_2+1) \cdot ... \cdot (\alpha_n+1) \tag{1.36}
$$

and

$$
\sigma(N) = \frac{p_1^{\alpha_1+1}-1}{p_1-1} \cdot \frac{p_2^{\alpha_2+1}-1}{p_2-1} \cdot ... \cdot \frac{p_n^{\alpha_n+1}-1}{p_n-1}, \tag{1.37}
$$

respectively.

Example. Consider $N = 6 = 2^1 \cdot 3^1$. We can see that this number has four divisors, $1, 2, 3, 6$. By formulas (1.36) and (1.37), $\tau(6) = (1+1)(1+1) = 4$ and the sum of all divisors is $\sigma(6) = \frac{2^2-1}{2-1} \cdot \frac{3^2-1}{3-1} = 3 \cdot 4 = 12$, that is $1+2+3+6 = 12$.

Both, $\tau(N)$ and $\sigma(N)$ are multiplicative functions in number theory.

Problem 45

Evaluate $\tau(\sigma(120))$.

Solution. First, we will evaluate

$$\sigma(120) = \sigma(2^3 \cdot 3 \cdot 5) = \frac{2^4 - 1}{2 - 1} \cdot \frac{3^2 - 1}{3 - 1} \cdot \frac{5^2 - 1}{5 - 1} = 2^3 \cdot 3^2 \cdot 5.$$

then

$$\tau(2^3 \cdot 3^2 \cdot 5) = (3 + 1)(2 + 1)(1 + 1) = 4 \cdot 3 \cdot 2 = 24$$

Answer. $\tau(\sigma(120)) = 24$.

Problem 46

Find a natural number, x, a multiple of 75, and with 505 total number of factors.

Solution. By the condition of the problem

$$\tau(x) = 505 = 5 \cdot 101,$$

which means that the prime factorization of x contains only two primes. Using formulas (1.36) and (1.37), we obtain
$x = p^\alpha \cdot q^\beta$ and $\tau(x) = (\alpha + 1)(\beta + 1) = 5 \cdot 101$. Hence $\alpha = 4$, $\beta = 100$.
    Thus, we have to solve the equation,

$$x = p^4 \cdot q^{100} = 75n = 5^2 \cdot 3 \cdot n$$

This case is possible if either

$$p = 5, q = 3, \ x = 5^4 \cdot 3^{100}$$

or

$$p = 3, q = 5, \ x = 3^4 \cdot 5^{100}$$

Answer. Either $x = 5^4 \cdot 3^{100}$ or $x = 3^4 \cdot 5^{100}$.

Problem 47

Given $\sigma(n) = 91$ and $\tau(n) = 9$. Find n.

Solution. Because $91 = 7 \cdot 13$, and $9 = 3 \cdot 3 = 1 \cdot 9$, the following factorization of n is possible:

$$n = p^{\alpha} \cdot q^{\beta}$$

Then

$$\tau(n) = (\alpha+1)(\beta+1) = 9, \Rightarrow \alpha = \beta = 2, \; n = p^2 \cdot q^2.$$

Next,

$$\sigma(n) = \frac{p^3 - 1}{p - 1} \cdot \frac{q^3 - 1}{q - 1} = 7 \cdot 13, \Rightarrow p = 2, \, q = 3.$$

Finally, $n = 2^2 \cdot 3^2 = 36$.
Answer. $n = 36$.

Problem 48

When a number is multiplied by 3 or by 13, the number of its divisors does not change. Find the number.

Solution. Assume that the number is x, then the problem is equivalent to solving the equation:

$$\tau(3x) = \tau(13x)$$

Let

$$x = p^{\alpha} \cdot q^{\beta},$$

then

$$3x = 3 \cdot p^{\alpha} \cdot q^{\beta}$$
$$\tau(3x) = 2 \cdot (\alpha+1) \cdot (\beta+1)$$

and if $p = 3$, then $\tau(3x) = (\alpha+2)(\beta+1)$

$$13x = 13 \cdot p^{\alpha} \cdot q^{\beta}$$
$$\tau(13x) = 2 \cdot (\alpha+1) \cdot (\beta+1)$$

and if $q = 13$, then $\tau(13x) = (\alpha+1) \cdot (\beta+2)$.
  Equating both formulas, we obtain that $\alpha = \beta$ and that

$$x = 3^{\alpha} \cdot 13^{\alpha}$$

so

$$3x = 3^{\alpha+1} \cdot 13^{\alpha}$$
$$13x = 3^{\alpha} \cdot 13^{\alpha+1}$$
$$\tau(3x) = \tau(13x) = (\alpha+2)(\alpha+1).$$

Answer. $x = 3^{\alpha} \cdot 13^{\alpha}$.

  Next, let us find the sum of all proper divisors.

Property 19

For any number $N = p^a \cdot q^b \cdot r^c$, the sum of all its proper divisors equals

$$T = \frac{p^{a+1} - 1}{p - 1} \cdot \frac{q^{b+1} - 1}{q - 1} \cdot \frac{r^{c+1} - 1}{c - 1} - p^a \cdot q^b \cdot r^c.$$

By the Pythagoras classification, $N = 10,125 < T = 8751$ so it is a deficient number. Can we find a perfect number? The formula for an even perfect number was derived by Euclid. Let us prove his formula!

Problem 49

Prove that any Euclidean even perfect number can be written as

$$N = 2^n \cdot (2^{n+1} - 1).$$

Proof. Since a Euclidean perfect number is even, it can be written as $N = p \cdot 2^n$, where its even part is contained in the second factor and p is prime. We can organize all factors of N as

$$1, \ 2, \ 2^2, \ ... \ 2^{n-1}, \quad 2^n$$
$$p, \ 2p, \ 2^2 p, \ ... 2^{n-1} p, \ 2^n p$$

Adding the factors by columns, we obtain

$$S = (1 + p) \cdot (1 + 2 + 4 + 8 + ... + 2^n) = (\frac{2^{n+1} - 1}{2 - 1})(1 + p) = (2^{n+1} - 1)(p + 1)$$

This sum S contains the number itself, hence, the sum of the proper divisors equals $T = S - N$ and must be equal to N. We have the equation to solve,

$$(2^{n+1} - 1)(p + 1) - N = N$$
$$(2^{n+1} - 1)(p + 1) = 2N$$
$$2^{n+1} + 2^{n+1} \cdot p - p - 1 = 2 \cdot 2^n \cdot p$$
$$2^{n+1} = 1 + p.$$

From which we obtain the formulas for p and N, respectively:

$$p = 2^{n+1} - 1, \ N = 2^n \cdot (2^{n+1} - 1).$$

which for $m = n + 1$ coincides with Euclid's formula.

Question. Are there any other forms of even perfect numbers? The answer, "no" was given by Euler. Let us follow Euler's proof by doing Problem 50.

> **Problem 50**
>
> Prove that there are no other forms of an even perfect numbers aside from the Euclidean form
> $$N = 2^{m-1} \cdot (2^m - 1).$$

Proof. We will prove this statement by contradiction. Assume that

$$N = 2^n \cdot r.$$

The prime factorization of N must contain some power of 2 because N is even. The second factor, r, by our assumption is not prime, so it is an odd number and hence it can be a product of different odd prime factors raised to some powers. Using similar arguments, we can rewrite the sum of all factors of N as

$$S = (2^{n+1} - 1) \cdot R,$$

where R represents the sum of all factors of r.

If N is a perfect number, then we have the chain of equations,

$$\begin{aligned}
(2^{n+1} - 1) \cdot R &= 2 \cdot N \\
(2^{n+1} - 1) \cdot R &= 2 \cdot 2^n \cdot r \\
2^{n+1} \cdot R - R &= 2^{n+1} \cdot r \\
2^{n+1} \cdot (R - r) &= R \\
2^{n+1} \cdot (R - r) &= (R - r) + r \\
(2^{n+1} - 1) \cdot (R - r) &= r.
\end{aligned}$$

In order that $(2^{n+1} - 1) \cdot (R - r) = r$ to have any integer solutions, the number r on the right-hand side must be divisible by R. However, R is the sum of all divisors of r. If r is not prime, but has, for example, a very simple form, such as $r = a \cdot b$, then the sum of all its divisors will be $R = 1 + a + b + ab$ and $R - r = 1 + a + b$. Clearly, $(R - r)$ must be equal to 1 and this happens only if r is prime of the form,

$$r = 2^{n+1} - 1.$$

Finally, by setting $m = n + 1$, we conclude that all even perfect numbers have the form given by Euclid,

$$N = 2^{m-1} \cdot (2^m - 1).$$

Pythagoras also introduced the so-called **amicable numbers**, such that each number of a pair equals the sum of the proper divisor of the other number and vice versa. Let us denote by $\sigma(a)$ the sum of all divisors of the number a and $\sigma(b)$ the sum of all divisors of b. Thus 220 and 284 are an amicable pair,

$$220 = 2^2 \cdot 5 \cdot 11; \sigma(220) = \frac{2^3-1}{2-1} \cdot (5+1) \cdot (11+1) = 7 \cdot 72 = 504; \ \sigma(220) - 220 = 284.$$

$$284 = 2^2 \cdot 71, \ \sigma(284) = \frac{2^3-1}{2-1} \cdot (71+1) = 504, \ \sigma(284) - 284 = 220.$$

For any amicable pair, (a,b) obviously, the following is true:

$$\sigma(a) - a = b$$
$$\sigma(b) - b = a$$

Hence, there is a relationship between the two amicable numbers that the sum of all divisors of either of two amicable numbers is the same and equals the sum of the pair,

$$\sigma(a) = \sigma(b) = a + b$$

1.6.2 Triangular and Square Numbers

The numbers $1, 3, 6, 10, 15, 21, 25, \ldots$ have a special name—they are **triangular numbers**. Ancient Greeks knew about these numbers and even gave them this name. The Greeks tried to solve problems geometrically. Imagine a triangle where each side is formed by n dots or n billiard balls. If we arrange four such triangular numbers as in Fig. 1.3, we can see how the number of the "balls" in each case denoted by $T(n)$ can be calculated. For example, we can add the balls by the rows.

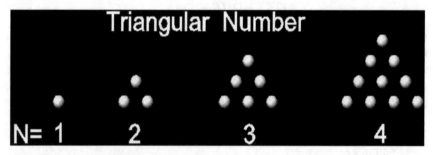

Fig. 1.3 Triangular Numbers

Let us solve Problem 51.

Problem 51

Find the formula for n^{th} term of a sequence

$$1,3,6,10,15,21,28,36,45,\ldots.$$

Solution. Notice that $1 = 1$, $3 = 1+2$, $6 = 1+2+3$, $10 = 1+2+3+4$, $15 = 1+2+3+4+5$, etc. By induction, the nth term is a sum of the first n natural numbers. Therefore, its formula can be found as $a_n = S_n = 1+2+3+\ldots+n = \frac{2\cdot1+\boxed{1}\cdot(n-1)}{2}\cdot n = \frac{n(n+1)}{2}$. It is easy to check that $a_7 = S_7 = 1+2+\ldots+7 = \frac{7\cdot8}{2} = 28$. We can evaluate the series using the formula for the sum of the arithmetic progression, emphasizing the value of the common difference.

Answer. $a_n = \frac{n(n+1)}{2}$.

Lemma 3

A triangular number can be evaluated as $T(n) = \frac{n(n+1)}{2}$

Proof. The way that each triangular number is constructed gives the formula,

$$T(n) = 1+2+3+\ldots+n = \frac{n(n+1)}{2}.$$

Definition. A **square number** or perfect square is the product of some integer with itself.

Square and triangular numbers are very connected. The ancient Greeks proved everything visually. For example, they noticed that each square number can be written as a sum of two consecutive triangular numbers. For example, $4 = 1+3$, $9 = 3+6$, $16 = 6+10$, etc. By playing with them we can find many different properties. For example, let us prove Lemma 4.

Lemma 4

The sum of two consecutive triangular numbers is a square number, i.e., $T(n-1)+T(n) = n^2$.

Proof. The picture or construction similar to our Figure 1.4 validates the statement.

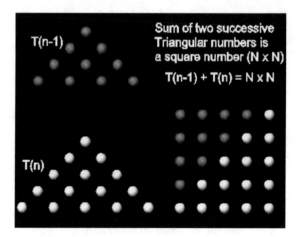

Fig. 1.4 Sum of two consecutive triangular numbers

This formula can be easily derived analytically as well. By adding numbers in each triangle by the rows, it is obvious that
$$T(1) = 1, T(2) = 1+2 = 3, T(3) = 1+2+3 = 6, T(4) = 1+2+3+4 = 10, \ldots.$$
Since each sum can be evaluated analytically, we have

$$T(n) = 1+2+3+\ldots+(n-1)+n = \frac{n(n+1)}{2}$$

and

$$T(n-1) = 1+2+3+\ldots+(n-1) = \frac{(n-1)n}{2}.$$

Therefore,

$$T(n-1)+T(n) = \frac{(n-1)n}{2} + \frac{n(n+1)}{2} = \frac{(n+1+n-1)n}{2} = n^2.$$

Remark. Additionally, problems on triangular and square numbers will be considered in Chapter 3 (Problems 181 and 182) and in Chapter 4 (Theorem 28).

1.7 Congruence and Divisibility

I think that it is time to introduce congruences to my reader. Although many number theory problems were stated circa 2000 BC in the era of Archimedes, the congruence notation was not introduced until 1801 in the work of the famous mathematician Carl Friedrich Gauss (1777–1855), only 24 year old at that time. His notation simplifies the proofs of many theorems. You will appreciate this soon. Congruences have a lot of applications and will be used in the following material. Our intention is to prepare you for solving a variety of number theory problems which can be sometimes difficult without knowledge of congruences.

We say that a is congruent to b modulo m (in symbols $a \equiv b \pmod{m}$) if and only if m divides $(a - b)$ assuming that $m > 0$. For example, $n \equiv 7 \pmod{8}$ means that a number n divided by 8 leaves the remainder of 7. This can be written as $n = 8k + 7$, which means that $n - 7 = 8k$. There are infinitely many numbers that divided by 8 leave a remainder of 7, such as $7, 15, 23, 31, 39, ..., 8k + 7, ...$ All these numbers are equivalent to each other under division by 8.

If we divide 15, 16, and 17 by 3 we can represent it as

$$15 = 3 \cdot 5 + 0 \Rightarrow 15 \equiv 0 \pmod{3},$$
$$16 = 3 \cdot 5 + 1 \Rightarrow 16 \equiv 1 \pmod{3},$$
$$17 = 3 \cdot 5 + 2 \Rightarrow 17 \equiv 2 \pmod{3}.$$

Theorem 2

Every integer is congruent $\pmod{m}$ to exactly one of: $0, 1, 2, \ldots (m - 1)$.

Theorem 2 has a simple explanation. As you remember, the entire set of natural numbers can be divided into m disjoint subsets with respect to their divisibility by m. Thus,

$$n \equiv 0 \pmod{m} \sim n = m \cdot k,$$
$$n \equiv 1 \pmod{m} \sim n = m \cdot k + 1,$$
$$n \equiv 2 \pmod{m} \sim n = m \cdot k + 2, \text{ etc.}$$

Theorem 3

$a \equiv b \pmod{m}$ if and only if a and b leave the same remainders on division by m.

Proof. Let us prove it by contradiction. Suppose that $a = mk_1 + r_1$ and $b = mk_2 + r_2$, where $r_1 \neq r_2$. Subtracting a and b we obtain

$$a - b = m(k_1 - k_2) + (r_1 - r_2). \tag{1.38}$$

Since $r_1 \neq r_2$, we have that (1.38) contradicts the condition of the problem, $a - b = km$. Therefore, a and b must leave the same remainders when divided by m.

Lemma 5

For integers a, b, c, and d:

 i $a \equiv a \pmod{m}$,
 ii if $a \equiv b \pmod{m}$, then $b \equiv a \pmod{m}$,
 iii if $a \equiv b \pmod{m}$ and $b \equiv c \pmod{m}$, then $a \equiv c \pmod{m}$,
 iv if $a \equiv b \pmod{m}$ and $c \equiv d \pmod{m}$, then $ac \equiv bd \pmod{m}$,
 v if $a \equiv b \pmod{m}$ and $c \equiv d \pmod{m}$, then $(a+c) \equiv (b+d) \pmod{m}$,
 vi if $a \equiv b \pmod{m}$, then $a^k \equiv b^k \pmod{m}$.

Many other important properties of congruence will be emphasized throughout the book. You can always obtain a correct congruence statement if you multiply both sides of it by the same factor or if you raise both sides to a power. In fact, you can either go through the procedure of consecutive squaring or doing other proper simplifications. Please read more about it in Section 3.3.

Problem 52

Find the remainder of 3^{89} divided by 7.

Solution.
Method 1. Notice that because $27 = -1 + 7 \cdot 4$, the following is true:

$$3^3 = 27 \equiv -1 \pmod{7}.$$

Since $89 = 84 + 5 = 3 \cdot 28 + 5$, we can raise both sides of the congruence to the 28th power and then multiply it with the congruence for the fifth power of 3,

$$3^{84} \equiv 1 \pmod{7}$$
$$3^2 = 9 \equiv 2 \pmod{7}$$
$$3^3 \equiv -1 \pmod{7}.$$

Finally we obtain
$$3^{89} \equiv -2 \pmod{7} \equiv 5 \pmod{7}.$$

We can state that 3^{89} divided by 7 leaves the remainder of 5.
   Method 2. By doing consecutive squaring,

$$3 \equiv 3 \quad (\mathrm{mod}\ 7)$$
$$3^2 = 9 \equiv 2 \quad (\mathrm{mod}\ 7)$$
$$3^4 \equiv 4 \quad (\mathrm{mod}\ 7)$$
$$3^8 \equiv 16 \quad (\mathrm{mod}\ 7) \equiv 2 \quad (\mathrm{mod}\ 7)$$
$$3^{16} \equiv 4 \quad (\mathrm{mod}\ 7)$$
$$3^{32} \equiv 16 \equiv 2 \quad (\mathrm{mod}\ 7)$$
$$3^{64} \equiv 4 \quad (\mathrm{mod}\ 7).$$

Now the power of 89 can be written as

$$89 = 64 + 16 + 8 + 1 = 2^6 + 2^4 + 2^3 + 1$$

and

$$3^{89} = 3^{64} \cdot 3^{16} \cdot 3^8 \cdot 3 \equiv 4 \cdot 4 \cdot 2 \cdot 3 \equiv 5 \quad (\mathrm{mod}\ 7).$$

We can see that the first method was faster because we found a power of 3 that leaves a remainder of 1 when divided by 7.
Answer. 5.

Problem 53

Prove that $2^{20} - 1$ is divisible by 41.

Proof. Let us start from the obvious true congruence,

$$2^5 = 32 \equiv -9 \quad (\mathrm{mod}\ 41)$$

Raising both sides of this congruence to the fourth power, we obtain

$$2^{20} \equiv 9^4 = 81 \cdot 81 \quad (\mathrm{mod}\ 41)$$

Because $81 \equiv (-1) \ (\mathrm{mod}\ 41)$, then $81 \cdot 81 \ (\mathrm{mod}\ 41) \equiv 1 \ (\mathrm{mod}\ 41)$. Then

$$2^{20} \equiv 1 \quad (\mathrm{mod}\ 41)$$
$$2^{20} - 1 \equiv 0 \quad (\mathrm{mod}\ 41)$$

The last congruence indicates that $2^{20} - 1$ is divisible by 41.

Theorem 4 (Chinese Remainder Theorem)

The system of congruences $x \equiv a_i \ (\mathrm{mod}\ m_i)$, $i = 1,2,3,\ldots,k$, where $(m_i, m_j) = 1$ and $i \neq j$ has a unique solution modulo $m_1 \cdot m_2 \cdots \cdots m_k$.

Here are other important congruence statements:

> **Lemma 6**
>
> If $a \equiv b \pmod{m}$ and d divides m then $a \equiv b \pmod{d}$.

Example. If $a \equiv b \pmod{1001}$, then $a \equiv b \pmod{7}$ because 7 divides 1001.

$$a = 101n + b = 7 \cdot 143 \cdot n + b = 7k + b \iff a \equiv b \pmod{7}.$$

> **Lemma 7**
>
> If $a \equiv b \pmod{r}$ and $a \equiv b \pmod{s}$, then
>
> $$a \equiv b \pmod{[s,r]},$$
>
> where $[s,r]$ is the least common multiple of s and r.

For example, if $a \equiv b \pmod{7}$, $a \equiv b \pmod{13}$, and $a \equiv b \pmod{11}$, then $a \equiv b \pmod{1001}$ ($lcm = 1001 = 7 \cdot 11 \cdot 13$).

Proof. If $a \equiv b \pmod{r}$, then

$$a = r \cdot k + b, \tag{1.39}$$

and if $a \equiv b \pmod{s}$ then

$$a = s \cdot l + b. \tag{1.40}$$

Combining (1.39) and (1.40) we obtain $r \cdot k = s \cdot l$ and

$$k = sm, \; l = kp. \tag{1.41}$$

Substituting (1.41) into (1.39) and (1.40) we finally have $a = r \cdot sm + b$ or

$$a \equiv b \pmod{r \cdot s}.$$

Hence, $a \equiv b \pmod{[r,s]}$. The proof is complete.

> **Problem 54** (Divisibility by 11)
>
> Prove divisibility by 11 using the congruence, 11 divides N if and only if 11 divides the alternating sum of the digits of N.

Proof. Since $10 \equiv -1 \pmod{11}$ then for any power r, we obtain

$$10^r \equiv (-1)^r \pmod{11}.$$

Then for number N we have

$$(a_n 10^n + a_{n-1} 10^{n-1} + \cdots + a_2 10^2 + a_1 10^1 + a_0) \equiv$$
$$(a_n(-1)^n + a_{n-1}(-1)^{n-1} + \cdots + a_2(-1)^2 + a_1(-1)^1 + a_0) \pmod{11}.$$

Even powers of -1 will become one and odd become negative one. The proof is complete. Thus, $37,594,953$ is divisible by 11 because $3 - 7 + 5 - 9 + 4 - 9 + 5 - 3 = -11$ is divisible by 11.

Problem 55

Show that every prime (except 2) is congruent to 1 or 3 $\pmod{4}$.

Solution. Any integer when divided by 4 leaves a remainder of 0 (multiples of 4), 1, 2, or 3. All numbers with remainder 2 are even ($4k + 2 = 2m$) and cannot be primes. Therefore, every prime is congruent to 1 or 3 $\pmod{4}$.

Congruence notation can be used for checking the calculation (multiplication) of large numbers. Consider a decimal notation of a natural number N,

$$N = a_n \cdot 10^n + a_{n-1} \cdot 10^{n-1} + \ldots + a_1 \cdot 10 + a_0 = (a_n a_{n-1} \ldots a_0)_{10}$$

and the sum of its digits

$$S_N = a_n + a_{n-1} + \ldots + a_0$$

Because each power of 10 divided by 9 leaves a remainder of 1,

$$1 \equiv 1 \pmod{9}$$
$$10 \equiv 1 \pmod{9}$$
$$10^2 \equiv 1 \pmod{9}$$
$$\ldots$$
$$10^n \equiv 1 \pmod{9}.$$

Hence,

$$N \equiv S_N \pmod{9}.$$

If we have two numbers N and M and S_N and S_M are the sum of their digits, then Lemma 8 is valid.

Lemma 8

$$S_N \pm S_M \equiv N \pm M \equiv S_{N\pm M} \pmod 9$$
$$S_N \cdot S_M \equiv N \cdot M \equiv S_{N \cdot M} \pmod 9$$

(1.42)

Let us apply this lemma to Problems 56 and 57.

Problem 56

Maria had to multiply $17,953$ and 5847 without a calculator and obtained $114,971,191$. Her teacher marked the problem wrong without multiplying the numbers. Please explain.

Solution. We know that a number divided by 9 leaves the same remainder as the sum of its digits divided by 9. If the calculations are correct, the remainders of the left and right sides must be the same. Using lemma 8 we obtain for the left-hand side:

$$N = 17953 \equiv 1+7+9+5+3 = 25 \equiv 7 \pmod 9$$
$$M = 5847 \equiv 5+8+4+7 = 24 \equiv 6 \pmod 9$$
$$N \cdot M \equiv 42 \equiv 6 \pmod 9$$

Now let us check the sum of digits of Maria's result.

$$114,971,191 \equiv 1+1+4+9+7+1+1+9+1 \equiv 34 \equiv 7 \pmod 9$$

Since $6 \neq 7$, her calculation was wrong.

Problem 57

In the multiplication $31,415 \cdot 92,653 = 2,910,X93,995$, one digit in the product is missing and replaced by X. Find the missing digit without doing multiplication.

Solution. On one hand, we have

$$(3+1+4+1+5)(9+2+6+5+3) \equiv (14 \cdot 25) \equiv (5 \cdot 7 = 35) \equiv 8 \pmod 9.$$

On the other hand, we see

$$(2+9+1+X+9+3+9+9+5 = (47+X) \equiv (2+X) \pmod 9.$$

Answer. $X = 6$.

Problem 58

What is the remainder when

$$1! + 2! + 3! + 4! + \ldots + 2017!$$

is divided by 12?

Solution. Starting from 4! each term of the finite series is divisible by 12, then we need to find the remainder of only the sum of the first three terms:

$$1! + 2! + 3! = 1 + 2 + 6 = 9.$$

Answer. 9 (mod 12).

Let us solve using congruence the following problem on proof.

Problem 59

Prove that $9^{2n+1} + 8^{n+2}$ is always divisible by 73.

Proof. Using properties of exponent, let us rewrite the given expression as

$$\begin{aligned}
9 \cdot 9^{2n} + 64 \cdot 8^n &= 9 \cdot 81^n + (73 - 9) \cdot 8^n \\
&= 9 \cdot (73 + 8)^n + (73 - 9) \cdot 8^n \\
&\equiv 9 \cdot 8^n - 9 \cdot 8^n \equiv 0 \pmod{73}.
\end{aligned}$$

The last row shows that the given expression is divisible by 73.

Problem 60

The three digit numbers abc, def, ghi represented in base 10 are divisible by 19. Prove that the determinant of 3×3 matrix

$$A_{3 \times 3} = \begin{pmatrix} a\ b\ c \\ d\ e\ f \\ g\ h\ i \end{pmatrix}$$

is always divisible by 19.

Proof. Denote by $\det A$ the determinant of matrix A. The given multiples of 19 can be written as follows:

$$abd = 100a + 10b + c = 19m$$
$$def = 100d + 10e + f = 19k$$
$$ghi = 100g + 10h + i = 19l$$

Using the properties of determinants, let us first multiply the first column of matrix A by 100 and the second column by 10, which will create a new matrix, B,

$$B = \begin{pmatrix} 100a & 10b & c \\ 100d & 10e & f \\ 100g & 10h & i \end{pmatrix}.$$

The determinants of matrices A and B are connected by the relationship,

$$\det B = 1000 \cdot \det A.$$

Next, we will make a new matrix, C, $C = B^T$, transpose of matrix B,

$$C = \begin{pmatrix} 100a & 100d & 100g \\ 10b & 10e & 10h \\ c & f & i \end{pmatrix}$$

By the properties of determinants,

$$\det C = \det B^T = \det B = 1000 \cdot \det A.$$

Let us take a close look at matrix C. If we replace its second row by the sum of the second row and the third row ($R_2 \rightarrow R_2 + R_3$) and then replace the first row by the sum of the new second row and the first row ($R_1 \rightarrow R_2 + R_1$), we will obtain a new matrix D as follows:

$$D = \begin{pmatrix} 100a + 10b + c & 100d + 10e + f & 100g + 10h + i \\ 10b + c & 10e + f & 10h + i \\ c & f & i \end{pmatrix}$$
$$= \begin{pmatrix} abc & def & ghi \\ bc & ef & hi \\ c & f & i \end{pmatrix}$$
$$= \begin{pmatrix} 19m & 19k & 19l \\ bc & ef & hi \\ c & f & i \end{pmatrix}$$

It follows from the properties of determinants that the determinant of matrix D equals determinant of matrix C. Additionally, because each number of the first row of matrix D is a multiple of 19, then its determinant is also divisible by 19. This can be written as

$$\det D = 19n = 1000 \cdot \det A.$$

Since the entries of matrix A are natural numbers, its determinant is also a natural number. Because 19 is prime and 1000 is not divisible by 19, then in order that this equation have an integer solution, the determinant of matrix A must satisfy the equation,

$$\det A = 19p, \ p \in \mathbb{N}.$$

This completes our proof.

Problem 61

Are there any natural numbers N, with 2017 digits, such that the number

$$M = N^N + (N+1)^{N+1}$$

is a composite number?

Solution. Let us consider even numbers in the form $N = 6n+4$. When such a number is divided by 3 it will leave a remainder of 1, so as its Nth power, which can be written as

$$N \equiv 1 \pmod{3}.$$

Obviously, another number $N+1$ divided by 3 would leave a remainder of 2 or -1,

$$N+1 \equiv -1 \pmod{3}.$$

Next, if N is even, then $N+1$ is odd and then the following is true:

$$N^N \equiv 1 \pmod{3}$$
$$(N+1)^{N+1} \equiv -1 \pmod{3}$$
$$M = N^N + (N+1)^{N+1} \equiv 0 \pmod{3}.$$

Therefore M is a multiple of 3 and it is not 3, and hence it is a composite number for $N = 6n+4$. There are infinite number of 2017 digit numbers of this type.
Answer. $N = 6n+4$.

Chapter 2
Further Study of Integers

People have been intrigued with problems involving numbers for as long as there is recorded history, although the intensity of the interest has ebbed and flowed. After great results obtained by the ancient Greeks, Indians, Babylonians, and Chinese, there was a period of relative quiet in the study of number theory until interest was refreshed by the works of Pierre de Fermat. Then again, the study of number theory declined at the end of the 17th and the beginning of the 18th centuries as there was more attention on the newly developed differential and integral calculus. But the work of Leonard Euler attracted many mathematicians back to number theory. In his work, *Theory of Comparisons*, Chebyshev described Euler's influence in number theory:

> Euler laid the beginning of all the investigations that make up the general part of the theory of numbers; In these investigations, Euler was preceded by Fermat; he first began to study the properties of numbers with respect to their ability to satisfy indefinite equations of one kind or another and the result of his research was the discovery of many general theorems of number theory. Euler proved many of the results of his predecessors in particular Fermat's Little Theorem. He successfully applying the means of arithmetic and algebra and method of infinite descent, created new analytical methods in number theory, and set a number of important tasks. Euler's contribution to the theory of numbers is so great that it is not possible to mention here even his main results set out in more than one hundred publications.

In this chapter, the reader will learn about the Euclidean algorithm and its application to reducing fractions, finding the greatest common divisor of two expressions in integers, representing a number in different bases, and representing a number by finite or infinite continued fractions. In addition, the theory of continued fractions, representation of irrational numbers by nonperiodic continued fractions, quadratic irrationalities and the Lagrange Theorem will be explored. The reader will learn how to determine the last digits of a number raised to a power using several methods including the use of Fermat's Little Theorem and Euler's formula. Continuing with the study of special numbers started in Chapter 1, we will find the formula for Catalan numbers. And finally, the chapter concludes with an introduction to basic methods of proof.

© Springer International Publishing AG, part of Springer Nature 2018   63
E. Grigorieva, *Methods of Solving Number Theory Problems*,
https://doi.org/10.1007/978-3-319-90915-8_2

2.1 Euclidean Algorithm

When two numbers a and b are less than 100 it is easy to find their greatest common divisor, $d = (a,b)$. Difficulties appear when a and b are comparatively big numbers, like 76,084 and 63,020. Let us find their *gcd*. For this we need to know the Euclidean algorithm, one of the basic methods of elementary number theory:

Assuming that $a \geq b$, we divide a by b with respect to the least positive remainder

$$a = q_1 b + r_1, \ \ 0 \leq r_1 < b,$$

next we divide b by r_1 and obtain some remainder r_2

$$b = q_2 r_1 + r_2, \ \ 0 \leq r_2 < r_1.$$

Then we divide r_1 by r_2 obtaining a new remainder r_3

$$r_1 = q_3 r_2 + r_3, \ \ 0 \leq r_3 < r_2.$$

Continue this procedure until a decreasing sequence of positive integer remainders, $r_1, r_2, r_3, ..., r_n$ arrives at a division for which $r_{n+1} = 0$.

$$r_2 = q_4 r_3 + r_4$$
$$r_3 = q_5 r_4 + r_5$$
$$.............................$$
$$r_{n-2} = q_n r_{n-1} + r_n$$
$$r_{n-1} = q_{n+1} r_n + 0.$$

From the last expression we obtain that the greatest common divisor of a and b is r_n, the last nonzero remainder.

2.1.1 Finding the gcd, lcm and Reducing Fractions

In order to practice the Euclidean algorithm let us try to solve Problem 62.

Problem 62

What is the greatest common divisor of 40,300 and 54,990?

Solution. We find that $54,990 > 40,300$, then $a = 54,990$ and $b = 40,300$. Hence,

$$54990 = 1 \cdot 40300 + 14690$$
$$40300 = 2 \cdot 14690 + 10920$$
$$14690 = 1 \cdot 10920 + 3770$$
$$10920 = 2 \cdot 3770 + 3380$$
$$3770 = 1 \cdot 3380 + 390 \tag{2.1}$$
$$3380 = 8 \cdot 390 + 260$$
$$390 = 1 \cdot 260 + 130$$
$$260 = 2 \cdot 130 + 0.$$

From the last row of (2.1), we conclude that $d = (54990, 40300) = 130$, the last nonzero remainder.

Answer. $d = 130$.

Obviously, using prime factorization, the given numbers can be factored as $54,990 = 130 \cdot 423$ and $40,300 = 130 \cdot 310$, which gives us the same answer $gcd = 130$. What if you need to find the greatest common divisor of two variable expressions, e.g., $(7n+1, 15n+4)$ or $(n^4 - 3 \cdot n^2 + 2, n^2 + 3)$?

The Euclidean algorithm will help us solve such problems as well. Consider division with a remainder. The following formulas are true:

$$a = b \cdot q + r$$
$$(a,b) = (b,a) = (a,r) = (b,r) \tag{2.2}$$
$$(a,b) = (a-b,b) = (a,b-a).$$

Let us examine this repeated subtraction procedure by solving the following problems.

Problem 63

What is the greatest common divisor of $(30n+2)$ and $(12n+1)$?

Solution.
Method 1. Let $a = 30n+2$, $b = 12n+1$, and $d = (a,b)$. Applying repeated subtractions by formula (2.2), we obtain the chain of operations,

$$(a,b) = (30n+2, 12n+1)$$
$$= (a-b,b)$$
$$= (30n+2-(12n+1), 12n+1)$$
$$= (18n+1, 12n+1)$$
$$= (6n, 12n+1)$$
$$= (12n+1, 6n)$$
$$= (6n+1, 6n)$$
$$= 1.$$

Therefore the given numbers are relatively prime and do not have any common divisor. You can see that our search of the *gcd* was reduced to finding the common divisor of two consecutive integers, $(6n+1)$ and $(6n)$ which are always relatively prime.

Method 2. Instead of repeated subtraction, we can simply divide the bigger number by the smaller number and apply the Euclidean algorithm,

$$30n+2 = (12n+1)\cdot 2+6n$$
$$12n+1 = 6n\cdot 2+1$$
$$r = 1,\ d = 1.$$

Answer. $d = 1$.

Try to use the second method (Euclidean algorithm) in solving the Problem 64.

Problem 64

Find the greatest common divisor of $(6n^4+n^2+3n)$ and $(2n^3+1)$.

Solution. Dividing two polynomial expressions and recording remainders we obtain

$$6n^4+n^2+3n = (2n^3+1)\cdot(3n)+n^2$$
$$2n^3+1 = n^2\cdot(2n)+1$$
$$n^2 = 1\cdot n^2+0.$$

The last nonzero remainder is 1. Therefore the two polynomial expressions do not have any common factors for any integer value of n.

Answer. $d = 1$.

Problem 65

Let m and d be the *lcm* and *gcd* of integer numbers x and y, respectively. Find the minimal value of $\frac{m}{d}$ if x and y satisfy the equation,

$$8x = 29+3y.$$

Solution. By the definition of the least common multiple and greatest common divisor,

$$m = \frac{xy}{d},$$

then the requested expression is

$$\frac{m}{d} = \frac{xy}{d^2} \tag{2.3}$$

and we need to minimize it subject to $8x - 3y = 29$.

Let us solve this equation. Notice that $x_0 = 1$ and $y_0 = -7$ are the solutions to this equation. Hence, all other possible integer solutions can be written as

$$x = 3n + 1, \ y = 8n - 7, \ n \in \mathbb{Z}.$$

The method of solving linear equations like this will be explained in more detail in Chapter 3. Let us find the greatest common divisor, $d = (x, y)$. Using consecutive subtractions, we obtain the chain of equations,

$$d = (8n - 7, 3n + 1) = (5n - 8, 3n + 1) = (2n - 9, 3n + 1) = (3n + 1, 2n - 9)$$
$$= (n + 10, 2n - 9) = (n + 10, n - 19)$$
$$= (29, n + 10).$$

Formula (2.3) can be rewritten now as

$$\frac{m}{d} = \frac{(8n - 7)(3n + 1)}{(29, n + 10)^2} \tag{2.4}$$

We have to consider two cases:

Case 1. If $n = 19$, then $d = (x, y) = 29$ and

Case 2. If $n \neq 19$, then $d = (x, y) = 1$.

Considering the numerator of the fraction in (2.4), and because it is quadratic function with positive leading coefficient, the minimal value of the numerator in this case will be at $n = 1$. Next, we will see which n, $n = 1$ or $n = 19$ would give us the smaller values of $\frac{l}{d}$ in (2.4).

Case 1. $d = 29$, $n = 19$, $\frac{(8 \cdot 19 - 7)(3 \cdot 19 + 1)}{29^2} = \frac{254}{29} > 4$

Case 2. $d = 1$, $n = 1$, $x = 4$, $y = 1$, $\frac{(8 \cdot 1 - 7)(3 \cdot 1 + 1)}{1} = 4$.

Answer. min $\frac{m}{d} = 4$.

In many problems, the *gcd* of two expressions is not 1 so these cases are more interesting. Often in contest problems instead of finding the greatest common divisor, you are asked to reduce a fraction. It is not difficult to work with numerical examples, such as

$$\frac{66}{12} = \frac{6 \cdot 11}{6 \cdot 2} = \frac{11}{2},$$

where we reduced $66/12$ to $11/2$ by canceling 6, the greatest common divisor of the numerator and denominator. Similarly, we will use the Euclidean algorithm in order to find the *gcd* of the numerator and denominator depending on n and reduce the fraction.

Problem 66

Reduce the fraction $\frac{16n + 60}{11n + 41}$.

Solution. Dividing two expressions and recording the remainders, we obtain

$$16n + 60 = (11n + 41) \cdot 1 + (5n + 19)$$
$$11n + 41 = (5n + 19) \cdot 2 + (n + 3)$$
$$5n + 19 = (n + 3) \cdot 5 + 4$$
$$n + 3 = 4 \cdot k + r$$
$$(n + 3, 4) = ?$$

How can we find $((n + 3), 4)$? Clearly, the answer will depend on n. Assume that

$$n + 3 = 4 \cdot k + r, \ r = 0, 1, 2, 3$$

and examine the possible remainders.

Case 1. Let $r = 0$, then $n + 3 = 4 \cdot k$ which means that $n + 3$ is divisible by 4 and that the last nonzero remainder of the division algorithm above equals 4. Then

$$n = 4 \cdot k - 3$$

This also can be rewritten as $n = 4t + 1$. Substituting, for example, the first formula into the fraction for n, we will be able to reduce it as follows:

$$\frac{16(4k - 3) + 60}{11(4k - 3) + 41} = \frac{64k + 12}{44k + 8} = \frac{4 \cdot (16k + 3)}{4 \cdot (11k + 2)} = \frac{16k + 3}{11k + 2}$$

For example, $n = 13 = 4 \cdot 4 - 3$, $k = 4$, then we can reduce the fraction as follows:

$$\frac{260}{184} = \frac{16 \cdot 13 + 60}{11 \cdot 13 + 41} = \frac{65}{46}.$$

Case 2. Let $r = 1$, then $n + 3 = 4 \cdot k + 1$ or $n = 4k - 2$, and 1 is the last nonzero remainder of the division algorithm,

$$16n + 60 = (11n + 41) \cdot 1 + (5n + 19)$$
$$11n + 41 = (5n + 19) \cdot 2 + (n + 3)$$
$$5n + 19 = (n + 3) \cdot 5 + 4$$
$$n + 3 = 4 \cdot k + 1$$
$$4 = 1 \cdot 4 + 0$$

In this case the numerator and denominator are relatively prime and the fraction $\frac{16n+60}{11n+41}$ cannot be reduced.

Case 3. Let $r = 2$, then $n + 3 = 4 \cdot k + 2$ or $n = 4k - 1$. In this case the last two rows of the division algorithm can be written as

$$n+3 = 4 \cdot k + 2$$
$$4 = 2 \cdot 2 + 0$$
$$(4k+2, 4) = 2.$$

Thus with the substitution $n = 4k - 1$, both fractions will have the common factor 2 and can be reduced as

$$\frac{16(4k-1)+60}{11(4k-1)+41} = \frac{64k+44}{44k+30} = \frac{2 \cdot (32k+22)}{2 \cdot (22k+15)} = \frac{32k+22}{22k+15}.$$

Finally, we need to investigate the last case.

Case 4. Let $r = 3$, then $n+3 = 4 \cdot k + 3$ or $n = 4k$. Consider the last two steps of the algorithm:

$$n+3 = 4 \cdot k + 3$$
$$4 = 3 \cdot 1 + 1$$
$$3 = 1 \cdot 3 + 0.$$

The last nonzero remainder of the Euclidean algorithm is 1, hence at $n = 4k$ the fraction $\frac{16n+60}{11n+41}$ is not reducible.

Answer. If $n = 4k - 3$, then $\frac{16n+60}{11n+41} = \frac{16k+3}{11k+2}$; if $n = 4k - 1$, then $\frac{16n+60}{11n+41} = \frac{32k+22}{22k+15}$.

Problem 67

Find the greatest common divisor of $(9n + 17)$ and $(7n + 11)$.

Solution. Applying formula (2.2) several times, we obtain

$$(9n+17, 7n+11) = (2n+6, 7n+11) = (5n+5, 2n+6)$$
$$= (3n-1, 2n+6) = (n-7, 2n+6) = (n+13, n-7) = (20, n-7).$$

Let $n = 20k + r$, $r = 0, 1, 2, ..., 19$. Substituting all possible remainders, we obtain that the remainder, $r = 7$. Hence, $n = 20k + 7$ and the greatest common divisor is 20. $(20, 20k) = 20$.

$$(9(20k+7)+17, 7(20k+7)+11) = (180+80, 140+60) = (260, 200) = 20.$$

Answer. $gcd = (9n+17, 7n+11) = 20$, $n = 20k + 7$.

Problem 68

Find all natural numbers n, for which the fraction $\frac{3n^3-8n^2+14n-8}{3n-5}$ a) can be reduced and b) is an integer.

Solution.

a) The given fraction can be rewritten as

$$q(n) = \frac{3n^3 - 8n^2 + 14n - 8}{3n - 5} = n^2 - n + \frac{9n - 8}{3n - 5}$$

Next, let us find the greatest common divisor of $(9n - 8)$ and $(3n - 5)$. Applying formula (2.2) several times, we obtain

$$\begin{aligned}(9n - 8, 3n - 5) &= (6n - 3, 3n - 5) \\ &= (3n + 2, 3n - 5) \\ &= (7, 3n - 5).\end{aligned}$$

Let $n = 7k + r$, $r = 0, 1, ..., 6$. and by substituting all possible remainders, we obtain that $r = 4$. Hence, if $n = 7k + 4$, the greatest common divisor of $(7, 3n - 5)$ is 7. At this value of n the given fraction $q(n)$ can be first reduced to

$$n^2 - n + 3 + \frac{7}{3n - 5}$$

and further to

$$(7k + 4)^2 - (7k + 4) + 3 + \frac{7}{3 \cdot (7k + 4) - 5} = 49k^2 + 49k + 15 + \frac{1}{3k + 1}.$$

b) In order for the given fraction to be an integer, let us divide the numerator $9n - 8$ by the denominator $3n - 5$.

$$9n - 8 = (3n - 5) \cdot 3 + 7;$$
$$\tfrac{9n-8}{3n-5} = 3 + \tfrac{7}{3n-5}.$$

From the last row above, we can see that condition b) holds in two cases, if $3n - 5 = 7$, $n = 4$ and if $3n - 5 = 1$, $n = 2$. The given fraction will be equal to 16 and to 12, respectively.

Answer. a) $49k^2 + 49k + 15 + \frac{1}{3k+1}$, $n = 7k + 4$, b) if $n = 4$, $q(4) = 16$, and if $n = 2$, $q(2) = 12$.

2.1.2 Other Bases

Ancient people represented numbers in a different base system than base 10. For example, ancient Babylonians used base 60 and now we are aware that ancient Egyptians knew that any natural number can be written as the sum of powers of 2, so they used system base 2. Moreover, the Egyptians knew that the representation of a number by some powers of 2 is unique.

Let us consider several numbers written in base 10 and decompose them into the sum of powers of 2, then rewrite them with the new base of 2:

$$17 = 2^0 + 2^4 = 10001_2$$

$$34 = 2^5 + 2^1 = 100010_2$$

$$119 = 2^0 + 118 = 2^0 + 2^6 + 54 = 2^0 + 2^6 + 2^5 + 22$$
$$= 2^0 + 2^6 + 2^5 + 2^4 + 6 = 2^6 + 2^5 + 2^4 + 2^2 + 2^1 + 2^0 = 1101110001_2.$$

In base 2 any number ending in 1 is odd and those ending in 0 is even. Next we will state Lemma 9.

Lemma 9

Every natural number can be written uniquely in the base $b \geq 2$.

$$n = d_0 + d_1 \cdot b^1 + d_2 \cdot b^2 + ... + d_k \cdot b^k$$
$$0 \leq d \leq b, \quad i = 0, 1, 2, ..., k.$$

Proof. The proof of this lemma is almost trivial. Assume the contradiction and that the same number can be written differently in the same base b.

$$n = s_0 + s_1 \cdot b^1 + s_2 \cdot b^2 + ... + s_k \cdot b^k$$
$$0 \leq s \leq b, \quad i = 0, 1, 2, ..., k.$$

Subtracting the left sides and the right sides of two representations, we will obtain

$$0 = (d_0 - s_0) + (d_1 - s_1) \cdot b^1 + (d_2 - s_2) \cdot b^2 + ... + (d_k - s_k) \cdot b^k$$

From the equation above it follows that the representations are the same and that

$$d_i - s_i = 0, \quad i = 0, 1, 2, ..., k.$$

In this text we will assume that any integer written without a subscript is in base 10. Let us do a little exercise.

Example. Rewrite 111_7 in base ten.

$$111_7 = 1 + 1 \cdot 7 + 1 \cdot 7^2 = 57_{10} = 57.$$

In order to find the representation of a base 10 integer in base b, we will use the Euclidean algorithm. Numbers in base 2 are used in computers. However, in order to read long numbers easily, we can use base 8 instead, because $8 = 2^3$. Let us see how some numbers can be represented in base 2 and then in base 8.

$$0 = 000, \quad 4 = 100,$$
$$1 = 001, \quad 5 = 101,$$
$$2 = 010, \quad 6 = 110,$$
$$3 = 011, \quad 7 = 111$$

$$1961 = 8 \cdot 245 + \boxed{1}$$
$$245 = 8 \cdot 30 + \boxed{5}$$
$$30 = 8 \cdot 3 + \boxed{6}$$
$$3 = 8 \cdot 0 + \boxed{3}$$
$$1961 = (011, 110, 101, 001) = (3, 6, 5, 1)_8$$

Let us rewrite 3241 in base 8 and then in base 60 like the Babylonians did. We will divide 3241 by 8 and record the remainder, then divide the quotient by 8 and again record the remainder, and so on until we obtain zero quotient.

$$3241 = (d_k \dots d_0)_8$$
$$3241 = 8 \cdot \underline{405} + \boxed{1}$$
$$405 = 8 \cdot \underline{50} + \boxed{5}$$
$$50 = 8 \cdot \underline{6} + \boxed{2}$$
$$6 = 8 \cdot \underline{0} + \boxed{6}$$

The number 3241 in base 8 will be

$$6251_8 = 1 \cdot 8^0 + 5 \cdot 8^1 + 2 \cdot 8^2 + 6 \cdot 8^3.$$

Similarly, in base 60,

$$3241 = 60 \cdot \underline{54} + \boxed{1}$$
$$54 = 60 \cdot \underline{0} + \boxed{54}$$
$$3241 = (54, 1)_{60} = 54 \cdot 60 + 1$$

and the same number in base 7,

$$3241 = 7 \cdot 463 + \boxed{0}$$
$$464 = 7 \cdot 66 + \boxed{1}$$
$$66 = 7 \cdot 9 + \boxed{3}$$
$$9 = 7 \cdot 1 + \boxed{2}$$
$$1 = 7 \cdot 0 + \boxed{1},$$

which can be written as

$$0 \cdot 7^0 + 1 \cdot 7^1 + 3 \cdot 7^2 + 2 \cdot 7^3 + 1 \cdot 7^4 = (1, 2, 3, 1, 0)_7.$$

And finally, in base 15,

$$3241 = 15 \cdot 216 + \boxed{1}$$
$$216 = 15 \cdot 14 + \boxed{6}$$
$$14 = 15 \cdot 0 + \boxed{14}.$$

Hence,

$$1 \cdot 15^0 + 6 \cdot 15^1 + 14 \cdot 15^2 = (14, 6, 1)_{15}.$$

I hope that you noticed that while the decimal representation of 3241 does not end in zero, its representation in base 7 ends in 0 and is an even number in this base. This should not be a surprise for you because the number is a multiple of 7. Please check that $3241 = 7 \cdot 463$. Additionally, if you add all the digits of 3241, you would obtain $3 + 2 + 4 + 1 = 10$, an even number. Does it mean that if the sum of the digits of a number in base ten is even number, then its representation in any odd base would be even? Is it true that if a number is divisible by 7, then it will end in zero in base 7? What if it is divisible by 7^n, will it end in n zeros? What if instead of N in decimal representation, we want to find the number of zeros of $N!$ in base d? Can we use Legendre's formula that we derived and used in Chapter 1? We will check our observations and answer the questions in solving or proving the next problems.

Example. Consider 1065. You can see that its decimal representation does not have zero in the end. However, this number is divisible by 15,

$$1065 = 15 \cdot 71.$$

Using the Euclidean algorithm, we can see that in base 15 it does end in zero, i.e.,

$$1065 = 15 \cdot 71 + 0$$
$$71 = 15 \cdot 4 + 11$$
$$4 = 15 \cdot 0 + 4$$
$$1065_{10} = (4, 11, 0)_{15}.$$

This happens because 1065 is not divisible by 2, just by 5. As you recall from Chapter 1, each ending zero can be obtained only as a product of 2 and 5. Additionally, please, notice that 1065 is an odd number in base ten and even number in base 15. This is not true for all odd numbers but just for such that the sum of digits of is an even number. In fact, $1 + 0 + 6 + 5 = 12$. The general case is proven in Problem 73.

This example and Problem 69 will help us develop Lemma 10.

Problem 69

How many zeros ending in 1065! written in base 15?

Solution. As we have learned, each product of 3 and 5 for a number written in base 10 will give us a zero in base 15. Since multiples of 3 appear in $n!$ more often than multiples of 5, the number of all zeros, k, can be calculated as

$$k = \left[\frac{1065}{5}\right] + \left[\frac{1065}{5^2}\right] + \left[\frac{1065}{5^3}\right] + \left[\frac{1065}{5^4}\right] = 213 + 42 + 8 + 1 = 264.$$

In this case, 1065! will have the same number of zeros in base 10.
Answer. 264 ending zeros.

The following statement is true.

Lemma 10

A number $N = n!$ written in base 15 has the same number of ending zeros
as this number written in base 10.

Proof. The number of ending zeros is determined by factors of 5 only in both bases.

Problem 70

3241! written in base 7 will end in how many zeros?

Solution. We have to add all factors of 3241 that are divisible by $7, 7^2, 7^3...$,

$$n = \left[\frac{3241}{7}\right] + \left[\frac{3241}{7^2}\right] + \left[\frac{3241}{7^3}\right] + \left[\frac{3241}{7^4}\right] = 463 + 66 + 9 + 1 = 539.$$

Answer. 3241! ends in 539 zeros.

Problem 71

Prime factorization of some number in base 10 is $N = a^n b^m$. Find the number of zeros of N written in base a, b, and ab.

Solution.

1. A number N will have n zeros in base a.
2. A number N will have m zeros in base b
3. It will have $p = \min(n, m)$ zeros in base ab.

Problem 72 (Dudley)

In which bases b, is $N = 1111_b$ divisible by 5?

Solution. Expanding 1111_b we obtain

$$N = 1 + b + b^2 + b^3 = (b+1)(b^2+1)$$

Assume that N is divisible by 5, then

$$(b+1)(b^2+1) = 5n.$$

Substituting for b with $b = 5k$, $b = 5k+1$, $b = 5k+2$, $b = 5k+3$ and $b = 5k+4$, we can see that only in the last three cases is the left side of the equation divisible by 5. For example,

$$b = 5k+2$$
$$(b+1)(b^2+1) = (5k+3)((5k+2)^2+1)$$
$$= (5k+3)(25k^2 + 10k + 5)$$
$$= 5n.$$

Similar results can be obtained for $b = 5k+3$ and $b = 5k+4$. For example, if $b = 13$ then $b = 5k+3$ and the number

$$1111_{13} = 1 + 1 \cdot 13 + 1 \cdot 13^2 + 1 \cdot 13^3 = 2380_{10} = 5n.$$

Answer. $b = 5k+2$, $b = 5k+3$, $b = 5k+4$.

When a number is written in a base different from base ten, i.e., in base 7, it is sometimes difficult to decide if the number is even or odd. We will establish a new rule by solving the Problem 73.

Problem 73

Show that an integer in any odd base is even, if and only if the sum of its digits is even. For example, 6217_7 is an even number and 5217_7 is an odd number.

Proof. Denote the number by N and assume that it is written in an odd base $b = 2n+1$,

$$N = d_0 + (d_1 \cdot b + d_2 \cdot b^2 + d_3 \cdot b^3 + \ldots + d_k \cdot b^k).$$

It was shown earlier in the book that an odd number raised to any integer power is an odd number. (For example, $7^2 = 49$ or $3^3 = 27, \ldots$) Let us subtract and add 1 to each term on the right-hand side, inside the parentheses:

$$N = d_0 + d_1(b - 1 + 1) + d_2 \cdot (b^2 - 1 + 1) + d_3 \cdot (b^3 - 1 + 1) + \ldots + d_k \cdot (b^k - 1 + 1)$$
$$= (d_1(b-1) + d_2 \cdot (b^2 - 1) + d_3 \cdot (b^3 - 1) + \ldots + d_k \cdot (b^k - 1)) + [d_0 + d_1 + d_2 + \ldots + d_k]$$

Since $b = 2n + 1$ is an odd number, then $b^k - 1 = 2m$ is an even number for any power of k, then the expression inside parentheses is always even, and then if the sum inside brackets is also even, then the entire number N will be even. Otherwise, it will be odd. Therefore, if

$$d_0 + d_1 + d_2 + \ldots + d_k = 2s \implies N = 2n.$$

For the same reason 6213_7 is even number because $6 + 2 + 1 + 3 = 12 = 2s$. Check $6213_7 = 3 + 1 \cdot 7 + 2 \cdot 7^2 + 6 \cdot 7^3 = 2166$. True.

> **Problem 74**
>
> Assume that you can have four weights to measure items. What weights should you choose if you must be able to weight all possible items with an integer weight from one to forty pounds $(1, 2, 3, 4, \ldots 39, 40$ lbs$)$?

Solution. Let us demonstrate that it is enough to have the weights in 1, 3, 9 and 27 pounds.

$$40 = 27 + 9 + 3 + 1 = 1 \cdot 3^0 + 1 \cdot 3^1 + 1 \cdot 3^2 + 1 \cdot 3^3 = 1111_3$$
$$39 = 27 + 9 + 3 = 0 \cdot 3^0 + 1 \cdot 3^1 + 1 \cdot 3^2 + 1 \cdot 3^3 = 1110_3$$
$$\ldots$$
$$30 = 27 + 3 = 1010_3$$
$$\ldots$$
$$✠ + 3 + 1 = 9$$
$$\ldots$$

For example, in order to weigh 5 lbs denoted by symbol ✠, we can put on the left this item and two weights of 1 and 3 lbs, then on the right side we will put 9 pounds, etc.

Answer. We need four weights of 1, 3, 9 and 27 pounds.

> **Problem 75** (Alfutova, Shestopal)
>
> We want to find a counterfeit coin among 12 coins. Find the solution if:
> a) It is known that a fake coin is either heavier or lighter than the other coins. Using only three weighings on a two-cup scale, find the counterfeit coin.
> b) It is unknown if the fake coin is lighter or heavier. Using only three weighings, find the fake coin and establish whether it is lighter or heavier than a real one.

Solution.
a) Without loss of generality we can assume that a counterfeit coin is lighter than other coins. Let us divide the coins into two groups and place six of them on either

scale bowl. If, for example, the left bowl is heavier, then the false coin is on the right, among those six. Now take those six coins of the light group and divide them into groups of three and again place on the scale. The counterfeit coin will be in the lighter group of three. Finally, take any two coins from that group and place each one in a separate bowl. If they are equal, then the remaining coin is the counterfeit coin. If one of the two coins is heavier, then the lighter coin is the false coin.

b) When we do not know anything about the weight of the counterfeit coin, the above procedure would not be appropriate.

Method 1. For convenience, we number the coins from 1 to 12. By the first weighing, let's compare two groups of four coins: 1, 2, 3, 4 and 5, 6, 7, 8.

Case 1. The first weighing showed equality in the two groups. If the scales show equality, then the counterfeit coin is among the remaining four coins. Then for the second weighing we will compare three coins, 9, 10, 11, with the obviously real coins 1, 2, 3. If this time the scales show equality again, then the counterfeit coin is number 12, and for the third weighing we will compare it with a real coin and find out whether it is lighter or heavier. If three coins 9, 10, 11 were lighter (heavier), then for the third weighing we will compare the coins 9 and 10 with each other. If they are equal, then coin 11 is fake, and it is lighter (heavier) than a real coin. Otherwise, we conclude that out of the two coins, 9 and 10, that one is fake, which is lighter (heavier) than the other.

Case 2. The first weighing showed an inequality. Now suppose that the first weighing showed that coins 1, 2, 3, 4 are heavier than 5, 6, 7, 8. The case when the first coins were lighter is symmetrical. In the second weighing, we put coins 1, 2, 5 on one cup of the scales, and coins 3, 4, 9 on the other (coin 9 is obviously real). If the second weighing showed equality, then we have three coins 6, 7, 8, one, of which is lighter than the rest. By the third weighing we compare coins 6 and 7. If they are equal, then the coin 8 is lighter than the rest. Otherwise, the one that is lighter than the other is fake.

Now, suppose that in the second weighing, the coins 1, 2, 5 were heavier than 3, 4, 9. This means that the counterfeit coin is among coins 1 and 2, and it is heavier than the rest. Comparing these two coins with each other in the third weighing, we will determine the counterfeit. Alternatively, suppose that in the second weighing, the coins 1, 2, 5 were lighter than 3, 4, 9. This means that either coin 5 is lighter than the others, or one of the coins 3 and 4 is heavier than the rest. The third weighing will compare the coins 3 and 4 with each other and find the answer.

Method 2. This is called Dyson's method and it is based on a representation of a number in base 3. Before we put any coins on the scale, let us first renumber them as follows:

$$001, \ 010, \ 011, \ 012, \ 112, \ 120, \ 121, \ 122, \ 200, \ 201, \ 202, \ 220.$$

For the first weighing, let's put on one scale those coins in which the highest digit is zero $(001, 010, 011, 012)$ and on the other a scale those coins where it is equal to two $(200, 201, 202, 220)$. If the cup with zero is overtaken, we will write the number zero on the piece of paper. If the cup with 2 is overtaken, we will write the number 2. If the cup with two weights will remain in equilibrium, we write one.

For the second weighing, on one bowl, we put coins with the second digit equal to zero, i.e., $001, 200, 201, 202$, and on the other bowl, those coins whose middle digit is two—$120, 121, 122, 220$. We will write the result of weighing in such a way that it was in the first weighing procedure. By the third weighing, we compare coins $010, 020, 200, 220$ with coins $012, 112, 122, 202$ (coins with the zeros and 2 as the last digit, respectively), and after weighting we write the third digit.

We got three figures, in other words, a three-digit number. If this number coincides with the number of some coin, then this coin is counterfeit and heavier than the rest, if not, then we replace all the zeros in this number by 2, and all the 2 by zeros, then it must coincide with the number of some coin and this coin is counterfeit and lighter than other coins.

2.1.3 Continued Fractions and Euclidean Algorithm

In this section we will learn how to rewrite any number as a finite or infinite continued fraction. Many math competitions offer a problem, such as

Determine the real number represented by the expression,

$$1 + \cfrac{1}{1 + \cfrac{1}{1 + \cfrac{1}{1 + \cfrac{1}{1 + \cdots}}}}. \tag{2.5}$$

The trick is to observe that if we take the reciprocal of this expression and add 1, then we get the same expression back. Denoting the unknown expression by x, then

$$x = 1 + \frac{1}{x}.$$

This equation can be easily solved as

$$x^2 - x - 1 = 0, \ x > 0$$
$$x = \frac{1+\sqrt{5}}{2}.$$

The positive root of a quadratic equation is a so-called **golden ratio**. This number was admired by the Ancient Babylonians and Greeks and plays a very important role in mathematics and art. The expression given by (2.5) is an example of an infinite continued fraction. These objects provide us with an alternative to decimal expansion for the approximation of irrational numbers and have very important number-theoretic applications.

2.1.3.1 Finite Continued Fractions

All rational numbers can be represented by **finite continued fractions**. For example, $5/3$ can be written as

$$\frac{5}{3} = 1 + \frac{2}{3} = 1 + \frac{1}{\left(\frac{3}{2}\right)} = 1 + \frac{1}{1 + \frac{1}{2}}.$$

We first extract the largest integer (1) and rewrite a number as sum of this integer and a fraction (2/3), then we rewrite this fraction as 1 divided by reciprocal of this fraction (3/2) and again decompose it as a number (1) and a fraction (1/2). We are done.

Alternatively, we could use the Euclidean algorithm,

$$5 = 3 \cdot \underline{1} + 2$$
$$3 = 2 \cdot \underline{1} + 1$$
$$2 = 1 \cdot \underline{2} + 0$$

Recording all underlined quotients (on the right), we can represent $5/3$ as a finite continued fraction (unique representation obtained above).

Next, consider the fractions: $\frac{33}{78}$, $\frac{87}{55}$ and $\frac{107}{37}$.

$$\frac{33}{78} = \frac{1}{\frac{78}{33}} = \frac{1}{2 + \frac{12}{33}} = \frac{1}{2 + \frac{1}{\frac{33}{12}}}$$

$$= \frac{1}{2 + \frac{1}{2 + \frac{9}{12}}} = \frac{1}{2 + \frac{1}{2 + \frac{1}{1 + \frac{1}{3}}}}$$

$$\frac{87}{55} = 1 + \frac{32}{55} = 1 + \frac{1}{\frac{55}{32}} = 1 + \frac{1}{1 + \frac{23}{32}}$$

$$= 1 + \frac{1}{1 + \frac{1}{\frac{32}{23}}} = 1 + \frac{1}{1 + \frac{1}{1 + \frac{9}{23}}}$$

$$= 1 + \frac{1}{1 + \frac{1}{1 + \frac{1}{2 + \frac{5}{9}}}}$$

$$= 1 + \frac{1}{1 + \frac{1}{1 + \frac{1}{2 + \frac{1}{1 + \frac{4}{5}}}}}$$

$$= 1 + \frac{1}{1 + \frac{1}{1 + \frac{1}{2 + \frac{1}{1 + \frac{1}{1 + \frac{1}{4}}}}}}.$$

If we evaluate $87/55$ on a calculator, we will get that

$$\frac{87}{55} = 1.5\overline{81} = 1.5818181...$$

Let us consider consecutive convergent fractions for this number,

$$\frac{P_0}{Q_0} = \frac{1}{1} < \frac{87}{55}$$

$$\frac{P_1}{Q_1} = \frac{2}{1} > \frac{87}{55}$$

$$\frac{P_2}{Q_2} = 1 + \frac{1}{1+1} = \frac{3}{2} < \frac{87}{55}$$

$$\frac{P_3}{Q_3} = 1 + \frac{1}{1+\frac{1}{1+\frac{1}{2}}} = \frac{8}{5} = 1.6 > \frac{87}{55}$$

$$\frac{P_4}{Q_4} = 1 + \frac{1}{1+\frac{1}{1+\frac{1}{2+1}}} = \frac{11}{7} \approx 1.5714 < \frac{87}{55}$$

$$\frac{P_5}{Q_5} = 1 + \frac{1}{1+\frac{1}{1+\frac{1}{2+\frac{1}{1+1}}}} = \frac{19}{12} \approx 1.5833 > \frac{87}{55}$$

$$\frac{P_6}{Q_6} = \frac{87}{55}.$$

Each such a fraction gives us an approximate value of $\frac{87}{55}$ that is either greater than the actual value or less than it. By moving from the first fraction to the last one, it happens in alternation but this approximation becomes more or more precise, finally getting exactly $\frac{87}{55}$ at the last step.

In general, for a reduced fraction $\frac{m}{a}$ we consider a sequence of the continued fractions

$$\frac{P_0}{Q_0}, \frac{P_1}{Q_1}, \frac{P_2}{Q_2}, \dots, \frac{P_{s-1}}{Q_{s-1}}, \frac{P_s}{Q_s} = \frac{m}{a}, \quad (m,a) = 1$$

for which

$$P_{s-1}Q_s - P_sQ_{s-1} = (-1)^s.$$

Thus, for fraction $\frac{87}{55}$ we obtain that

$$P_5Q_6 - P_6Q_5 = (-1)^6 = 1,$$

or

$$19 \cdot 55 - 87 \cdot 12 = 1045 - 1044 = 1.$$

This can be used to solve a linear equation, e.g.,

$$55x + 87y = 1.$$

Take the initial solutions $x_0 = 19$ and $y_0 = -12$ from

$$55 \cdot 19 + 87 \cdot (-12) = 1045 - 1044 = 1,$$

and then it can be shown that a general solution to the linear equation has the form, $x = 19 + t$, $y = -12 + t$. Next, looking at the fraction $\frac{107}{37} \approx 2.891891892$,

$$\frac{107}{37} = 2 + \frac{33}{37} = 2 + \frac{1}{\frac{37}{33}} = 2 + \frac{1}{1 + \frac{4}{33}} = 2 + \frac{1}{1 + \frac{1}{\frac{33}{4}}} = 2 + \frac{1}{1 + \frac{1}{8 + \frac{1}{4}}}.$$

Here,

$$\frac{P_0}{Q_0} = \frac{2}{1} < \frac{107}{37},$$
$$\frac{P_1}{Q_1} = 2 + \frac{1}{1} = \frac{3}{1} > \frac{107}{37},$$
$$\frac{P_2}{Q_2} = 2 + \frac{1}{1 + \frac{1}{8}} = \frac{26}{9} \approx 2.88888... < \frac{107}{37},$$
$$\frac{P_3}{Q_3} = \frac{107}{37},$$

and

$$P_2 Q_3 - P_3 Q_2 = (-1)^3 = -1,$$

or

$$26 \cdot 37 - 107 \cdot 9 = 962 - 963 = -1,$$

from which we obtain a particular solution to the linear equation

$$37x - 107y = -1,$$

such as $x_0 = 26$, $y_0 = 9$.

We know that $\frac{87}{55}$ is a **rational number**. The proof of this fact also follows from the Euclidean algorithm. If we apply this to 87 and 55, we eventually stop because of a zero remainder.

$$87 = 55 \cdot \underline{1} + 32$$
$$55 = 32 \cdot \underline{1} + 23$$
$$32 = 23 \cdot \underline{1} + 9$$
$$23 = 9 \cdot \underline{2} + 5$$
$$9 = 5 \cdot \underline{1} + 4$$
$$5 = 4 \cdot \underline{1} + 1$$
$$4 = 1 \cdot \underline{4} + 0.$$

By listing all the quotients underlined above, we can rewrite a fraction of $\frac{87}{55}$ as

$$1 + \cfrac{1}{1 + \cfrac{1}{1 + \cfrac{1}{2 + \cfrac{1}{1 + \cfrac{1}{1 + \frac{1}{4}}}}}},$$

or as

$$\frac{87}{55} = [1, 1, 1, 2, 1, 1, 4]$$

we have

$$\alpha_0 = \tfrac{87}{55} \approx 1.58$$
$$\alpha_1 = \tfrac{55}{32} \approx 1.72$$
$$\alpha_2 = \tfrac{32}{23} \approx 1.39$$
$$\alpha_3 = \tfrac{23}{9} \approx 2.56$$
$$\alpha_4 = \tfrac{9}{5} = 1.8$$
$$\alpha_5 = \tfrac{5}{4} = 1.25$$
$$\alpha_6 = \tfrac{4}{4} = 1 \text{ (an integer!)}.$$

Let us show that rewriting a fraction $\frac{a}{b}$ as a continued fraction can be explained by the Euclidean algorithm. Assuming that $a \geq b$, we will divide a by b with respect to the least positive remainder,

$$a = q_1 b + r_1, \ 0 \leq r_1 < b, \qquad \alpha_0 = \frac{a}{b} = q_1 + \frac{r_1}{b}$$

$$b = q_2 r_1 + r_2, \ 0 \leq r_2 < r_1, \quad \alpha_1 = \frac{b}{r_1} = q_2 + \frac{r_2}{r_1}.$$

Then divide r_1 by r_2 obtaining a new remainder r_3 and

$$r_1 = q_3 r_2 + r_3, \ 0 \leq r_3 < r_2, \quad \alpha_2 = \frac{r_1}{r_2} = q_3 + \frac{r_3}{r_2}.$$

After some steps, we will have

$$r_{n-1} = q_{n+1} r_n + r_{n+1}, \ 0 \leq r_{n+1} < r_n, \quad \alpha_n = \frac{r_{n-1}}{r_n} = q_{n+1} + \frac{r_{n+1}}{r_n}.$$

Therefore, we obtained the following continued fraction:

$$\frac{a}{b} = q_1 + \cfrac{1}{q_2 + \cfrac{1}{q_3 + \cfrac{1}{q_4 + \cdots + \frac{1}{q_n}}}}.$$

Denoting by $[\alpha_k]$ the greatest integer $\leq \alpha_k$. Since,

$$q_{k+1} = [\alpha_k] \Rightarrow \alpha_k = [\alpha_k] + \frac{1}{\alpha_{k+1}},$$

we obtain that

$$\alpha_{k+1} = \frac{1}{\alpha_k - [\alpha_k]}. \tag{2.6}$$

Problem 76

Solve in natural numbers

$$x + \cfrac{1}{y + \frac{1}{z}} = \frac{10}{7}.$$

Solution. Rewriting 10/7 as a finite continued fraction we obtain

$$\frac{10}{7} = 1 + \frac{3}{7} = 1 + \frac{1}{\frac{7}{3}} = 1 + \frac{1}{2 + \frac{1}{3}}.$$

Therefore, $x = 1$, $y = 2$, $z = 3$.
Answer. $(x, y, z) = (1, 2, 3)$.

Finite continued fractions have an application in **physics**. For example, the theory of circuit construction is based on two laws:

1. If we connect resistors in series, then the total resistance will be the sum of all resistances:

$$R = R_1 + R_2 + \ldots + R_n.$$

2. If we connect the same resistors in parallel, the total resistance is given by formula

$$\frac{1}{R} = \frac{1}{R_1} + \frac{1}{R_2} + \ldots + \frac{1}{R_n}$$

or

$$R = \frac{1}{\frac{1}{R_1} + \frac{1}{R_2} + \ldots + \frac{1}{R_n}}.$$

Using only unit resistors, we can construct an electrical circuit with a given resistance, for example, with the total resistance of 19/15. First, let's connect three resistors in parallel and add one resistor to this block, we will obtain the electrical circuit with resistance 4/3.

$$1 + \frac{1}{1 + 1 + 1} = 1 + \frac{1}{3} = \frac{4}{3}.$$

Now, if we connect this obtained circuit in parallel with the block that consists of three unit resistors connected in parallel, we will get a circuit with common resistance of 4 /15.

$$\frac{1}{\frac{1}{\frac{1}{3}} + \frac{1}{\frac{4}{3}}} = \frac{1}{3 + \frac{3}{4}} = \frac{4}{15}$$

Next, we will connect to this circuit one single unit resistor,

$$1 + \frac{4}{15} = \frac{19}{15} = 1 + \frac{1}{3 + \frac{1}{1 + \frac{1}{3}}}.$$

So, we have constructed a circuit of eight single resistors, the total resistance of which is 19/15.

2.1.3.2 Infinite Continued Fraction Representation of Irrational Numbers

We can use formula (2.6) for representing irrational numbers as infinite continued fractions. Moreover, for each irrational number such representation is unique. Let us demonstrate that $\sqrt{11}$ is an irrational number,

$$\alpha_0 = \frac{\sqrt{11}}{1}, \quad [\alpha_0] = 3, \quad \alpha_0 = 3 + \frac{1}{\alpha_1}$$

$$\alpha_1 = \frac{1}{\alpha_0 - [\alpha_0]} = \frac{1}{\sqrt{11} - 3} = \frac{\sqrt{11} + 3}{2}, \quad [\alpha_1] = 3, \quad \alpha_1 = 3 + \frac{1}{\alpha_2}$$

$$\alpha_2 = \frac{1}{\alpha_1 - [\alpha_1]} = \frac{1}{\frac{\sqrt{11}+3}{2} - 3} = \frac{2}{\sqrt{11} - 3} = \sqrt{11} + 3, \quad [\alpha_2] = 6, \quad \alpha_2 = 6 + \frac{1}{\alpha_3}$$

$$\alpha_3 = \frac{1}{\alpha_2 - [\alpha_2]} = \frac{1}{\sqrt{11} + 3 - 6} = \frac{1}{\sqrt{11} - 3} = \frac{\sqrt{11} + 3}{2} = \alpha_1.$$

Since

$$\alpha_{k+1} = \frac{1}{\alpha_k - [\alpha_k]}$$

and $\alpha_1 = \alpha_3$, then

$$\alpha_2 = \alpha_4, \quad \alpha_3 = \alpha_5, \text{ etc.}$$

However, $\alpha_1 = \frac{\sqrt{11}+3}{2}$ is irrational, then $\sqrt{11}$ cannot be written as convergent continued fraction, only as periodic continued infinite fraction as

$$\sqrt{11} = [3, \overline{363}]$$

$$\sqrt{11} = 3 + \cfrac{1}{3 + \cfrac{1}{6 + \cfrac{1}{3 + \dots}}}.$$

We can show that the right-hand side equals $\sqrt{11}$. Let $\sqrt{11} = N$. From the right-hand side we can see that $N = x + 3$, where

$$x = \frac{1}{3 + \frac{1}{6+x}}$$
$$x = \frac{6+x}{3x+19}$$
$$3x^2 + 19x = x + 6$$
$$x^2 + 6x - 2 = 0$$
$$x = -3 + \sqrt{11}.$$

Finally, $N = \sqrt{11}$.

The representation above can be used for approximation of an irrational number by a closest rational number. Similar to what we did for the approximation of $87/55$, we can get a finer approximation of the actual value of $\sqrt{11}$ by cutting some terms of the infinite fraction. Using a calculator we can see that

$$\sqrt{11} \approx 3.31662479$$

Let us see how to consecutively improve the approximations for this number,

$$\frac{P_0}{Q_0} = 3 < \sqrt{11}$$

$$\frac{P_1}{Q_1} = 3 + \frac{1}{3} = 3.\overline{3} > \sqrt{11}$$

$$\frac{P_2}{Q_2} = 3 + \frac{1}{3 + \frac{1}{6}} = \frac{63}{19} \approx 3.315789474 < \sqrt{11}$$

$$\frac{P_3}{Q_3} = 3 + \frac{1}{3 + \frac{1}{6 + \frac{1}{3}}} = \frac{199}{60} \approx 3.31666.. > \sqrt{11}.$$

By doing this we could get the integer part of the number, its first, second, third, and more digits precisely. We can see that the last approximation allowed us to get a very good approximation for $\sqrt{11}$ by getting a number with the correct four digits after the decimal point. This method was very popular when calculators and computer software did not exist. Cutting an infinite fraction at certain number of terms we find a rational approximation of a given irrational number with the necessary precision. Moreover, if we substitute the numerator and denominator of each such a fraction into the expression,

$$x^2 - 11 \cdot y^2,$$

we will obtain

$$3^2 - 11 \cdot 1^2 = -2$$
$$10^2 - 11 \cdot 3^2 = 1$$
$$63^2 - 11 \cdot 19^2 = -21$$
$$199^2 - 11 \cdot 60^2 = 1...$$

and would see that each such pair is a solution to the Pell type equation $x^2 - 11 \cdot y^2 = \pm M$ that will be discussed in Chapter 3.

2.2 Representations of Irrational Numbers by Infinite Fractions

Definition. An infinite continued fraction is the expression of type,

$$a_0 + \cfrac{1}{a_1 + \cfrac{1}{a_2 + \cfrac{1}{\ddots + \cfrac{1}{a_s + \cfrac{1}{\ddots}}}}} = [a_0; a_1, a_2, ..., a_s...], \qquad (2.7)$$

where $a_0 \in Z, a_i \in N, i = \overline{1,\infty}$.

Definition. A fraction $\frac{P_n}{Q_n}$ approaching an infinite fraction (2.7) is called **convergent** if it can be written as a finite continued fraction

$$\frac{P_n}{Q_n} = [a_0; a_1, a_2, ..., a_n]. \qquad (2.8)$$

Definition. An infinite fraction (2.7) is called a convergent continued fraction if the limit of its approaching fractions exists, i.e., $\lim\limits_{n\to\infty} \frac{P_n}{Q_n} = \alpha$.

Theorem 5

Any infinite continued fraction is convergent.

Theorem 6

Let $\alpha = [a_0; a_1, a_2, .., a_n, \alpha_{n+1}]$, α_{n+1} be full quotients in the decomposition of α, then

$$\alpha = \frac{P_n \cdot \alpha_{n+1} + P_{n-1}}{Q_n \cdot \alpha_{n+1} + Q_{n-1}},$$
$$\alpha_{n+1} = \frac{P_{n-1} - \alpha Q_{n-1}}{\alpha Q_n - P_n}. \qquad (2.9)$$

Next, we consider the decomposition of an irrational number into continued fraction.

Definition. Decomposition of a real number α into a continued fraction $\alpha = [a_0; a_1, a_2, ..., a_n, ...]$, where $a_0, a_1, a_2, ...$ is a finite or infinite sequence of natural numbers such that for $k \geq 1$ all $a_k \geq 1$. In case of a finite fraction, the last element of the decomposition is greater than one ($a_k > 1$).

Theorem 7

Any real number can be represented by a continued fraction and its decomposition is unique.

Proof. Consider the case when α is irrational number, because we have proven earlier this for a rational number. Denote by a_0 the integer part α, and by α_1, the reciprocal of its fractional part, α, that is $\alpha_1 = \frac{1}{\alpha - \alpha_0}$, so that $\alpha = a_0 + \frac{1}{\alpha_1}$. Because α is irrational, $a_0 \neq \alpha$ and α_1 is also irrational number, such that $\alpha_1 > 1$. In this way for α_1 the numbers $a_1 = [\alpha_1]$ and $\alpha_2 > 1$, etc., we obtain,

$$\alpha = a_0 + \frac{1}{\alpha_1}, a_0 = [\alpha],$$
$$\alpha_1 = a_1 + \frac{1}{\alpha_2}, a_1 = [\alpha_1],$$
$$\dots \tag{2.10}$$
$$\alpha_n = a_n + \frac{1}{\alpha_{n+1}}, a_n = [\alpha_n],$$
$$\dots$$

where for $n \geq 1$ all irrational numbers are greater than unity ($\alpha_n > 1$) and hence for all such numbers n the numbers $a_n = [\alpha_n] \geq 1$. The numbers $a_0, a_1, a_2, \dots$ form an infinite sequence of natural numbers. Because for all $n \geq 1$ the inequality $\alpha_n \geq 1$ is valid, we can take these numbers as elements to make an infinite continued fraction $[a_0; a_1, a_2, \dots]$, which by Theorem 5 converges.

Let us prove that the value of this continued fraction equals our original number α. Indeed, from the inequality (2.10) we obtain that $\alpha = [a_0; a_1, a_2, \dots, a_n, \alpha_{n+1}]$, then in regard to (2.9) we obtain

$$\alpha = \frac{P_n \cdot \alpha_{n+1} + P_{n-1}}{Q_n \cdot \alpha_{n+1} + Q_{n-1}}$$

and

$$\left| \alpha - \frac{P_n}{Q_n} \right| = \left| \frac{P_n \alpha_{n+1} + P_{n-1}}{Q_n \alpha_{n+1} + Q_{n-1}} - \frac{P_n}{Q_n} \right| = \frac{1}{(Q_n \alpha_{n+1} + Q_{n-1})Q_n} < \frac{1}{Q_n^2 \alpha_{n+1}} < \frac{1}{Q_n^2}.$$

Since $Q_n \to \infty$, then $\lim_{n \to \infty} \frac{P_n}{Q_n} = \alpha$. The proof of the uniqueness of this decomposition can be done using contradiction.

Therefore, for a given irrational number α we have an algorithm that allows us to create an infinite continued fraction equal to α.

Problem 77

Find the irrational number that can be written as $\alpha = [(1, 3)]$.

Solution. $\alpha = [1, 3, \alpha] = 1 + \frac{1}{3 + \frac{1}{\alpha}} = 1 + \frac{\alpha}{3\alpha + 1} = \frac{4\alpha + 1}{3\alpha + 1}$, from which we obtain a quadratic equation $3\alpha^2 - 3\alpha - 1 = 0$. Find its roots $\alpha_{1,2} = \frac{3 \pm \sqrt{21}}{6}$. Since $\alpha > 1$, we will take only a positive root of a quadratic equation $\alpha = \frac{3 + \sqrt{21}}{6}$.

Answer. $\alpha = \frac{3 + \sqrt{21}}{6}$.

2.2.1 Quadratic Irrationalities

Among all irrational numbers, the numbers obtained as solutions of quadratic equations with integer coefficients are considered to be the simplest irrational numbers.

Definition. The number α is called a quadratic irrationality if α is an irrational root of a quadratic equation,

$$ax^2 + bx + c = 0 \tag{2.11}$$

with integer, not simultaneously zero, coefficients. At this α, we would have $a \neq 0, c \neq 0$. Coefficients of (2.11), can be taken, as relatively prime and in this case the discriminant of the equation $D = b^2 - 4ac$ will be also called the discriminant of α.

The roots of equation (2.11) equal $\frac{-b \pm \sqrt{b^2 - 4ac}}{2a}$, so any quadratic irrationality α can be written as $\alpha = \frac{P + \sqrt{D}}{Q}$, where P and Q are integers and D $(D > 1)$ is an integer but not a perfect square. The second root of the equation, $\alpha' = \frac{P - \sqrt{D}}{Q}$ is called the irrational conjugate to α.

Example.
a) $\sqrt{5}$ is a quadratic irrationality because $\sqrt{5}$ is a solution to $x^2 - 5 = 0$.
b) $\alpha = \frac{1 - \sqrt{5}}{3}$ is also a quadratic irrationality because α is the irrational root of the equation $9x^2 - 6x - 4 = 0$.
c) $\sqrt[5]{2}$ is not a quadratic irrationality.

Definition. A continued fraction $\alpha = [a_0; a_1, a_2, ...]$ is called periodic if the sequence of the elements $a_0, a_1, a_2, ...$ is periodic. If the sequence is periodic, then the continued fraction is truly periodic. The length of the period of the sequence of the elements is called the length of the period of continued fraction.

Theorem 8

A continued fraction $\alpha = [a_0; a_1, a_2, ...]$ is periodic fraction with period k if and only if at some s, $\alpha_{s+k} = \alpha_s$.

Considering the values of periodic continued fractions, we obtain some of the real numbers. It turns out that at first glance it seems unexpected that the set of such numbers coincides with the set of quadratic irrationalities. This remarkable result was obtained for the first time in 1770 by Lagrange. The fact that the value of any periodic continued fraction is a quadratic irrationality can be proven quite simply. It is more difficult to prove that any quadratic irrationality decomposes into a periodic continued fraction; this fact is usually called the Lagrange theorem.

Problem 78

Find the value of the following continued fractions:
a) $\alpha = [5; (1,3)]$, b) $\alpha = [1; 4, 1, 4, 1, 4, ...]$, c) $\alpha = [2; 2, 2, 1, 2, 2, 2, 1, ...]$

Solution.
a) $\alpha = 5 + \frac{1}{[(1,3)]} = 5 + \frac{1}{3 + \frac{\sqrt{21}}{6}} = \frac{7 + \sqrt{21}}{2}$.

b) $\alpha = 1 + \frac{1}{4 + \frac{1}{\alpha}} = 1 + \frac{1}{4\alpha + 1} = \frac{4\alpha + 2}{4\alpha + 1}$, obtain the equation $4\alpha^2 - 4\alpha - 1 = 0$, with the roots: $\alpha_{1,2} = \frac{1 \pm \sqrt{2}}{2}$, hence, $\alpha = \frac{1 + \sqrt{2}}{2}$.

c) $\alpha = [2; 2, 2, 1, \alpha]$, $\alpha = \frac{P_3 \alpha + P_2}{Q_3 \alpha + Q_2}$, find the numerators and denominators of the convergent fractions, obtaining $\alpha = \frac{17\alpha + 12}{7\alpha + 5}$, from which the quadratic equation is $7\alpha^2 - 12\alpha - 12 = 0$ and its positive root is $\alpha > 0$ $\alpha = \frac{6 + 2\sqrt{30}}{7}$.

Before proceeding to the Lagrange theorem, we will prove the following auxiliary theorem.

Theorem 9

If a quadratic irrationality is represented by the sequence $\alpha = [a_0; a_1, a_2, ..., a_{n-1}, \alpha_n]$, where all a_i are integers, then α_n is also quadratic irrationality with the same discriminant as α.

Proof. Let α be the root of the quadratic equation $A\alpha^2 + B\alpha + C = 0$, where A, B, C are integer numbers. Substituting $\alpha = a_0 + \frac{1}{\alpha_1}$, we obtain,

$$A\left(a_0 + \frac{1}{\alpha_1}\right)^2 + B\left(a_0 + \frac{1}{\alpha_1}\right) + C = 0,$$

or

$$(Aa_0^2 + ba_0 + C)\alpha_1^2 + (2Aa_0 + B)\alpha_1 + A = 0,$$

It means that α_1 is the root of the equation $A_1\alpha_1^2 + B_1\alpha_1 + C_1 = 0$, with integer coefficients the discriminant of which equals

$$(2Aa_0 + B)^2 - 4(Aa_0^2 + ba_0 + C)A = B^2 - 4AC,$$

where $C_1 = A \neq 0$. Replacing in the previous quadratic equation α_1 by $a_1 + \frac{1}{\alpha_2}$, by analogy we obtain that α_2 is the root of a quadratic equation with integer coefficients with the same discriminant.

Next, we can state Theorem 10.

Theorem 10 (Lagrange)

Any quadratic irrationality can be written as periodic continued fraction.

Proof. Let α be a quadratic irrationality, i.e., α is an irrational number, the positive root of a quadratic polynomial

$$f(x) = Ax^2 + Bx + C$$

with integer coefficients. Substituting it into equation $A\alpha^2 + B\alpha + C = 0$

$$\alpha = \frac{P_{n-1} \cdot \alpha_n + P_{n-2}}{Q_{n-1} \cdot \alpha_n + Q_{n-2}}$$

And putting the fractions over the common denominator,

$$A(P_{n-1}\alpha_n + P_{n-2})^2 + B(P_{n-1}\alpha_n + P_{n-2})(Q_{n-1}\alpha_n + Q_{n-2}) + C(Q_{n-1}\alpha_n + Q_{n-2})^2 = 0,$$

that can be written as

$$A_n\alpha_n^2 + B_n\alpha_n + C_n = 0 \qquad (2.12)$$

where the coefficients

$$A_n = AP_{n-1}^2 + BP_{n-1}Q_{n-1} + CQ_{n-1}^2 = Q_{n-1}^2 f\left(\frac{P_{n-1}}{Q_{n-1}}\right),$$
$$C_n = AP_{n-2}^2 + BP_{n-2}Q_{n-2} + CQ_{n-2}^2 = Q_{n-2}^2 f\left(\frac{P_{n-2}}{Q_{n-2}}\right),$$
$$B_n = 2AP_{n-1}P_{n-2} + B(P_{n-1}Q_{n-2} + P_{n-2}Q_{n-1}) + 2CQ_{n-1}Q_{n-2}$$

are integer numbers.

Regarding the previous theorem, the discriminant of equation (2.12) can be written as

$$B_n^2 - 4A_nC_n = B^2 - 4AC, \qquad (2.13)$$

and it does not change with the change of n.

First, let us prove that A_n and C_n for big n have opposite signs, and second, using true equality (2.13), we will prove that the values of B_n are bounded.

The values of $\frac{P_{n-1}}{Q_{n-1}}$ and $\frac{P_{n-2}}{Q_{n-2}}$ are on the opposite sides of α, moreover, when n is big, they become very close to α. Further, we know that $f(\alpha) = 0$, but because α is irrational, then $f'(\alpha) = 2A\alpha + B \neq 0$. Otherwise, we would obtain that $\alpha = -\frac{B}{2A}$ is rational. Therefore, α a simple root of $f(x) = 0$.

It is known that in a sufficiently small neighborhood to the left and to the right of the simple root, the values of a continuous function, in this case of the polynomial $f(x) = Ax^2 + Bx + C$, have opposite signs. (See, for example, my book *Methods of Solving Nonstandard Problems*, page 70.) This means that as n increases and becomes large, the values of

$$A_n = Q_{n-1}^2 f\left(\frac{P_{n-1}}{Q_{n-1}}\right), \quad C_n = Q_{n-2}^2 f\left(\frac{P_{n-2}}{Q_{n-2}}\right)$$

become opposite in sign. Moreover, the values of $f\left(\frac{P_{n-1}}{Q_{n-1}}\right)$ and $f\left(\frac{P_{n-2}}{Q_{n-2}}\right)$ and hence, A_n and C_n are not equal to zero. Therefore, for big n, the product $A_n \cdot C_n < 0$. Since $-4A_nC_n > 0$, $B_n^2 \geq 0$, we have,

$$0 \leq B_n^2 < B_n^2 - 4A_nC_n = B^2 - 4AC, \quad 0 < -4A_nC_n \leq B_n^2 - 4A_nC_n = B^2 - 4AC,$$

Therefore, the values B_n^2 and $(-4A_nC_n)$ are bounded. It follows from the boundedness of these values the boundedness of $|A_n|$, $|B_n|$, $|C_n|$. Because they are integers, then among the equations (2.12) at increasing of n there exist only finite number of different equations. Each quadratic equation has only two roots, so among the roots of equations (2.12) there are a finite number of different, and hence among the quantities: $\alpha = \alpha_0$, α_1, α_2, ... there are only a finite number of different values. Hence, there can be found such α_k, that $\alpha_k = \alpha_{k+n}$, which means that the continued fraction decomposition of α is periodic.

Problem 79

Rewrite by continued fraction $\alpha = \frac{1+\sqrt{23}}{3}$.

Solution. Find consecutive terms:

$$\alpha_0 = [\alpha_0] + \frac{1}{\alpha_1}, \quad [\alpha_0] = \left[\frac{1+\sqrt{23}}{3}\right] = 1$$

$$\alpha_0 = 1 + \frac{1}{\alpha_1} \Rightarrow \alpha_1 = \frac{1}{\alpha_0 - 1} = \frac{1}{\frac{1+\sqrt{23}}{3} - 1} = \frac{3}{\sqrt{23} - 2} = \frac{3(\sqrt{23}+2)}{19}$$

$$[\alpha_1] = 1, \quad \alpha_1 = 1 + \frac{1}{\alpha_2} \Rightarrow$$

$$\alpha_2 = \frac{1}{\alpha_1 - 1} = \frac{1}{\frac{3(\sqrt{23}+2)}{19} - 1} = \frac{19}{3\sqrt{23} - 13} = \frac{19(3\sqrt{23}+13)}{9 \cdot 23 - 13^2}$$

We continue as

$$\alpha_2 = \frac{3\sqrt{23}+13}{2} \Rightarrow [\alpha_2] = 13, \quad \alpha_2 = 13 + \frac{1}{\alpha_3}$$

$$\alpha_3 = \frac{3\sqrt{23}+13}{19} = 1 + \frac{1}{\alpha_4}$$

Further, we obtain the following:

$$\alpha_4 = \frac{\sqrt{23}+2}{3} = 2 + \frac{1}{\alpha_5}, \quad \alpha_5 = \frac{3}{7}(\sqrt{23}+4) = 3 + \frac{1}{\alpha_6}, \quad \alpha_6 = \frac{\sqrt{23}+3}{6} = 1 + \frac{1}{\alpha_7},$$

$\alpha_7 = \frac{3}{7}(\sqrt{23}+3) = 3 + \frac{1}{\alpha_8}$, $\alpha_8 = \frac{\sqrt{23}+4}{3} = 2 + \frac{1}{\alpha_9}$, $\alpha_9 = \frac{3}{19}(\sqrt{23}+2)$, i.e., $\alpha_9 = \alpha_1$.

Finally, $\frac{1+\sqrt{23}}{3} = [1;(1,13,1,2,3,1,3,2)]$.

Answer. $\frac{1+\sqrt{23}}{3} = [1;(1,13,1,2,3,1,3,2)]$.

Since we now know that any quadratic irrationality can be represented by a periodic continued fraction, it is natural to find out for which quadratic irrationalities such an expansion will be purely periodic. The Theorem 11 gives an answer to this question.

Theorem 11

A quadratic irrationality $\alpha = \frac{P+\sqrt{D}}{Q}$, where P, Q and D $(D > 1)$ are integers, can be written by periodic continued fraction if and only if $\alpha > 1$ and its conjugate irrationality $\alpha' = \frac{P-\sqrt{D}}{Q}$ belongs to the interval $(-1, 0)$.

Example.

1) $\alpha = \frac{1+\sqrt{13}}{3}$ can be written as a pure periodic continued fraction because $\alpha > 1$ and $\alpha' = \frac{1-\sqrt{13}}{3} \in (-1, 0)$. Indeed, $\alpha = [(1; 1, 1, 6, 1)]$.

2) $\frac{2+\sqrt{3}}{4}$ can be written by mixed periodic continued fraction because its conjugate, $\frac{2-\sqrt{3}}{4}$ is greater than zero.

Theorem 12

Let D not be a perfect square, Q an integer, and $D > Q^2 > 0$, then the decomposition $\frac{\sqrt{D}}{Q}$ into continued fraction has the type,

$$\frac{\sqrt{D}}{Q} = [a_0; (a_1, a_2, ..., a_{k-1}, 2a_0)].$$

Proof. If $D > Q^2 > 0$, then the number $\alpha = a_0 + \frac{\sqrt{D}}{Q}$, where $a_0 = \left[\frac{\sqrt{D}}{Q}\right]$, will have a purely periodic continued fraction type. Indeed, $\alpha > 1$ and $\alpha' = a_0 - \frac{\sqrt{D}}{Q} \in (-1, 0)$, $\left[a_0 + \frac{\sqrt{D}}{Q}\right] = 2a_0$, then $\alpha = a_0 + \frac{\sqrt{D}}{Q} = [(2a_0; a_1, ..., a_{k-1})]$. These facts imply the validity of the theorem.

Example.

1) $\sqrt{7} = [2; (1, 1, 1, 4)]$ $(a_0 = 2)$. 2) $\sqrt{53} = [7; (3, 1, 1, 3, 14)]$ $(a_0 = 7)$. 3) $\frac{\sqrt{11}}{3} = [1; (9, 2)]$ $(a_0 = 1)$.

You just learned how to represent an irrational number as an infinite continued fraction. At the beginning of this section I demonstrated how infinite continued fractions can be replaced by an irrational number. Let me offer you the following problem.

Problem 80

Simplify the following

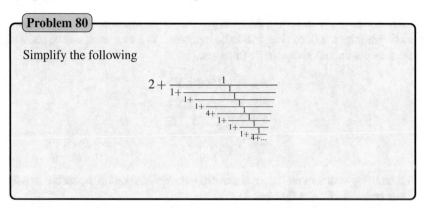

Solution. It follows from the expression above that in the infinite fraction representation of some irrational number, digits $1, 1, 1, 4$ are repeated infinitely many times. Since the integer part of the number is 2, then the number is greater then 2 but less than 3. If we omit the integer part, and denote

$$x = \cfrac{1}{1 + \cfrac{1}{1 + \cfrac{1}{1 + \frac{1}{4 + \dots}}}},$$

Hence, our variable x must satisfy the equation,

$$x = \cfrac{1}{1 + \cfrac{1}{1 + \cfrac{1}{1 + \frac{1}{4 + x}}}}$$

Working backward, we will obtain a quadratic equation in x,

$$1 + \frac{1}{4+x} = \frac{5+x}{4+x}$$

$$1 + \frac{1}{\frac{5+x}{4+x}} = 1 + \frac{4+x}{5+x} = \frac{9+2x}{5+x}$$

$$1 + \frac{1}{\frac{9+2x}{5+x}} = 1 + \frac{5+x}{9+2x} = \frac{14+3x}{9+2x}$$

$$x = \frac{9+2x}{14+3x}$$

$$3x^2 + 12x - 9 = 0$$

$$x^2 + 4x - 3 = 0$$

$$x = -2 \pm \sqrt{7}$$

Taking only the positive root, we obtain that $x = \sqrt{7} - 2$. The original problem is greater by 2, $b = 2 + x = \sqrt{7}$.
Answer. $\sqrt{7}$.

While irrational numbers in the form $a + \sqrt{b}$ can be represented by a continuous periodic infinite fraction, other irrational numbers, such as e or π, can be similarly written as continuous nonperiodic fractions.

$$\pi = 3.14159265358...$$
$$\pi = 3 + \cfrac{1}{7.0625133064}$$
$$\pi = 3 + \cfrac{1}{7 + \cfrac{1}{15.996594286}}$$
$$\pi = 3 + \cfrac{1}{7 + \cfrac{1}{15 + ...}} ...$$

An interesting problem on the representation of irrational numbers, such as e using infinite series is given in Problem 81.

Problem 81

Prove that

$$\frac{e+1}{e-1} = [2, 6, 10, ..., 4n+2, ...]. \qquad (2.14)$$

Proof. Recall decomposition of e into infinite Taylor's series. Define a function $f_n(x)$ $(n = 0, 1, 2, ...)$, as sum of the series:

$$f_n(x) = \frac{n!}{(2n)!} + \frac{(n+1)!}{1!(2n+2)!}x^2 + \frac{(n+2)!}{2!(2n+4)!}x^4 + ... = \sum_{s=0}^{\infty} \frac{(n+s)!}{s!(2n+2s)!}x^{2s}.$$

This series converges for any value of x; so we consider, for example, the values of x, in the interval $(0; 1)$. It is easy to see that the following is true:

$$f_n(x) - (4n+2) \cdot f_{n+1}(x) = 4x^2 f_{n+2}(x) \qquad (2.15)$$

Denote

$$\alpha_n = \frac{f_n\left(\frac{1}{2}\right)}{f_{n+1}\left(\frac{1}{2}\right)}.$$

Since

$$f_0(x) = 1 + \frac{x^2}{2!} + \frac{x^4}{4!} + ... = \frac{1}{2}(e^x + e^{-x}), f_1(x) = \frac{1}{2x}\left(x + \frac{x^3}{3!} + \frac{x^5}{5!} + ...\right) = \frac{1}{4x}(e^x - e^{-x}),$$

then

$$\alpha_0 = \frac{f_0\left(\frac{1}{2}\right)}{f_1\left(\frac{1}{2}\right)} = \frac{e^{\frac{1}{2}} + e^{-\frac{1}{2}}}{e^{\frac{1}{2}} - e^{-\frac{1}{2}}} = \frac{e+1}{e-1}.$$

Dividing the both sides of the equality (2.15) by $f_{n+1}(x)$ at $x = \frac{1}{2}$, we obtain

$$\alpha_n = (4n+2) + \frac{1}{\alpha_{n+1}} \qquad (2.16)$$

Since α_{n+1} is positive, the equality (2.16) shows that for all n, $\alpha_n > 4n+2 > 1$, $\frac{1}{\alpha_{n+1}} < 1$, i.e., $4n+2 = [\alpha_n]$ and the sequence (2.16) at $n = 0, 1, 2, \ldots$

$$\alpha_0 = 2 + \frac{1}{\alpha_1}, \alpha_1 = 6 + \frac{1}{\alpha_2}, \alpha_2 = 10 + \frac{1}{\alpha_3}, \ldots$$

Gives decomposition of α_0 into continues fraction.

The proof is complete.

You are probably wondering "What is to be gained by the expression of a number as a continued fraction besides solving math competition problems?" The answer will impress you a bit. (See more information on this in Chapter 3 of the book.)

2.3 Division with a Remainder

Returning to the decimal representation of a natural number, we can state Property 20.

> **Property 20**
>
> A number (every integer) divided by 3 or 9 gives the same remainder as the sum of its digits.

Proof.

$$N = a_n \cdot \underbrace{999\ldots99}_{n \text{ digits}} + a_{n-1} \cdot \underbrace{999\ldots99}_{(n-1) \text{ digits}} + \cdots + a_2 \cdot 99 + a_1 \cdot 9 + (a_n + a_{n-1} + \cdots + a_2 + a_1 + a_0).$$

Since the first n terms are multiples of 3 or 9 then a remainder when divided by 3 or 9 is determined by the sum inside parentheses (the sum of the digits).

> **Problem 82**
>
> Prove that if the sum of digits of a number equals 15, then such a number cannot be a perfect square.

Solution.
Method 1. One should just sum the digits again and note that the number is divisible by 3 but not 3. Number 15 is a multiple of 3 but not a multiple of 9. Assuming that such number N exists, it should satisfy the equation:

$$N = 3k = m^2. \tag{2.17}$$

A number m, in turn, can be either a multiple of 3 or not a multiple of 3 (giving remainders of 1 or 2). Let us consider these three possible situations for equation (2.17):

1. $m = 3n$—a multiple of 3, then $N = 3k = 9n^2$ that is impossible because N is not a multiple of 9. We obtained a contradiction.
2. $m = 3n + 1$, then $N = m^2 = 9n^2 + 6n + 1$. This yields that N is not a multiple of 3 that is false again.
3. $m = 3n + 2$, then $N = (3n+2)^2 = 9n^2 + 12n + 4 = \underbrace{9n^2 + 12n + 3}_{\div 3} + 1$. This means

   that N is not a multiple of 3, which is another contradiction. We have proven that number N cannot be a perfect square.

Method 2. We argue by contradiction. Assume that such a number exists, then $N = 3k = m^2$. In order to have solution in integers m must be divisible by 3. Let $m = 3n$. We obtain

$$3k = (3n)^2$$
$$3k = 9n^2$$
$$k = 3n^2$$
$$k = 3p.$$

Then $N = 9p$ and must be divisible by 9. However, the sum of its digits is divisible by 15, not by 9. Therefore, we obtain a contradiction and such number as N does not exist.

Now, if we want to know the remainder of some number divided by 3 or 9, we can add all its digits and look at the remainder of this number. For example, the number 124 can be written as $124 = 3 \cdot 41 + 1$ (1 is the remainder). Further, the sum of its digits is $1 + 2 + 4 = 7$, and $7 = 3 \cdot 2 + 1$. We see that both 124 and 7 divided by 3 give the same remainder of 1. Let us divide 124 by 9. The result can be written as $124 = 9 \cdot 13 + 7$ (7 is a remainder). Of course, 7 itself is less than 9, then $7 = 9 \cdot 0 + 7$, and 7 is a remainder. Property 20 is useful when we should find a remainder of very big numbers such as α^b divided by 3 and 9.

Problem 83

A number $\underbrace{252525\ldots2525}_{\text{2018 digits}}$ is divided by 9. What is the remainder?

Solution. The given number is huge and has the number 25 within it repeated $2018/2 = 1009$ times. We cannot solve the problem on a calculator, but are able to do it applying Property 20.

We can find the sum of its digits that is $(2+5) \cdot 1009 = 7063$. We divide 7063 by 9 and look for the remainder: $7063 = 9 \cdot 784 + 7$.

Answer. The given number divided by 9 leaves a remainder of 7.

> **Property 21**
>
> A number divided by 5 and 10 gives the same remainder as its last digit.

The proof of this statement follows from the divisibility by 5 or 10 (See Section 1.3). Using the digital form of a number N, we can rewrite it as $N = 10k + a_0$. This will allow us to find the last digit of any integer raised to a power in the future. Thus,

$$7 = 5 \cdot 1 + 2$$
$$57 = 5 \cdot 11 + 2.$$

> **Property 22**
>
> A number divided by 4, 25, 50 and 100 leaves the same remainder as a number written by the last two digits (by tens and ones).

Proof. In order to prove this let us consider any n digits number as

$$N = a_n a_{n-1} \ldots a_2 a_1 a_0 = (a_n 10^n + a_{n-1} 10^{n-1} + \ldots a_2 10^2) + (a_1 10^1 + a_0).$$

Since the first part of it is divisible by 4, 25, 50, and 100, and 100 is a multiple of 4, 25, 50, and 100, then the number can be written as $N = 100k + a_1 a_0$. Therefore, the remainder of N when divided by these numbers is determined by a number written by the last two digits. Note that we will use this property in finding the last two digits of some "big" power of a number.

> **Problem 84**
>
> Find all remainders of $9,451,174,652$ divided by 4, 5, 25, 50, and 100.

Solution.

1. Divisibility by the first four numbers is united in one answer following by Property 22. Number $9,451,174,652$ divided by any of these numbers (4, 25, 50, and 100) will give the same remainder as 52,

$$52 = 4 \cdot 13 + 0$$
$$52 = 25 \cdot 2 + 2$$
$$52 = 50 \cdot 1 + 2$$
$$52 = 100 \cdot 0 + 52.$$

2. The last digit of $9,451,174,652$ is 2, then divided by 5 its remainder equals 2.

Answer. A number $9,451,174,652$ divided by $4, 5, 25, 50$, and 100 gives remainders of $0, 2, 2, 2$, and 52, respectively.

Property 23

A number divided by 125 leaves the same remainder as a number written by its last three digits (by hundreds, tens, and ones).

Proof. The proof of this is clear,

$$N = a_n a_{n-1} \ldots a_2 a_1 a_0$$
$$= (a_n 10^n + a_{n-1} 10^{n-1} + \ldots a_3 10^3) + (a_2 10^2 + a_1 10^1 + a_0),$$

where the first term is divisible by 125.

Property 24

A number divided by 625 leaves the same remainder as a number written by its last four digits.

The following statement generalizes our ideas and knowledge.

Property 25

All arithmetic operations with numbers (except division) repeat operations with their remainders. Thus, adding numbers we add their remainders, raising to a power we raise a remainder to this power, etc.

This powerful property has many applications. Solving the problems below, we will demonstrate this.

Problem 85

Numbers m and n divided by 3 leave remainders of 1 and 2, respectively. Dividing by 3 what is a remainder of

(a) their sum $(m+n)$,
(b) their product (mn)?

Solution. Using this problem let us show that Property 25 works. The numbers m and n can be written as

$$\begin{cases} m = 3k+1 \\ n = 3l+2. \end{cases}$$

And their sum $m + n = 3(k+l) + 3 = 3(k+l+1)$ is divisible by 3, and has a remainder of 0. Can we find the same result using Property 25? Yes, we can. Adding remainders $1 + 2 = 3$ we notice that because 3 is a multiple of 3, then its remainder is 0, so as a remainder of $(m+n)$.

Let us consider a product mn.

$$mn = (3k+1)(3l+2) = 9kl + 6k + 3l + 2 = 3(3kl + 2k + l) + 2$$

A product of m and n divided by 3 gives a remainder of 2. Again, let us show that the same answer we can get considering only remainders of m and n. We have $1 \cdot 2 = 3 \cdot 0 + 2 = 2$ remainder equals 2.
Answer. a) 0, b) 2.

The following problem is also very interesting and can be solved using division with a remainder.

Problem 86

Solve $2^x - 15 = y^2$ in natural numbers.

Solution.

1. Let $x = 2k+1$, be an odd number. After substitution into the equation

$$2^{2k+1} = 2 \cdot 4^k = 2(3+1)^k = 2(3m+1) = 3l+2.$$

Since 15 is divisible by 3, then

$$2^x - 15 = 3r+2.$$

Hence the left side divided by 3 leaves a remainder of 2. On the other hand, the right-hand side, y^2, as a square a natural number, either leaves a remainder 1 or 0. Therefore, if x is an odd, then the given equation has no solutions.

2. Let $x = 2k$, an even number. After substitution we obtain

$$2^{2k} - 15 = y^2.$$

Moving 15 to the right and factoring the difference of squares,

$$(2^k - y)(2^k + y) = 15$$

Obviously that factor $(2^k + y)$ is positive, then $(2^k - y)$ must be also positive and we now have to solve two systems of equations:

$$\begin{cases} 2^k - y = 1 \\ 2^k + y = 15, \end{cases}$$

or

$$\begin{cases} 2^k - y = 3 \\ 2^k + y = 5. \end{cases}$$

Finally, we obtain two solutions: $(4, 1)$ and $(6, 7)$.
Answer. $\{(x, y) : (4, 1), (6, 7)\}$.

Problem 87

What is the remainder of 10^{2018} divided by 7?

Solution.
Method 1. Earlier we found that $1001 = 7 \cdot 11 \cdot 13$. Because $1000 = 1001 - 1$, then the following is true

$$10^3 = 7k - 1$$

Using properties of the exponent and the fact that 2016 is a multiple of 3, the given number can be written by the following chain of correct expressions:

$$10^{2018} = 10^{2016} \cdot 100 = (7k - 1)^{\frac{2016}{3}} \cdot 100$$
$$= (7k - 1)^{672} \cdot 100 = (7m + 1) \cdot (7 \cdot 14 + 2) = 7n + 2.$$

Note that in the long formula above, we used the Newton Binomial formula and the fact that $100 = 7 \cdot 14 + 2$. Finally, $10^{2018} = 7n + 2$.
Method 2. It is easy to see that $10^2 = 100 = 98 + 2 = 7 \cdot 14 + 2$, then

$$10^2 = 7k + 2$$

How can we represent 10^{2018}?

$$10^{2018} = (10^2)^{1009} = (7k + 2)^{1009} = 7n + 2^{1009}.$$

What is the remainder of 2^{2009} divided by 7? Notice that

$$8 = 2^3 = 7 + 1$$

Next, 2^{1009} can be written as

$$2^{1009} = 2^{1008} \cdot 2 = (2^3)^{336} \cdot 2 = (7m + 1) \cdot 2 = 7p + 2.$$

Therefore, $10^{2018} = 7p + 2$.

Answer. 10^{2018} divided by 7 leaves a remainder of 2.

Remark. Using the problem above, we can answer the following question: What is the largest multiple of 7 closest to 10^{2018}? Answer: Because $10^{2018} = \underbrace{100\ldots00}_{\text{2018 digits}}$,

then its closest multiple of 7 is $N = \underbrace{999\ldots98}_{\text{2017 digits}}$.

2.4 Finding the Last Digits of a Number Raised to a Power

Problems on finding the last digits are sometimes offered on SAT, AP exam or on contests, such as AIME. Let us try to find the remainder of 3^{1995} divided by 5 or 2^{2001} divided by 10. Guess what: If we knew the last digit of 3^{1995} or 2^{2001}, we could apply Property 20 and find the remainders. Can we find the last digit? Not yet!

This section will teach you how to find the last digits of some very big numbers, which will help you to solve the problems above. Because we need to know the last digit or the last two digits of the numbers for problems stated above, let us try to find a general idea for solving such problems.

First, we will use pattern recognition.

Problem 88

Find the last digit of the following numbers:

1. $(a)\ 6^{1971}$ $(b)\ 9^{1971}$ $(c)\ 3^{1971}$ $(d)\ 2^{1971}$
2. $(a)\ 6^{1999}$ $(b)\ 9^{1999}$ $(c)\ 3^{1999}$ $(d)\ 2^{1999}$.

Solution. All numbers in this problem are too big to be evaluated on calculators. In order to find the last digit of some big powers of 6 we can consider the series,

$$6^1 = 6$$
$$6^2 = 36$$
$$6^3 = 216$$
$$6^4 = 1296$$
$$6^5 = 7776$$
$$6^6 = 46656$$
$$6^7 = 279936.$$

We can notice that whatever n is, 6^n always ends in the digit 6. Then $6^{1971} = \ldots\ldots 6$ as well, as does $6^{1999} = \ldots\ldots 6$.

Let us try to find the last digit of 9^{1971}, 9^{1999} or 9^{2000}. Note that

$$9^1 = 9$$
$$9^2 = 81$$
$$9^3 = 729$$
$$9^4 = 6561$$
$$9^5 = 59049$$
$$9^6 = 531441.$$

We can conclude that 9^n ends in digit 9 if n is an odd number, and ends in digit 1 if n is even. Then 9^{1999} and 9^{1971} will end with a 9, and 9^{2000} will end with a 1.

Let us find the last digit of 3^{1971} and 3^{1999}. Writing down the first nine powers of 3,

$$3^1 = 3$$
$$3^2 = 9$$
$$3^3 = 27$$
$$3^4 = 81$$
$$3^5 = 243$$
$$3^6 = 729$$
$$3^7 = 2187$$
$$3^8 = 6561$$
$$3^9 = 19683.$$

We can unite all these in a small table where the numbers in the columns are powers of n for each last digit of 3^n.

The last digit is 3	The last digit is 9	The last digit is 7	The last digit is 1
1	2	3	4
5	6	7	8
9	10	11	12
13	14	15	16
17	18	19	20
$4k+1$	$4k+2$	$4k+3$	$4k$

Table 2.1 Last digit of 3^n

From Table 2.1 we notice that if the degree of 3 is 1, 5, 9, 13, 17 and so on that can be written as $(4k+1)$ for any $k=0,1,2,\dots$, then a number 3^n ends in digit of 3. These are shown in the first column. If the degree of 3 is 2, 6, 10, 14, 18 or $(4k+2)$, where $k=0,1,2,\dots$, then the number 3^n ends in 9 (second column). If the degree of 3 is 3, 7, 11, 15 or $(4k+3)$, where $k=0,1,2,\dots$, then 3^n ends in a 7 (third column), and if the degree of 3 is 4, 8, 12, 16 or $4k$, where $k=0,1,2,\dots$, then the last digit is 1.

Now we are ready to find the last digit of (a) 3^{1971}, (b) 3^{1999}, and (c) 3^{2000}. Actually, we are able to find the last digit of 3 raised to any power n. Following our new algorithm, we just divide a power of three by four and the remainder (0, 1, 2 or 3) will reveal the last digit (1, 3, 9, or 7, respectively).

1. Rewriting 1971 as $1971=4\cdot492+3=4k+3$, we find that 3^{1971} will end in a 7.
2. $1999=4\cdot496+3=4m+3$, thus 3^{1999} has a last digit of 7.
3. $2000=4\cdot500+0=4p$, and 3^{2000} ends with a 1.

Let us try the same technique with powers of 2. We have

$$
\begin{aligned}
2^1 &= 2\\
2^2 &= 4\\
2^3 &= 3\\
2^4 &= 16\\
2^5 &= 32\\
2^6 &= 64\\
2^7 &= 128\\
2^8 &= 256\\
2^9 &= 512\\
2^{10} &= 1024.
\end{aligned}
$$

Note that powers of 2 can end in 2, 4, 6 or 8, so we create Table 2.2.

The last digit is 2	The last digit is 4	The last digit is 8	The last digit is 6
1	2	3	4
5	6	7	8
9	10	11	12
13	14	15	16
17	18	19	20
$4k+1$	$4k+2$	$4k+3$	$4k$

Table 2.2 Last digit of 2^n

Using the Table 2.2, let us find the last digit of (a) 2^{1971}, (b) 2^{1999}, and (c) 2^{2000}.

(a) $1971=4\cdot492+3=4k+3$, and 2^{1971} ends in the digit 8.
(b) $1999=4\cdot499+3$, and 2^{1999} ends with an 8 again.
(c) $2000=4\cdot500+0=4k$, then 2^{2000} has a last digit of 6.

Remark. You probably noticed that each power of 2 or 3 was divided by 4 and depending on a remainder that a power of 2 or 3 leaves, we judged about the last digit of a number. This "magic" number 4 will be explained in Section 2.5.1.

Now we can solve the problems that started this section.

Problem 89

What is the remainder of

(a) 3^{1995} divided by 5?
(b) 2^{2001} divided by 10?

Solution. First, we find the last digit of expression 3^{1995}. Since, $1995 = 4 \cdot 498 + 3$, then 3^{1995} ends in a digit of 7. The last digit of 2^{2001} is 2, because $2001 = 4 \cdot 500 + 1 = 4k + 1$ (see Table 2.2). By Properties 20 and 21 of the previous section, 3^{1995} divided by 5 will give a remainder of 2, because $7 = 1 \cdot 5 + 2$. In turn, 2^{2001} divided by 3 gives a remainder of 2, because $2 = 0 \cdot 3 + 2$.

We have some ideas on how to find the last digit of big numbers, but it would be interesting to be able to find the last two digits of some big numbers. Let us try it now.

Problem 90

Find the last two digits of the number $99^{99} - 51^{51}$.

Solution. We are not able to evaluate this number on a calculator. However, using the previous techniques we can consider the first powers of 99 and 51.

$$99^1 = 99$$
$$99^2 = 9801$$
$$99^3 = 970299$$
$$99^4 = 96059601.$$

We notice that every even degree of 99 ends in 01, and every odd degree in 99. Then,

$$51^1 = 51$$
$$51^2 = 2601$$
$$51^3 = 132651$$
$$51^4 = 6765201.$$

We see that every even degree of 51 ends in a 01, and every odd degree in a 51. Powers, 99 and 51 are odd numbers, then $99^{99} - 51^{51}$ will end in $99 - 51 = 48$. Eureka!

Answer. 48.

Let us consider the problem-solving which we can use the properties of the last digit.

Problem 91

Solve the equation $2^n + 7 = k^2$.

Solution.
1) If $k = 2t$ (even) then the right side is even but the left side is always odd and there will be no solutions for even k. Alternatively, for all $n > 1$, then $k^2 = 4t^2$ divided by 4 gives a remainder of 0. On the other hand, the left side divided by 4 leaves a remainder of 3, because for $n > 1$, $2^n = 4m$ and $7 = 4 + 3$, then $2^n + 7 = 4p + 3$. Therefore, there are no solutions for even values of k.
2) If $k = 2t + 1$, an odd number and $n > 1$, then

$$k^2 = (2t + 1)^2 = 4k^2 + 4k + 1 = 4l + 1.$$

Hence, the right side divided by 4 leaves a remainder 1.

Can the odd number on the left side give the same remainder? If $n > 1$ and $n = 2p$, then $2^{2p} = 4k$, and $2^{2p} + 7 = 4m + 3$. If $n = 2p + 1$, then $22p + 1 + 7 = 2 \cdot 4^p + 7 = 4m + 3$. There is no solution.

Finally, if $n = 1$, then the left side is $2 + 7 = 9$ and the right side equals 9 if $k = 3$.
Answer. $n = 1$, $k = 3$.

Problem 92

Find the last two digits of $9999^{999^{99^9}}$ $_{..}$

Solution. Consider a few powers of 9999, then decide if a power of 999^{99^9} is an even or odd. From the first powers of 9999 we have

$$9999^1 = 9999$$
$$9999^2 = 99980001$$
$$9999^3 = \ldots\ldots\ldots 99$$
$$9999^4 = \ldots\ldots\ldots\ldots 01.$$

We see that every odd degree of 9999 will end in 99, and every even – in 01. To check this we can multiply 99980001 by 9999 by hand (a calculator will round the answer for such big numbers, but we want to know the exact last two digits).

Now we have to decide whether a number 999^{99^9} is an even or odd. Obviously, 999 is odd so as any of its power. It is useful to know that, if a number has an even last digit, or 0, such a number is an even. Therefore, we have to justify only the last digit of a number 999^{99^9}. As we know, a number 999 raised to any odd power ends in 99 and to any even power ends in 01. Neither one combination gives us an even last digit, then we conclude that 999^{99^9} is an odd number, and $9999^{999^{99^9}}$ ends in 99.

Answer. 99.

After solving this problem we can formulate the following lemma.

Lemma 11

Any number ending in 99 raised to an odd power ends in 99 and raised to an even power ends in 01.

Proof. Any l digits number ending in 99 can be written as

$$N = a_l a_{l-1}...a_2 99 = [a_l \cdot 10^l + a_{l-1} \cdot 10^{l-1} + ... + a_2 \cdot 10^2] + 9 \cdot 10 + 9$$
$$= 100k + 99 = 100k + 100 - 1 = 100s - 1.$$

Its odd power is

$$N^{2n+1} = (100s - 1)^{2n+1} = 100m + (-1)^{2n+1} = 100m - 1 = 100(m-1) + 99 = abc...99.$$

its even power is

$$N^{2n} = (100s - 1)^{2n} = 100m + (-1)^{2n} = 100m + 01 = abc...01.$$

Problem 93

Find the last digit of $2^{3^{4^5}}$.

Solution. In the previous problem, the answer (99) did not depend on the way we evaluate the number $9999^{999^{99^9}}$, and the process of finding the last two digits was reduced to finding the power (even or odd) of a number itself always ending in 99. In this problem, the answer does depend on the way we evaluate the given number.

First, let us find out what is $2^{3^{4^5}}$? Can we evaluate the power of 2? Is the given number unique? If we think about this number a little more, then we will see that the number can be evaluated differently depending on how we put parentheses between

2, 3, 4 and 5. It is interesting that if we had to evaluate the product of the same numbers instead, $2 \cdot 3 \cdot 4 \cdot 5$, then the answer would be always 120 despite the fact how we group the numbers for multiplication:

$$(2 \cdot 3) \cdot (4 \cdot 5) = 2 \cdot (3 \cdot 4) \cdot 5 = 2 \cdot (3 \cdot (4 \cdot 5)) = ... = 120.$$

For simplicity, let us use the following notation for $2^{3^{4^5}} = 2 * 3 * 4 * 5$, where $*$ means power. There are five possible cases and hence, five different answers:

1. First, we can raise 4 to the 5 power, then raise 3 to that power and then raise 2 to the obtained power:

$$2 * (3 * (4 * 5)) = 2^{(3^{(4^5)})} = 2^{3^{1024}}$$

This number is very big, however, we can still calculate its last digit, because we just need to figure out what remainder 3^{1024} leaves when divided by 4. Rewriting it as $(4 - 1)^{1024}$ and applying Newton's Binomial theorem, we can see that $3^{1024} = 4k + 1$. Hence, 2 raised to this power ends in 2.

2. We can raise 3 to the 4th power and then this number raise to the 5th power, then finally raise 2 to this power,

$$2 * ((3 * 4) * 5) = 2^{3^{20}}.$$

Similarly, the power of 2 is $3^{20} = (4 - 1)^{20} = 4m + 1$, and this means that the last digit of the number is also 2. (Using Maple, we obtain $3^{20} = 3,486,784,401$.)

3. If we place parentheses this way $(2 * 3) * (4 * 5) = (2^3)^{(4^5)}$, then we have to raise 2 to the third power and then the obtained number to the power of 4 raised to the fifth power, obtaining $(2^3)^{1024} = 2^{3072}$. Since 3072 is a multiple of 4, then in this case 2^{3072} ends in 6.

4. Next option is $((2 * 3) * 4) * 5 = ((2^3)^4)^5 = 2^{3 \cdot 4 \cdot 5} = 2^{60}$. This number is also ending in 6 because 60 is a multiple of 4. It is interesting that using Maple we can calculate that $2^{60} = 1152921504606846976$.

5. Finally, we have one more possibility for evaluating the given number,

$$(2 * (3 * 4)) * 5 = (2^{3^4})^5 = (2^{81})^5 = 2^{243}.$$

$243 = 4 \cdot 60 + 3 = 4t + 3$, therefore this time the number ends in 8.

Answer. Depending evaluation of this number, the last digit can be 2, 6 or 8.

Remark. By solving the last problem in this section we found three different answers and five different ways of placing parentheses between numbers 2, 3, 4, 5, without changing the order of the numbers. What if we had more than four numbers? For example, five or six, or even more. The number of the ways to place parentheses would increase. Of course, sometimes we can count all possible cases, but it is easy to miscalculate and make a mistake. In general, this problem of putting parentheses between n numbers is similar to the problem of finding the nth Catalan number, and the formula is derived in this chapter. Some of the problems in this section are solved by pattern recognition rather than method. In order to learn how further knowledge of number theory can justify our answers, please continue reading this chapter.

2.5 Fermat's Little Theorem

Many contest problems ask to find the ending digits of a number raised to a large power or to find a remainder of a big number divided by another number (prime or not prime). Some of such problems can be solved using information of this Section. Let's take a look at the "Little" Theorem by Fermat.

> **Theorem 13** (Fermat's Little Theorem)
>
> If p is prime, $a \in Z \backslash \{0\}$, and $(a, p) = 1$, then
>
> $$a^{p-1} \equiv 1 \pmod{p} \tag{2.18}$$

Proof. If a number is not divisible by p, then it leaves a remainder of 1, 2, 3, ..., $(p-1)$, when divided by p.

Let $(a, p) = 1$. Then

$$
\begin{aligned}
a & \equiv r_1 \pmod{p} \\
2a & \equiv r_2 \pmod{p} \\
3a & \equiv r_3 \pmod{p} \\
& \cdots\cdots\cdots\cdots \\
(p-2)a & \equiv r_{p-2} \pmod{p} \\
(p-1)a & \equiv r_{p-1} \pmod{p},
\end{aligned}
\tag{2.19}
$$

where $1 \leq r_i \leq (p-1)$, $i = 1, 2, \ldots, (p-1)$.

Let us now multiply the left-hand sides and the right-hand sides of the relationships (2.19). (We can do it because they have the same modulo.) Hence, the product is

$$(1 \cdot a \cdot 2a \cdot 3a \cdots (p-2)a \cdot (p-1)a) \equiv (r_1 \cdot r_2 \cdots r_{p-1}) \pmod{p}.$$

Putting remainders in the order on the right-hand side and applying factorial formula on the left we obtain,

$$
\begin{aligned}
(a^{p-1}(p-1)!) & \equiv (1 \cdot 2 \cdot 3 \cdots (p-1) \pmod{p} \\
\Rightarrow (a^{p-1}(p-1)!) & \equiv (p-1)! \pmod{p}.
\end{aligned}
$$

Canceling of factorials on both sides leads to the formula,

$$a^{p-1} \equiv 1 \pmod{p},$$

which completes the proof. For better understanding of the proof of Theorem 13, let us choose $p = 7$ and a can be any number such that $(a, 7) = 1$. (It means they are relatively prime and have no common divisors.) Such a number a, when divided by $p = 7$ would give the following possible remainders: 1, 2, 3, 4, 5 or 6, which also can be written as

$$a = 7n + 1 \sim a \equiv 1 \quad (\text{mod } 7)$$
$$a = 7n + 2 \sim a \equiv 2 \quad (\text{mod } 7)$$
$$\cdots\cdots\cdots\cdots\cdots\cdots\cdots$$
$$a = 7n + 6 \sim a \equiv 6 \quad (\text{mod } 7).$$

If a is not divisible by p, then any multiple of a, ka is not divisible by p if $k = 1, 2, 3, \ldots, (p - 1)$.

Example. If $a = 8$ is not divisible by $p = 7$, then

$$1 \cdot 8, \; 2 \cdot 8, \; 3 \cdot 8, \; 4 \cdot 8, \; \ldots, \; 6 \cdot 8$$

are not divisible by 7.

The multiples of a: a, $2a$, $3a$, ..., $(p - 1)a$ when divided by p will give remainders of 1, 2, 3, ..., $(p - 1)$ but in a random order. (For example, $4a$ may give a remainder of 1 and $7a$ may give a remainder of 3, etc.)

Regarding our example we have the following:

$$1a, \; 2a, \; 3a, \; 4a, \; 5a, \; 6a$$
$$8, \; \; 16, \; 24, \; 32, \; 40, \; 48$$

When divided by 7,

$$8 \equiv 1 \quad (\text{mod } 7)$$
$$16 \equiv 2 \quad (\text{mod } 7)$$
$$24 \equiv 3 \quad (\text{mod } 7)$$
$$32 \equiv 4 \quad (\text{mod } 7)$$
$$40 \equiv 5 \quad (\text{mod } 7)$$
$$48 \equiv 6 \quad (\text{mod } 7).$$

Remainders happened to be in ascending order. However, if we take $a = 4$ and $p = 7$, $(4, 7) = 1$, we will obtain that

$$1a, \; 2a, \; 3a, \; 4a, \; 5a, \; 6a$$
$$4, \; \; 8, \; \; 12, \; 16, \; 20, \; 24$$

When divided by 7,

$$4 \equiv 4 \quad (\text{mod } 7)$$
$$8 \equiv 1 \quad (\text{mod } 7)$$
$$12 \equiv 5 \quad (\text{mod } 7)$$
$$16 \equiv 2 \quad (\text{mod } 7)$$
$$20 \equiv 6 \quad (\text{mod } 7)$$
$$24 \equiv 3 \quad (\text{mod } 7).$$

The remainders are 1, 2, 3, 4, 5, and 6 but not in the order.

When Fermat stated this Theorem in 1640, the proof was missing. The proof of this theorem was given by Leonard Euler in 18th century. Using Theorem 13 many proofs can be done. For example, if N is not prime, then there exists such prime number, $a < N$, that the congruence

$$a^{N-1} \equiv 1 \pmod{N}$$

is false.

In order to demonstrate how useful this theorem can be for us, let us first solve Problem 87 again and find the remainder of 10^{2018} divided by 7. We can reformulate it as follows.

Find x that satisfies the congruence:

$$10^{2018} \equiv x \pmod{7}.$$

Because 7 is prime and $(10, 7) = 1$, it follows from Fermat's Little Theorem that

$$10^6 \equiv 1 \pmod{7}.$$

If we raise 10 to any multiple of 6 power, then it will be also congruent to 1 $\pmod{7}$. Since

$$2018 = 2016 + 2 = 6 \cdot 336 + 2,$$

then

$$10^{2018} = 10^{2016} \cdot 10^2 = (10^6)^{336} \cdot 100 \equiv 1 \cdot 100 \pmod{7} \equiv 2 \pmod{7}.$$

Therefore the remainder is 2.

Let us practice application of Fermat's Little Theorem by solving the following problems.

Problem 94

Solve the equation $(222^{333} - 333^{222}) \equiv x \pmod{7}$.

Solution. Using Theorem 13 we have

$$\begin{aligned} 222^6 &\equiv 1 \pmod{7} \\ (222^6)^{55} &= 222^{330} \equiv 1 \pmod{7}. \end{aligned} \tag{2.20}$$

On the other hand,

$$222^3 \equiv 6 \pmod{7}. \tag{2.21}$$

Combining (2.20) and (2.21) we obtain:

$$222^{333} \equiv 6 \pmod{7}. \tag{2.22}$$

Similarly, we get

$$333^6 \equiv 1 \pmod{7}$$
$$(333^6)^{37} \equiv 1 \pmod{7} \tag{2.23}$$
$$333^{222} \equiv 1 \pmod{7}.$$

Subtracting (2.22) and (2.23). Finally the following is valid:

$$(222^{333} - 333^{222}) \equiv (6 - 1)(\mod 7) \equiv 5 \pmod{7}.$$

Answer. Remainder is 5.

Problem 95

Solve the equation $314^{164} \equiv x \pmod{165}$.

Solution.
Method 1. The number 165 is not prime ($165 = 3 \cdot 5 \cdot 11$). We can represent $314^{164} = 165k + r$. It will leave the same remainder when divided by 3, 5, or 11 as its remainder, r, divided by 3, 5, or 11.

First, we will find remainder of the number when it is divided by 5:

$$314^{164} \equiv r_1 \pmod{5}.$$

By Theorem 13,

$$314^4 \equiv 1 \pmod{5}$$
$$314^{164} \equiv 1 \pmod{5}.$$

Second, we will find its remainder when divided by 3:

$$314^2 \equiv 1 \pmod{3}$$
$$(314^2)^{82} = 314^{164} \equiv 1 \pmod{3}.$$

Third, we will find the remainder of 314^{165} when divided by 11:

$$314^{10} \equiv 1 \pmod{11}$$
$$(314^{10})^{16} = 314^{160} \equiv 1 \pmod{11}$$

and

$$314^4 \equiv 9 \pmod{11}$$
$$314^{160} \cdot 314^4 = 314^{164} \equiv 9 \pmod{11}$$

Now we need to find such a number that satisfies the following relationships simultaneously:

$$x = r \equiv 1 \pmod{5} \equiv 1 \pmod{3} \equiv 9 \pmod{11}.$$

Such a number exists. It can be found using Chinese remainder theorem. It is 31.
Method 2. Consider the first five powers of 314 and its remainders:

$$314^2 \equiv 91 \quad (\text{mod } 165)$$
$$314^3 \equiv 29 \quad (\text{mod } 165)$$
$$314^4 \equiv 31 \quad (\text{mod } 165)$$
$$314^5 \equiv 164 \quad (\text{mod } 165) \equiv (-1) \quad (\text{mod } 165)$$

Raising the last congruence to 32 power, we obtain

$$314^{160} = (314^5)^{32} \equiv (-1)^{32} \equiv 1 \quad (\text{mod } 165).$$

Since $164 = 160 + 4$, we finally obtain the same answer, 31.

$$314^{160} \cdot 314^4 \equiv 31 \quad (\text{mod } 165)$$

Answer. $x = 31$.

Problem 96

Solve the equation $2^{340} \equiv x \ (\text{mod } 341)$.

Solution. Number 341 is not prime, but it's a product of two primes, $341 = 11 \cdot 31$.
From Theorem 13 it follows for division by 11 and 31, respectively:

$$\begin{aligned} 2^{10} &\equiv 1 \quad (\text{mod } 11) \\ 2^{340} &\equiv 1 \quad (\text{mod } 11). \end{aligned} \tag{2.24}$$

$$\begin{aligned} 2^{30} &\equiv 1 \quad (\text{mod } 31) \\ 2^{330} &\equiv 1 \quad (\text{mod } 31). \end{aligned} \tag{2.25}$$

$$\begin{aligned} 2^{5} &\equiv 1 \quad (\text{mod } 31) \\ 2^{10} &\equiv 1 \quad (\text{mod } 31). \end{aligned} \tag{2.26}$$

Multiplying the second equations of (2.25) and (2.26) we get

$$2^{340} \equiv 1 \quad (\text{mod } 31). \tag{2.27}$$

Considering (2.24) and (2.27) we conclude that the number leaves the same remainder when divided by 11 or 31, then from Lemma 7 it follows that the number will leave the same remainder 1 when divided by the least common multiple (*LCM*) of 11 and 31 (341). Therefore, $x = 1$.
Answer. $x = 1$.

Problem 97

For any odd number n find the least prime that can divide $n^8 + 1$ but not $n^{16} + 1$.

Solution.

1. If n is odd, then it is easy to show that both $(2k+1)^8 + 1 = 2m$ and $(2k+1)^{16} + 1 = 2k$ are even, and hence they are divisible by 2.

2. In order to find other possible prime factors, assume that $n^8 + 1$ is divisible by prime number p. The following must be true

$$n^8 + 1 \equiv 0 \pmod{p},$$

which can be rewritten as

$$n^8 \equiv -1 \pmod{p}.$$

Squaring both sides of the congruence we will obtain a new congruence:

$$n^{16} \equiv 1 \pmod{p}$$

Obviously if we raise this to any power k, the right side will be the same, i.e.,

$$n^{16k} \equiv 1 \pmod{p}$$

On the other hand, applying Fermat's Little Theorem, we have

$$n^{p-1} \equiv 1 \pmod{p}$$

Comparing two congruences we can state that prime factor must satisfy

$$p = 16k + 1.$$

Finally, besides 2, the other prime factors of $n^8 + 1$ must be odd and of the type $p = 16k + 1$. For example, we can check possible factors of $n^8 + 1$ among

$$p = 16k + 1: \quad 17, 97, 113, 193, \ldots$$

Similarly, any odd factors of $n^{16} + 1$ must be of the type $p = 32m + 1$, with the smallest factor of 97:

$$n^{16} + 1 \equiv 0 \pmod{p}$$
$$n^{16} \equiv -1 \pmod{p}$$
$$n^{32} \equiv 1 \pmod{p}$$
$$n^{32m} \equiv 1 \pmod{p}$$
$$n^{p-1} \equiv 1 \pmod{p}$$
$$p = 32m + 1.$$

Hence any odd prime factors of $n^{16} + 1$ must be of the type $p = 32m + 1$. Thus we can check possible factors of $n^{16} + 1$ among

$$p = 32m + 1: \ 97, 193, 257, \ldots$$

Therefore 17 is the least odd prime factor that can divide $n^8 + 1$ but not $n^{16} + 1$.
Answer. The least prime number that can divide $n^8 + 1$ but not $n^{16} + 1$ is $p = 17$.

Example.
Let us have an example for a small n, so $n^8 + 1$ can be easily evaluated and factored that will confirm our result.

$$3^8 + 1 = 6562 = 2 \cdot 17 \cdot 193.$$

Each prime factor, $17 = 16 \cdot 1 + 1$ and $193 = 16 \cdot 12 + 1$ is of the type $p = 16k + 1$.

Remark. The result of Problem 97 can help us in finding possible odd factors for "big" numbers. Consider $2017^8 + 1$. It is divisible by 2. We know that all other prime factors must fit the formula above. Let us check 17 as a start.

$$
\begin{aligned}
2017 &\equiv 11 \quad (\text{mod } 17) \\
2017^2 &\equiv 121 \quad (\text{mod } 17) \equiv 2 \quad (\text{mod } 17) \\
2017^4 &\equiv 4 \quad (\text{mod } 17) \\
2017^8 &\equiv 16 \quad (\text{mod } 17) \equiv -1 \quad (\text{mod } 17) \\
2017^8 + 1 &\equiv 0 \quad (\text{mod } 17) \\
2017^8 + 1 &= 2 \cdot 17 \cdot m \\
2017^{16} &\equiv 1 \quad (\text{mod } 17) \\
17 &\nmid 2017^{16} + 1.
\end{aligned}
$$

We found that 17 in fact divides $2017^8 + 1$ but not $2017^{16} + 1$. Please note that 17 is not always an odd prime factor of $n^8 + 1$.

Problem 98

Suppose that p is an odd prime, prove that

$$1^p + 2^p + 3^p + \cdots + (p-1)^p \equiv 0 \quad (\text{mod } p).$$

Proof. It is clear that numbers 1, 2, 3, ..., $(p-1)$ are relatively prime to p. By Theorem 13, the following statements are valid:

$$
\begin{aligned}
a^{p-1} &\equiv 1 \quad (\text{mod } p) \\
a^p &\equiv a \quad (\text{mod } p).
\end{aligned}
$$

Moreover, we also have

$$1^p \equiv 1 \quad (\mathrm{mod}\ p)$$
$$2^p \equiv 2 \quad (\mathrm{mod}\ p)$$
$$3^p \equiv 3 \quad (\mathrm{mod}\ p)$$

·····················

$$(p-1)^p \equiv (p-1) \quad (\mathrm{mod}\ p).$$

Adding all terms on the left and right sides we obtain

$$1^p + 2^p + 3^p + \cdots + (p-1)^p \equiv (1+2+3+\cdots+(p-1)) \quad (\mathrm{mod}\ p).$$

The sum of the first $(p-1)$ integers on the right can be evaluated as

$$\frac{1+(p-1)}{2} \cdot (p-1) = p \cdot \frac{p-1}{2}.$$

By the condition of the problem p is odd, then $(p-1)$ is an even number (multiple of 2), and $(p-1)/2$ is an integer. Therefore, we obtain

$$1^p + 2^p + 3^p + \cdots + (p-1)^p \equiv pk(\mod p) \equiv 0 \quad (\mathrm{mod}\ p).$$

2.5.1 Application of Fermat's Little Theorem to Finding the Last Digit of a^b

If you remember, we found the last digits in Section 2.4, but we did it empirically. Here we will apply Fermat's Little Theorem to finding the last digit or the last two digits of "big" numbers.

| Problem 99 |

What is the last digit of 2^{1995}?

Solution. We will use two important facts. The first is that the number divided by 5 or 10 leaves the same remainder as its last digit divided by 5 or 10. The second fact is Theorem 13, from which we have $a^{p-1} \equiv 1(\mod p)$. Therefore, selecting $a = 2$ and $p = 5$ we have

$$2^4 \equiv 1 \quad (\mathrm{mod}\ 5)$$
$$(2^4)^{498} = 2^{1995-3} \equiv 1 \quad (\mathrm{mod}\ 5).$$

Since $1995 = 4 \cdot 498 + 3 = 4k + 3$, we need to find out what remainder 2^3 will leave when divided by 5,

$$(2^4)^{498} \equiv 1 \quad (\mathrm{mod}\ 5), \tag{2.28}$$

$$8 = 2^3 \equiv 3 \quad (\text{mod } 5). \tag{2.29}$$

Multiplying the corresponding sides of (2.28) and (2.29) we obtain

$$2^{1995} \equiv 3 \quad (\text{mod } 5),$$

which means the that last digit of 2^{1995} divided by 5 leaves a remainder of 3. There are two possible numbers (less than or equal to 9 and greater or equal to 0), 3 and 8 that satisfy this condition. However, the last digit of an even number 2^{1995} must be even. Therefore correct answer is 8.

Answer. The last digit is 8.

Remark. We now can explain our empirical approach. Remember, we rewrote a power of 2 or 3 as $(4k + r)$ and depending on its remainder we made our conclusions about the last digit of 2^n or 3^n. This "magic" divisibility by 4 has a simple explanation: we can state that $2^4 \equiv 1(\mod 5)$ and $3^4 \equiv 1(\mod 5)$ (it also follows from Theorem 13).

If the power n of 2 or 3 is multiple of 4 then the remainder of 2^n or 3^n when divided by 5 will be also 1. If $n = 4k + r$, then remainder of 2^n or 3^n will depend on a remainder that 2^r or 3^r will leave when divided by 5.

Problem 100

What is the last digit of 3^{1995}?

Solution. Since $(3,5) = 1$, then solution to this problem is very similar to Problem 97:

$$(3^4)^{498} \equiv 1 \quad (\text{mod } 5), \tag{2.30}$$

$$3^3 = 27 \equiv 2 \quad (\text{mod } 5). \tag{2.31}$$

The product of (2.30) and (2.31) gives us

$$3^{1995} \equiv 2 \quad (\text{mod } 5),$$

which means that the last digit of 3^{1995} when divided by 5 leaves a remainder of 2. Since 3^{1995} is odd, then the last digit is 7.

Answer. The last digit is 7.

Remark. If p is not prime then Theorem 13 is not correct. Thus,

$$5^5 \equiv 5 \quad (\text{mod } 6)$$
$$2^8 \equiv 4 \quad (\text{mod } 9).$$

Question. Can we find the last two digits of the numbers above? Answer is simple! We want to use Theorem 13: a number divided by 4, 25, 50 or 100 leaves the same remainder as the number written by its last two digits. However, each number in the list $(4, 25, 50, 100)$ is not prime, so the Fermat's Little Theorem (Theorem 13) cannot be used!

It is time for us to learn a new theorem.

2.6 Euler's Formula

We have the following statement.

Theorem 14 (Euler's Formula)

Suppose that $m \in \mathbb{N}$, and $a \in \mathbb{Z} \backslash \{0\}$, where $(a, m) = 1$. Then

$$a^{\phi(m)} \equiv 1 \pmod{m}, \tag{2.32}$$

where $\phi(m)$ is the number of integers between 0 and m that are relatively prime to m, i.e., $\phi(m) = \{a : 1 \leq a \leq m, \ GCD(a, m) = 1\}$.

Thus, for $m = 10$ there are four integers between 0 and 10 that are relatively prime to 10: $m = 10$, $\phi(10) = 1, 3, 7, 9$. On the other hand, for $m = 9$ there are exactly 6 integers: $m = 9$, $\phi(9) = 1, 2, 4, 5, 7, 8$. If p is prime, then $\phi(p) = p - 1$.

We can make the following table.

m	1	2	3	4	5	6	7	8	9	10	11
$\phi(m)$	1	1	2	2	4	2	6	4	6	4	10

Table 2.3 Relatively prime

By this Table 2.3, the formula

$$7^4 \equiv 1 \pmod{10}$$

is valid. For us this can be interpreted as follows. The last digit of 7^4 when divided by 10 gives remainder of 1. It means that it is 1. We can see that it is true because $7^4 = 2401$.

In general, how do we compute $\phi(m)$? Let $m = p^k$ (a power of some prime, like $25 = 5^2$). $\phi(m)$ can be evaluated by subtracting integers that are the multiples of p:

$$1p, \ 2p, \ 3p, \ 4p, \ldots, (p^{k-1}-2)p, \ (p^{k-1}-1)p, \ p^k.$$

Then $\phi(p^k) = p^k - p^{k-1}$. For example,

$$\begin{aligned}
\phi(25) &= 5^2 - 5^1 &= 20 \\
\phi(121) &= 11^2 - 11^1 &= 110 \\
\phi(343) &= 7^3 - 7^2 &= 294.
\end{aligned}$$

Let us consider $m = 25$. We will record every number less than or equal to 25,

$$1,2,3,4,5,6,7,8,9,10,11,12,13,14,15,16,17,18,19,20,21,22,23,24,25.$$

There exactly 5 multiples of 25, and $25 - 5 = 20$ its relatively prime.
Considering $m = 343 = 7^3$. We have

$$\begin{aligned}
&1,2,3,4,5,6,7,8,9,10,11,12,13,14,15,16,17,18,19, \\
&20,21,22,23,24,25,26,27,28,29,\ldots,35,36,\ldots, \\
&42,43,\ldots,49,50,\ldots,56,57,\ldots,63,64,\ldots,70,71, \\
&\ldots,77,\ldots,84,85,\ldots,91,\ldots,342,343.
\end{aligned}$$

Formulas for $\phi(m)$ are as follows:

1. If $m = p^k$ is prime and $k \geq 1$, then

$$\phi(p^k) = p^k - p^{k-1} \tag{2.33}$$

2. If $m = p_1^{k_1} \cdot p_2^{k_2} \cdots \cdot p_r^{k_r}$, then

$$\phi(m) = \left(p_1^{k_1} - p_1^{k_1-1} \right) \left(p_2^{k_2} - p_2^{k_2-1} \right) \cdots \left(p_r^{k_r} - p_r^{k_r-1} \right) \tag{2.34}$$

3. If $GCD(m,n) = 1$, then

$$\phi(mn) = \phi(m)\phi(n) \tag{2.35}$$

4. If $p_1, \ p_2, \ \ldots, p_r$ are distinct primes that divide m, then

$$\phi(m) = m \left(1 - \frac{1}{p_1} \right) \left(1 - \frac{1}{p_2} \right) \cdots \left(1 - \frac{1}{p_r} \right). \tag{2.36}$$

Problem 101

Solve the equation $314^{164} \equiv x \pmod{165}$ using Euler's formula.

Solution. As you remember, we solved this problem in the previous section. Using Euler's formula from Theorem 14 gives the answer faster.

We have $165 = 3 \cdot 5 \cdot 11$, then $\phi(165) = (3-1) \cdot (5-1) \cdot (11-1) = 80$. Then,

$$314^{80} \equiv 1 \pmod{165}$$
$$(314^{80})^2 = 314^{160} \equiv 1 \pmod{165}$$
$$314^4 \equiv 31 \pmod{165}$$
$$314^{164} \equiv 31 \pmod{165}.$$

From this we find the answer.

Answer. $x = 31$.

The last problem of this section will demonstrate how to solve the congruence $a^n \equiv x$ $\pmod{m}$ when $(a, m) \neq 1$.

Problem 102

Find the remainder of $2^{7^{2018}}$ divided by 352.

Solution. This problem can be rewritten using congruence as

$$2^{7^{2018}} \equiv x \pmod{352}, \quad 0 \leq x < 352. \tag{2.37}$$

Using prime factorization of $a = 2^{7^{2018}}$ and $m = 352$, we can see that they are not relatively prime:

$$\left.\begin{array}{l} 352 = 2^5 \cdot 11 \\ 2^{7^{2018}} = 2^5 \cdot 2^{7^{2018}-5} \end{array}\right\} \Rightarrow x = 2^5 \cdot x_1$$

By dividing all terms of the original congruence by gcd=$(a, m) = 2^5$, we can rewrite it in a reduced form:

$$2^{7^{2018}-5} \equiv x_1 \pmod{11} \tag{2.38}$$

We know from Fermat's Little Theorem that

$$2^{10} \equiv 1 \pmod{11} \tag{2.39}$$

In order to use (2.39) we need to know the remainder of $7^{2018} - 5$ when divided by 10. Assume that such number, y, is found and that

$$7^{2018} - 5 \equiv y \pmod{10}, \quad 0 \leq y < 10.$$

Hence,

$$7^{2018} - 5 = 10k + y$$

and then

$$2^{7^{2018}-5} \equiv 2^y \pmod{11} \Leftrightarrow 2^y \equiv x_1 \pmod{11}$$

Next, we can find unknown y applying Euler's formula.
Because $(7, 10) = 1$, then

$$\varphi(10) = 2 \cdot 5 = (2-1) \cdot (5-1) = 4$$
$$7^4 \equiv 1 \,(mod\,10)$$
$$2018 = 4 \cdot 504 + 2$$
$$7^{2018} - 5 \equiv \left(7^4\right)^{504} \cdot 7^2 - 5$$
$$\equiv 7^2 - 5 \equiv 9 - 5 \equiv 4 \,(mod\,10)$$

Finally, because $y = 4$, the following is true.

$$2^4 = 16 \equiv 5 \,(mod\,11) \Rightarrow x_1 = 5$$
$$x = 2^5 \cdot x_1 = 32 \cdot 5 = 160.$$

Answer. $x = 160$.

2.6.1 Application of the Euler's Formula to Finding the Last Digits of a^b

Problem 103

What are the last two digits of 2^{2009}?

Solution. First we will find the last digits of the numbers using Theorem 13:

$$2^4 \equiv 1 \quad (\text{mod } 5)$$
$$(2^4)^{502} \equiv 1 \quad (\text{mod } 5)$$
$$2^1 \equiv 2 \quad (\text{mod } 5)$$
$$2^{2009} \equiv 2 \quad (\text{mod } 5).$$

Hence, the last digit of 2^{2009} is 2.

Next we will find the last two digits of the number, using the following statements:

1. A number divided by 4, 25, 50 or 100 leaves the same remainder as the number written by its last two digits.
2. Euler's Formula $a^{\phi(m)} \equiv 1 \pmod{m}$.

For $m = 25$ we know that $(2, 25) = 1$ and $\phi(25) = 20$, then the following is true:

$$2^{20} \equiv 1 \quad (\text{mod } 25)$$
$$2^{2000} = (2^{20})^{100} \equiv 1 \quad (\text{mod } 25)$$
$$2^9 = 512 \equiv 12 \quad (\text{mod } 25)$$
$$2^{2009} \equiv 12 \quad (\text{mod } 25).$$

Now we are looking for such a number that divided by 25 leaves a remainder of 12 and ends in 2. It can be either 12 or 62. Which number is correct? Since $2^{2009} =$

$4 \cdot 2^{2007}$, then the number written by the last two digits of 2^{2009} must be a multiple of 4. Only 12 is divisible by 4.

Answer. 12.

Problem 104

What are the last two digits of 3^{2009}?

Solution. Applying Theorem 13,

$$3^4 \equiv 1 \pmod{5}$$
$$(3^4)^{502} \equiv 1 \pmod{5}$$
$$3^1 \equiv 3 \pmod{5}$$
$$3^{2009} \equiv 3 \pmod{5},$$

we can state that the last digit of 3^{2009} is 3.

Next, we will find the last two digits of the number, using the following statements:

1. A number divided by 4, 25, 50 or 100 leaves the same remainder as the number written by its last two digits.
2. Euler's Formula $a^{\phi(m)} \equiv 1 \pmod{m}$.

For $m = 25$ we know that $(3, 25) = 1$ and $\phi(25) = 20$, then

$$3^{20} \equiv 1 \pmod{25}$$
$$(3^{20})^{100} \equiv 1 \pmod{25}$$
$$3^9 = 19683 \equiv 8 \pmod{25}$$
$$3^{2009} \equiv 8 \pmod{25}.$$

Now we are looking for such a number that divided by 25 leaves a remainder of 8 and ends in 3. The answer is not unique: it can be either 33 or 83. Does it mean that we cannot find the last two digits of 3^{2009}? Since the $GCD(3, 100) = 1$, we can use divisibility of a number by 100. (Recall that a number divided by 100 gives the same remainder as the number written by its last two digits.) Consider

$$100 = 2^2 \cdot 5^2 \Rightarrow \phi(100) = (2^2 - 2^1)(5^2 - 5^1) = 40$$
$$3^{40} \equiv 1 \pmod{100}$$
$$(3^{40})^{50} \equiv 1 \pmod{100}$$
$$3^9 = 19683 \equiv 83 \pmod{100}$$
$$3^{2009} \equiv 83 \pmod{100},$$

a number 83 is the only two digits number satisfying our problem. It is unique.

Answer. 83.

By solving the following problem, you will learn how to find the last three digits. Should you have your own ideas then do not look at my solution but rather do it by yourself.

Problem 105

Find the last three digits of 3^{2009}.

Solution. In order to find the last 3 digits of a number, we can divide it by 125:

$$3^{5^3 - 5^2} \equiv 1 (mod\,125)$$
$$\left(3^{100}\right)^{20} \equiv 1 (mod\,125)$$
$$19683 = 3^9 \equiv 58 (mod\,125)$$
$$3^{2009} \equiv 58 (mod\,125)$$

Unfortunately, among all possible "candidates" there are two odd numbers: 183 and 683, which end in 83.

Next, we will use division by 1000, and

$$1000 = 2^3 \cdot 5^3 \Rightarrow \phi(1000) = (2^3 - 2^2)(5^3 - 5^2) = 400$$
$$3^{400} \equiv 1 \quad (mod\ 1000)$$
$$(3^{400})^5 \equiv 1 \quad (mod\ 1000)$$
$$3^9 = 19683 \equiv 683 \quad (mod\ 1000)$$
$$3^{2009} \equiv 683 \quad (mod\ 1000).$$

We can see that the last three digits of the number are unique. It is 683.
Answer. 683.

Problem 106

Find the last three digits of 557^{2012}.

Solution. First we will find the last digit. Applying Fermat's Little Theorem, we can state that

$$557^4 \equiv 1\,(mod\,5)$$

Raising both sides of this congruence to the power of 503, we obtain $\left(557^4\right)^{503} = 557^{2012} \equiv 1\,(mod\,5)$, which indicates that the last digit is 1. Next, we have to find the last two digits. Let us use Euler's formula and evaluate

$$\varphi(100) = \varphi\left(2^2 \cdot 5^2\right) = \left(2^2 - 2^1\right)\left(5^2 - 5^1\right) = 40.$$

Now we obtain a new congruence based on Euler's formula,

$$557^{40} \equiv 1 \,(mod\,100)$$

Since $2012 = 40 \cdot 50 + 12$, then we can state that $557^{2000} \equiv 1 \,(mod\,100)$ and we need to solve

$$557^{12} \equiv x\,(mod\,100).$$

We will find the answer starting from the congruence for the first power of 557 and by continuing to square it,

$$557 \equiv 157 \,(mod\,100)$$
$$557^2 \equiv 49 \,(mod\,100)$$
$$557^4 \equiv (50-1)^2 \,(mod\,100)$$

where we replaced 49 by the difference of 50 and 1. Now if we square this difference, we can rewrite it as follows $557^4 \equiv 1 \,(mod\,100)$ and then raising both sides to the cube,

$$557^{12} \equiv 557^{2012} \equiv 1 \,(mod\,100).$$

Now we know that the last two digits are 01. Finally we will check congruence modulo 1000.

$$\varphi(1000) = \varphi\left(2^3 \cdot 5^3\right) = \left(2^3 - 2^2\right)\left(5^3 - 5^2\right) = 400.$$

$$557^{2000} \equiv 557^{400} \equiv 1 \,(mod\,1000)$$
$$557^{12} \equiv x\,(mod\,1000)$$

Considering powers of 557 modulo 1000.

$$557 \equiv 557 \,(mod\,1000)$$
$$557^2 \equiv 249 \,(mod\,1000)$$
$$557^4 \equiv 62001 \,(mod\,1000) \equiv 001 \,(mod\,1000)$$
$$557^{2012} \equiv 001 \,(mod\,1000).$$

Answer. Last three digits are 001.

Now you can use similar approaches in finding the last digits of "big" numbers and apply them to other problems yourself.

2.7 Methods of Proof

In this book, we have already had many problems on proofs. We proved statements directly, indirectly, by contradiction, and mentioned Proof by Induction. Proofs have been conducted by people since Ancient time. Ancient Babylonians, Egyptians and Greeks wanted to know if this or that property was valid in a general case, so they could use it in building pyramids or construction of palaces for their pharaohs. At that time the apparatus of algebra was not developed and hence, ancient people tried to prove things geometrically. For example, ancient people that lived 2000 BC knew many summation formulas that we use in mathematics every day. They knew how to evaluate the sum of the first n consecutive natural numbers, the sum of their squares or cubes. The difference between us and them now is that we write summations formulas using, for example, sigma notation, the basic property of which are listed below.

Important Properties of Sigma Notation

1. $\sum_{k=1}^{n} a \cdot b_k = a \cdot \sum_{k=1}^{n} b_k$ (a constant can be put before the summation)
2. $\sum_{m}^{n} a_k \pm \sum_{m}^{n} b_k = \sum_{m}^{n} (a_k \pm b_k)$
3. $\sum_{1}^{n} (b_k \pm m) = \sum_{1}^{n} b_k \pm n \cdot m$ (because the number m appears n times)
4. $\sum_{k=1}^{n} a_k = \sum_{k=2}^{n+1} a_{k-1} = \sum_{k=m+1}^{n+m} a_{k-m}$

using sigma notation we can write down some very useful summation formulas.

$$\sum_{k=1}^{n} k = \frac{n(n+1)}{2} \tag{2.40}$$

$$\sum_{k=1}^{n} k^2 = \frac{n(n+1)(2n+1)}{6} \tag{2.41}$$

$$\sum_{k=1}^{n} k^3 = \left(\sum_{k=1}^{n} k \right)^2 = \frac{n^2(n+1)^2}{4} \tag{2.42}$$

Example.

$$\sum_{1}^{100} (n^2 + 5) = \sum_{1}^{100} n^2 + 100 \cdot 5 = 500 + \sum_{1}^{100} n^2$$
$$= 500 + \frac{100 \cdot 101 \cdot (2 \cdot 100 + 1)}{6} = 338,850.$$

Some of these formulas we will prove in this section.

2.7.1 Geometric Proof by Ancient Babylonians and Greeks

Ancient people knew all these summation formulas and proved them using a geometric approach. The formulae list could be continued to demonstrate, for example, the sum of the first fourth or the first fifth consecutive powers of natural numbers, which we cannot find in the work of ancient mathematicians. It is understood, that there was a restriction for such proofs because of dimensionality and the lack of knowledge of irrational numbers, etc. However, an interest in geometric proofs recently has been refreshed and because of visual effect associated with such proofs, many students like and understand geometric proofs better than a "standard", algebraic approach. In my three previous books I demonstrated geometric proofs a lot, so in this book I will just mention some interesting ideas developed by Ancient mathematicians, some of the ideas are very useful in modern time and even gave a start to new fields of mathematics.

Problem 107

Prove that the sum of all natural numbers from 1 to N equals

$$1+2+3+4+5+6+\ldots+N = \frac{N(N+1)}{2}.$$

Proof. (Approach known to ancient Greeks) Such a construction can be reproduced using billiard balls. Imagine a right triangle with the legs of length 6 made by the white balls. Make a similar right triangle out of red balls and assuming that such a creation keeps its shape, we can stick two triangles together as shown in Fig. 2.1. It is clear that two triangles together form a rectangle with one (vertical) side of 6

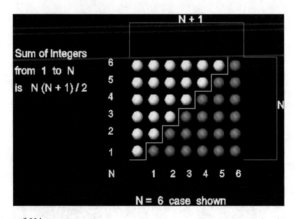

Fig. 2.1 The sum of N integers

and the other (horizontal) side of 7. The entire rectangle of the billiard balls now has $6 \cdot 7 = 42$ balls. If we look closely at this construction, we can see that starting from the very left corner (one white ball) and by moving up to 2 balls, 3 balls,..., 6 balls, we can this way add all the balls $1+2+3+4+5+6$ inside the white triangle. On the other hand, the same answer can be obtained by dividing 42 by 2. If instead of 6 we have N rows, then the answer for the sum of all natural numbers from 1 to N is $\frac{N(N+1)}{2}$. Additionally, it is known that ancient Greeks also geometrically proved a modification of (2.40) as stated in the following problem.

Problem 108

Prove that the sum of n consecutive odd numbers is a perfect square.

Proof.

The successive numbers added to 1 are 3, 5, 7, ..., (Fig. 2.2) that is to say, the successive odd numbers. The method of construction shows that the sum of any number of consecutive terms of the series of the odd numbers 1, 3, 5, 7 ....(starting from 1) is a square, and in fact $1+3+5+...+(2n-1)=n^2$, while the addition of the next odd number $(2n+1)$ makes the next higher square, $(n+1)^2$, e.g.,

$$1+3 = 4 = 2^2$$

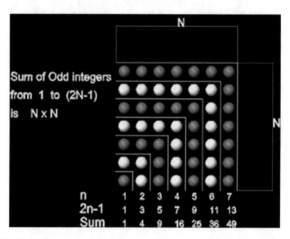

Fig. 2.2 Sum of odd numbers

Remark. An algebraic proof of this formula can be obtained in several ways, for example, using sigma notation,

$$1+3+5+...+2n-1 = \sum_{k=1}^{n}(2k-1) = 2\cdot\frac{n(n+1)}{2}-n = n^2.$$

2.7.2 Direct Proof

In this book, we have already had many problems on proofs. Most of such problems demanded proof that some number or expression dependent on a whole number n is divisible or not divisible by another number k or by an expression in variable $n \in \mathbb{N}$. Solving such problems can be done directly by decomposition of the given expression into a few others, the divisibility of which is obvious. In such case, we often used the fact that a product of k consecutive numbers is divisible by k. Usually direct proof requires creativity.

Please see the following problems on divisibility and summation.

Problem 109

Prove that if $(3x + 7y)$ is divisible by 19 for any integers x and y, then $(43x + 75y)$ is also divisible by 19.

Proof. Let us rewrite $(43x + 75y)$ as follows:

$$43x + 75y = 76x - 33x + 152y - 77y = 19 \cdot (4x + 8y) - 11 \cdot (3x + 7y) = 19 \cdot k.$$

The proof is complete.

Problem 110

Prove that the sum of all natural numbers from 1 to $(n - 1)$ is divisible by n.

Proof. Consider $1 + 2 + 3 + \cdots + (n - 1)$. How do we find this sum? Some of you may remember the story of how 10-year-old Carl Friedrich Gauss in math class added all natural numbers from 1 to 100. He wrote them in two rows, like

$$1 + 2 + 3 + \cdots + 98 + 99 + 100, \text{ ascending order,}$$
$$100 + 99 + 98 + \cdots + 3 + 2 + 1, \text{ descending order.}$$

He noticed that the sum of each column is $1 + 100 = 101$. There are 100 columns and the sum of each row is the same—his unknown. So he multiplied 101 by 100 than divided by 2 because we need only a single sum not a double.

$$S_{100} = \frac{100(1 + 100)}{2} = 50 \cdot 101 = 5050$$

Following Gauss's idea, we can derive the formula,

$$1 + 2 + 3 + \ldots + (n-1), \quad \text{ascending order,}$$
$$(n-1) + 99 + 98 + \ldots + \; 1, \text{ descending order.}$$

The sum in each column is n, we have exactly $(n-1)$ such columns. The answer is $S_n = \frac{n(n-1)}{2}$. This is a multiple of n.

Problem 111

Let $\{a_n\}$ be an arithmetic progression of integer numbers with first term a_1 and common difference d. Prove that the sum of its first n terms can be evaluated as

$$S_n = \frac{(2a_1 + (n-1)d)}{2} \cdot n$$

Proof. Let us rewrite S_n in two different ways:

$$S_n = a_1 + (a_1 + d) + (a_1 + 2d) + \ldots + (a_1 + (n-2)d) + (a_1 + (n-1)d)$$
$$S_n = a_n + (a_n - d) + (a_n - 2d) + \ldots + (a_n - (n-2)d) + (a_n - (n-1)d).$$

Again if we add elements by columns, in each column we will get $(a_1 + a_n)$. There are n such columns, so $2S_n = n(a_1 + a_n)$ or dividing both sides by 2,

$$S_n = \frac{(a_1 + a_n)}{2} \cdot n.$$

Replacing the nth term, a_n, in terms of a_1 and d we obtain another form for S_n,

$$S_n = \frac{(2a_1 + (n-1)d)}{2} \cdot n.$$

We leave it to the reader to obtain this form on her own. Notice that this formula can be more useful, because a_n is usually unknown.

Problem 112

Prove that the sum of the squares of the first $(n-1)$ natural numbers is a multiple of n.

Proof. We need to prove that the following relationship is true:

$$N = 1^2 + 2^2 + 3^2 + 4^2 + \cdots + (n-2)^2 + (n-1)^2 = n \cdot k.$$

Arranging sums in ascending and descending order does not help. We need to find a different approach. Let us consider the difference of two consecutive cubes:

$$
\begin{aligned}
1^3 - 0^3 &= 3 \cdot 1^2 - 3 \cdot 1 + 1 \\
2^3 - 1^3 &= 3 \cdot 2^2 - 3 \cdot 2 + 1 \\
3^3 - 2^3 &= 3 \cdot 3^2 - 3 \cdot 3 + 1 \\
&\cdots\cdots\cdots\cdots\cdots \\
(n-2)^3 - (n-3)^3 &= 3 \cdot (n-2)^2 - 3 \cdot (n-2) + 1 \\
(n-1)^3 - (n-2)^3 &= 3 \cdot (n-1)^2 - 3 \cdot (n-1) + 1.
\end{aligned}
$$

Adding the left and the right sides we obtain that

$$(n-1)^3 = 3[1^2 + 2^2 + 3^2 + \cdots + (n-1)^2] - 3(1 + 2 + 3 + \cdots + (n-1)) + 1(n-1).$$

Our unknown N is inside the brackets. Evaluating the sum of the $(n-1)$ natural numbers from 1 to $n-1$ as $\frac{(1+n-1)\cdot(n-1)}{2}$ and after simplification, this equation can be written as

$$(n-1)^3 = 3 \cdot N - 3 \cdot \left(\frac{(n) \cdot (n-1)}{2} \right) + (n-1).$$

Solving this for N we obtain,

$$
\begin{aligned}
N &= \frac{(n-1)^3 - (n-1) - \frac{3n(n-1)}{2}}{3} \\
&= \frac{(n-1)(2(n-1)^2 - 2 - 3n)}{6} \\
&= \frac{(n-1)n(2n-7)}{6} = n \cdot k.
\end{aligned}
$$

Therefore the sum of the first $(n-1)$ squares of natural numbers is always divisible by n.

2.7.3 Proofs by Contradiction

Suppose we need to prove that an equation does not have solution in integers or that some statement is not true for any values of the given variables. Such proofs are usually done by contradiction.

> **Problem 113**
>
> Proof that numbers $\sqrt{2}, \sqrt{3}, \sqrt{5}$ cannot be the terms of an arithmetic sequence.

When I give this problem to my students they very often say that irrational numbers $\sqrt{2}, \sqrt{3}, \sqrt{5}$ cannot differ from each other by the same number. Others take calculators and try to validate this statement by estimation: $\sqrt{3} - \sqrt{2} \approx 0.318$,

$\sqrt{5} - \sqrt{3} \approx 0.514$. However, estimation on a calculator cannot be considered as rigorous proof. Moreover, even if we accept the fact that the consecutive differences are not the same, we still have to prove that the numbers cannot be just three, not necessary neighboring terms of the arithmetic sequence.

Let us provide a correct proof by contradiction.

Proof. Assume that $\sqrt{2}$, $\sqrt{3}$, $\sqrt{5}$ are the kth, the mth and the nth terms of the arithmetic sequence with the first term of a_1 and common difference of d:

$$\begin{aligned}
\sqrt{2} &= a_k = a_1 + (k-1)d \\
\sqrt{3} &= a_m = a_1 + (m-1)d \\
\sqrt{5} &= a_n = a_1 + (n-1)d.
\end{aligned}$$

Subtracting the second and the first equations, and then the third and the second. Finally, dividing the results of subtractions we obtain:

$$\frac{\sqrt{3} - \sqrt{2}}{\sqrt{5} - \sqrt{3}} = \frac{m-k}{n-m}. \tag{2.43}$$

The right side of (2.43) is a rational number, because m, k, n are natural numbers. Denote this number by r:

$$r = \frac{\sqrt{3} - \sqrt{2}}{\sqrt{5} - \sqrt{3}}. \tag{2.44}$$

Expression (2.44) can be written as

$$r(\sqrt{5} - \sqrt{3}) = \sqrt{3} - \sqrt{2}. \tag{2.45}$$

Squaring both sides of (2.45) and collecting radicals on the left-hand side, we obtain:

$$r^2\sqrt{15} - \sqrt{6} = \frac{8r^2 - 5}{2}. \tag{2.46}$$

The right side of (2.46) is again a rational number and we can denote it by s:

$$r^2\sqrt{15} - \sqrt{6} = s.$$

Squaring both sides again, after simplification we have:

$$\sqrt{10} = \frac{15r^4 - s^2 + 6}{6r^2}. \tag{2.47}$$

Expression (2.47) indicates that $\sqrt{10}$ is a rational number. However, $\sqrt{10}$ is irrational. Therefore, we obtained a contradiction. Our original assumption was wrong and $\sqrt{2}$, $\sqrt{3}$, $\sqrt{5}$ cannot be terms of the same arithmetic sequence. The following problem involve prime numbers.

> ### Problem 114
>
> Prove that there is no infinite arithmetic progression of only prime numbers.

Proof. Assume that such progression exists and has the first term $a \neq 1$ and common difference d. Then the nth term of this progression can be written as

$$a_n = a + (n-1)d.$$

Clearly, if

$$n = a+1 \;\Rightarrow\; a_n = a + a \cdot d = a(d+1).$$

Thus, the first and $(a+1)$st term, a_{a+1}, of such arithmetic progression are not relatively prime, and this fact does not depend on the value of the common difference. Moreover, in such infinite progression all terms sitting in the positions of $n = a+1,\ 2a+1,\ 3a+1,\ 4a+1,\ldots$ will be multiples of the first term, a.

For example, in the progression $\underline{3},\ 7,\ 11,\ \underline{15},\ 19, 23, \underline{27}, 31, 35, \underline{39}$, there are infinitely many members divisible by 3, we underlined some of them. All of them are in the positions $4,\ 7,\ 10,\ \ldots (3k+1),\ldots.$ We obtained the contradiction.

Remark. Because our proof was based on the assumption that the first term of a progression is not one, a reasonable question is what if the first term of an infinite progression equals 1? Can such progression consist of only primes?

The answer is also "no" and the proof of this fact is very similar to the proof above. We just for any given progression start our arguments from the second term. Thus, infinite arithmetic progression $\{a_n\} : 1,\ 1+d,\ 1+2d,\ 1+3d,\ldots$ contains progression $\{b_n\} : 1+d,\ 1+2d,\ 1+3d,\ldots$ the first term of which equals the second term of the first progression, and then again prove that there are infinitely many terms divisible by $(1+d)$.

Remark. Any infinite arithmetic progression with natural members will have infinitely many multiples of the first, second, third or any other term and the location of such multiples will depend only on the value of the selected term of a progression. Suppose a number $b \in \mathbb{N}$ is a term of an infinite arithmetic progression, then there are infinitely many terms divisible by b in the relative location $n = b+1,\ 2b+1,\ 3b+1,\ldots.$ Thus if b is the kth term of the given progression, then all terms divisible by it will have positions of $k,\ k+b,\ k+2b,\ k+3b,\ \ldots.$

For example, since 11 is the third term of the given infinite progression,

$$3,\ 7,\ 11,\ 15,\ 19,\ 23,\ 27,\ 31,\ 35,\ 39,\ 43,\ 47,\ 51,\ 55,\ 59,\ 63,\ \ldots,$$

the terms divisible by 11 will appear at the positions

$$3,\ 3+11 = 14,\ 3+2\cdot 11 = 25,\ 3+3\cdot 11 = 36,\ \ldots,\ 3+(m-1)\cdot 11,\ \ldots,$$

where m represents the mth consecutive multiple of 11. You can see it yourself, 11 is the first multiple of 11, the second is 55, which is 14th term of the given progression, then the third consecutive multiple of 11 in the progression will correspond to the index $n = 25$ and will be evaluated as $3 + (25 - 1) \cdot 4 = 99$, etc.

We just proved that there is no infinite arithmetic progression that consists of only primes. Is this statement also true for a finite arithmetic progression? The shortest sequence of primes must contain three terms. We can see that the first three terms of the infinite progression discussed above, $\{3, 7, \text{and } 11\}$ are in arithmetic progression given by formula $a_n = 4n - 1$, $n = 1, 2, 3$. Are there arithmetic progressions with precisely 5, 10 or N prime numbers? The answer is yes, such progressions exist but it is hard to find them. You can find more information in my book "Methods of Solving Sequence and Series Problems", page 116–120 or in the book of Sierpinski([20]).

2.7.4 Mathematical Induction

The premise behind this method is intuitive. If a particular statement S_n (mathematical equation or inequality involving variable n) is true for $n = 1$ and if S_n is true for $n = k$ implying that S_n is true for $n = k + 1$, then S_n is true for every positive integer n. In some way the Ancient people's geometric proof could be classified as inductive proof, just sometimes not rigorous enough. The Method of Mathematical induction is very helpful in providing many statements about positive integers. According to this method, a mathematical statement involving the variable n can be shown to be true for any positive integer n by proving the following two statements:

1. The statement is true for $n = 1$.
2. If the statement is true for any positive integer k, then it is also true for $(k + 1)$.

Let us learn how to apply this method to the problems below.

Problem 115

Prove that $5^n - 3^n + 2n$ is divisible by 4 for any natural number n $(n \in \mathbb{N})$.

Proof. Let us check that the statement is true for $n = 1$, $5^1 - 3^1 + 2 \cdot 1 = 4$. Yes, 4 is divisible by 4. Assuming that expression $S(k) = 5^k - 3^k + 2k$ is divisible by 4 for $n = k$, let us show that $S(n)$ is true for $n = k + 1$. Thus, $S(k + 1) = 5^{k+1} - 3^{k+1} + 2(k + 1) = 4 \cdot m$. Now we are going to rewrite $S(k + 1)$ as a sum of $S(k)$ (a multiple of 4) and some other terms, and then we will show that other terms are divisible by 4 as well. We have:

$$\begin{aligned} S(k+1) &= 5(5^k - 3^k + 2k) + 2 \cdot 3^k - 8k + 2 \\ &= 5(5^k - 3^k + 2k) + 2(3^k + 1) - 8k. \end{aligned} \tag{2.48}$$

From (2.48) we notice that $S(k+1)$ consists of $5S(k)$, a multiple of 4, a term $(-8k)$ that is obviously divisible by 4, and a term $2(3^k+1)$, whose divisibility by 4 is not clear at first glance. However, in the previous sections we showed that every number 3^k+1 is always an even number, then it is at least a multiple of 2, therefore, $2(3^k+1)$ is divisible by 4. Since $S(k+1)$ is divisible by 4, then $5^n - 3^n + 2n$ is divisible by 4 for all $n \in \mathbb{N}$.

Problem 116

Prove that $7^n + 12n + 17$ is divisible by 18 for all $n \in \mathbb{N}$.

Proof. If $n = 1$ then $7^1 + 12 \cdot 1 + 17 = 36$, yes it is divisible by 18, and $S(1)$ is true. Suppose that $S(k) = 7^k + 12k + 17$ is divisible by 18. Let us show that $S(k+1)$ is divisible by 18 as well. We have

$$S(k+1) = 7^{k+1} + 12(k+1) + 17 = 7 \cdot 7^k + 12k + 29$$
$$= 7(7^k + 12k + 17) - 72k - 90. \tag{2.49}$$

Noticing that expression (2.49) consists of three terms; each of them a multiple of 18, we conclude that $S(k+1)$ is divisible by 18, and $7^n + 12n + 17$ is divisible by 18 for all $n \in \mathbb{N}$. Having solved the last two problems, I hope you understand the Method of Mathematical Induction and how to use it. We applied this method and it worked very well. However, it is sometimes difficult to decide whether to apply mathematical induction or to prove a statement directly using divisibility of integers. There is no written rule on which method is better, but practicing more you will be able to choose the right and easiest one. My own experience advises to use mathematical induction if it is needed to prove divisibility of some expression containing n within exponents, and to use direct proof for most polynomial expressions in n.

Let us prove Problem 5 using Mathematical Induction.

Problem 117

Prove that $n^3 + 6n^2 - 4n + 3$ is divisible by 3 for any natural number n.

Proof. We have

1. If $n = 1$, then $S(1) = 1 + 6 - 4 + 3 = 6$ is divisible by 3. True.
2. Assuming that $S(k) = k^3 + 6k^2 - 4k + 3$ is divisible by 3, let us show that $S(k+1)$ is divisible by 3 as well. Thus,

$$S(k+1) = (k+1)^3 + 6(k+1)^2 - 4(k+1) + 3$$
$$= k^3 + 3k^2 + 3k + 1 + 6(k^2 + 2k + 1) - 4k - 4 + 3$$
$$= k^3 + 3k^2 + 3k + 1 + 6k^2 + 12k + 6 - 4k - 4 + 3$$
$$= k^3 + 6k^2 - 4k + 3 + 3k^2 + 3k + 12k + 6$$
$$= S(k) + 3k^2 + 15k + 6$$

All terms of $S(k+1)$ are multiples of 3, then $S(k+1)$ is divisible by 3, and $n^3 + 6n^2 - 4n + 3$ is divisible by 3 for any natural number n.

Considering both proofs, direct and math induction, I think the first one was more elegant, but this is up to you.

Let us now prove by Induction the formula for the sum of the first n cubes.

Problem 118

Prove that $1^3 + 2^3 + ... + n^3 = (1 + 2 + ... + n)^2$ or that

$$\sum_{k=1}^{n} k^3 = \left(\sum_{k=1}^{n} k \right)^2 = \frac{n^2(n+1)^2}{4}.$$

Proof. Step 1. Replacing n by 1 in the above equality gives
$1^3 = 1^2$ which is true, so $n = 1$ satisfies the equation.
Step 2. Assume that the equality is true at $n = k$ and let us show that it will be true at $n = k + 1$:
If $1^3 + 2^3 + 3^3 + ... + k^3 = (1 + 2 + 3 + ... + k)^2 = \frac{k^2(k+1)^2}{4}$ is true, then let us show that for $n=k+1$ the left side of the given equality equals

$$(1 + 2 + 3 + ... + k + 1)^2 = \frac{(k+1)^2(k+2)^2}{4}.$$

We can state that

$$1^3 + 2^3 + 3^3 + ... + k^3 + (k+1)^3 = (1 + 2 + 3 + ... + k)^2 + (k+1)^3.$$

Replacing the right-hand side, putting fractions over the common denominator and factoring, we obtain the required formula,

$$\frac{k^2(k+1)^2}{4} + (k+1)^3 = \frac{k^2(k+1)^2 + 4(k+1)^3}{4}$$
$$= \frac{(k+1)^2(k^2 + 4k + 4)}{4}$$
$$= \frac{(k+1)^2(k+2)^2}{4}$$

The final equality proves that the equation is true for $n = k + 1$, assuming that it is true for $n = k$. Using the principle of mathematical induction, we have completed our proof.

You can use Math Induction to prove other formulas of summation as a homework exercise.

2.7.5 Using Analysis and Generating Functions

This method is well explained in my book 3 "Methods of Solving Sequence and Series Problems". It is very useful in deriving combinatorial formulas or some formulas for the nth term obtained recursively. Remember when we solved Problem 93 of this chapter we found that

$$2^{3^{4^5}}$$

can be evaluated in five different ways. Assume that we want to find in how many ways we can evaluate similar expression if there are not four but 23 consecutive numbers from 2 to 24. Next, we will derive the formula for the nth Catalan number by solving the following problem on proof.

Problem 119

Given $x_1 * x_2 * x_3 * \ldots * x_{n-1} * x_n$, where $*$ represents an operation, for example, raising to a power. Prove that the number of ways to place parentheses between these numbers equals $k_n = \frac{1}{n} \cdot C_{2n-2}^{n-1}$, where $C_n^k = \frac{n!}{k!(n-k)!}$.

Proof. Let us make a table and organize our possible cases by putting a major operation (a star) between two factors. For example, we can imagine the given expression as the first term multiplied by the combination of the remaining $(n-1)$ terms. In the second column we will express the number of such cases as the product of two Catalan numbers, k_1 for the first factor and k_{n-1} for all possibilities of placing parentheses inside the second factor (it is ok that we do not know it yet). The second row represents the case when we group the first two numbers and separately the remaining $(n-2)$ terms, which on the right can be written as the product of k_2 and k_{n-2}, etc.

Obviously, the left column of the Table 2.4 describes all possible cases of putting parentheses between n terms. Then the nth Catalan number is the sum of all numbers in the right column of the Table 2.4. It can be written recurrently as follows

$$k_n = k_1 k_{n-1} + k_2 k_{n-2} + \ldots + k_m k_{n-m} + \ldots + k_{n-1} k_1. \tag{2.50}$$

Using this formula we can find as many consecutive Catalan numbers as we wish. Thus, six of them are

Case	The number of ways to calculate Catalan number
$x_1 \star (x_2 * x_3 * \ldots * x_{n-1} * x_n)$	$k_1 \cdot k_{n-1}$
$(x_1 * x_2) \star (x_3 * \ldots * x_{n-1} * x_n)$	$k_2 \cdot k_{n-2}$
$(x_1 * x_2 * x_3) \star (x_4 * \ldots x_{n-1} * x_n)$	$k_3 \cdot k_{n-3}$
...	...
$(x_1 * x_2 \ldots * x_m) \star (x_{m+1} * \ldots * x_n)$	$k_m \cdot k_{n-m}$
...	...
$(x_1 * x_2 * x_3 * \ldots * x_{n-1}) \star (x_n)$	$k_{n-1} \cdot k_1$

Table 2.4 Calculation of Catalan numbers

$$
\begin{aligned}
k_1 &= 1, \\
k_2 &= k_1 \cdot k_1 = 1 \cdot 1 = 1, \\
k_3 &= k_1 \cdot k_2 + k_2 \cdot k_1 = 1 \cdot 1 + 1 \cdot 1 = 2, \\
k_4 &= k_1 \cdot k_3 + k_2 \cdot k_2 + k_3 \cdot k_1 = 1 \cdot 2 + 1 \cdot 1 + 2 \cdot 1 = 5, \\
k_5 &= k_1 \cdot k_4 + k_2 \cdot k_3 + k_3 \cdot k_2 + k_4 \cdot k_1 = 1 \cdot 5 + 1 \cdot 2 + 2 \cdot 1 + 5 \cdot 1 = 14 \\
k_6 &= \ldots = 1 \cdot 14 + 1 \cdot 5 + 2 \cdot 2 + 5 \cdot 1 + 14 \cdot 1 = 42.
\end{aligned}
$$

Recursions are very useful in order to evaluate the numbers using computers. However, if we want to know 50th or 125th Catalan number by hand, recursive formulas are not very efficient. In order to derive the formula for the nth Catalan number, consider an infinite series representing a function

$$
f(x) = k_1 x + k_2 x^2 + k_3 x^3 + k_4 x^4 + k_5 x^5 + \ldots, \tag{2.51}
$$

where k_i is the ith Catalan number. This series is convergent when $x \longrightarrow 0$. Without loss of generality (WLG) we can set that $x < 1/4$. If the series is convergent then its sum is bounded and we can multiply this series, which will be also bounded. Consider the square of $f(x)$,

$$
\begin{aligned}
f^2(x) &= (k_1 x + k_2 x^2 + k_3 x^3 + \ldots) \cdot (k_1 x + k_2 x^2 + k_3 x^3 + \ldots) \\
&= k_1 k_1 x^2 + (k_1 k_2 + k_2 k_1) x^3 + (k_1 k_3 + k_2 k_2 + k_3 k_1) x^4 \\
&\quad + (k_1 k_{n-1} + k_2 k_{n-2} + \ldots + k_{n-1} k_1) x^n + \ldots
\end{aligned} \tag{2.52}
$$

Let us now compare the coefficients of the powers of x with the recursive formula above. You can notice that each coefficient is the corresponding ith Catalan number! Let us rewrite the square of the function as follows

$$
f^2(x) = k_2 x^2 + k_3 x^3 + k_4 x^4 + \ldots + k_n x^n + \ldots \tag{2.53}
$$

Comparing two formulas for $f(x)$ and $f^2(x)$, we can see that they differ by only one term, $k_1 x$. Since $k_1 = 1$, the following equation is valid:

$$
f^2(x) = f(x) - x, \tag{2.54}
$$

which can be solved using a quadratic formula. The solution is

$$f^2(x) - f(x) + x = 0$$
$$f_{1,2}(x) = \frac{1 \pm \sqrt{1-4x}}{2}.$$

Our function must be given by a unique formula. In order to find the correct formula (with plus or minus) let us look at $f(x)$ given by infinite series again. Since $f(0) = 0$, then the correct solution is one with the minus sign between two terms!

$$f(x) = \frac{1 - \sqrt{1-4x}}{2} = \frac{1}{2} - \frac{1}{2} \cdot (1-4x)^{\frac{1}{2}}. \tag{2.55}$$

Next, we will obtain all the coefficients of (2.51) (Catalan numbers) by taking consecutive derivatives of (2.51) and (2.55), and by equating the corresponding expressions at $x = 0$. The first derivative of $f(x)$ using (2.55) on one hand equals

$$f'(x) = -\frac{1}{2} \cdot \frac{-4}{2 \cdot \sqrt{1-4x}} = \frac{1}{\sqrt{1-4x}} = (1-4x)^{-\frac{1}{2}}.$$

then at zero it is equal to

$$f'(0) = 1. \tag{2.56}$$

Differentiating (2.51) (infinite series), we obtain that

$$f'(x) = k_1 + 2k_2 x + 3k_3 x^2 + 4k_4 x^3 + \ldots$$

and at zero

$$f'(0) = k_1 \tag{2.57}$$

which gives us correct formula

$$k_1 = 1.$$

Continue differentiating and evaluate the second derivatives of (2.51) and (2.55):

$$f''(x) = (-1/2)(1-4x)^{-3/2}(-4) = 2(1-4x)^{-\frac{3}{2}},$$

which at zero equals

$$f''(0) = 2. \tag{2.58}$$

Differentiating the first derivative of the infinite series:

$$f''(x) = 2k_2 + 3 \cdot 2k_3 \cdot x + 4 \cdot 3k_4 \cdot x^2 + \ldots$$

which at zero is

$$f''(0) = 2k_2. \tag{2.59}$$

Equating both values of the second derivatives we get

$$k_2 = 1. \tag{2.60}$$

which is obviously true because there is only one operation possible for two terms $(a \star b)$. Finding the 3rd derivative results in

$$f'''(x) = 2(-\frac{3}{2})(1-4x)^{-\frac{5}{2}}(-4) = 2 \cdot 3 \cdot 2(1-4x)^{-\frac{5}{2}}.$$

And its value at zero is

$$f'''(0) = 2 \cdot 3 \cdot 2. \tag{2.61}$$

Differentiating the second derivative of the infinite series we obtain

$$f'''(x) = 3 \cdot 2k_3 + 4 \cdot 3 \cdot 2 \cdot k_4 \cdot x + ...$$

and its value at zero is

$$f'''(0) = 3 \cdot 2 \cdot k_3. \tag{2.62}$$

After equating two formulas we obtain the third Catalan number as

$$k_3 = 2, \tag{2.63}$$

which is also true (for three terms we have only two options, either $a \star (b \star c)$ or $(a \star b) \star c$. Let us continue our procedure and evaluate the 4th derivative, the fifth, and so on. The fourth derivative is

$$f^{(4)}(x) = 2 \cdot 3 \cdot 2 \cdot 5 \cdot 2 \cdot (1-4x)^{-\frac{7}{2}},$$

Its value at zero is

$$f^{(4)}(0) = 2 \cdot 3 \cdot 2 \cdot 5 \cdot 2. \tag{2.64}$$

The corresponding fourth derivative of the infinite series function is

$$f^{(4)}(x) = 4 \cdot 3 \cdot 2 \cdot k_4 + ...$$

and its value at zero is

$$f^{(4)}(0) = 4 \cdot 3 \cdot 2 \cdot k_4. \tag{2.65}$$

Equating the values of the fourth derivatives we obtain that

$$2 \cdot 3 \cdot 2 \cdot 5 \cdot 2 = 4 \cdot 3 \cdot 2 \cdot k_4$$

Hence

$$k_4 = 5.$$

Using **inductive** thinking, we can write down the nth derivative of function given by formula (2.51) as

$$f^{(n)}(x) = 2 \cdot 3 \cdot 2 \cdot 5 \cdot 2 \cdot 7 \cdot ...(2n-3) \cdot 2 \cdot (1-4x)^{-(\frac{2n-1}{2})}$$

and its value at zero as

$$f^{(n)}(0) = 2 \cdot 3 \cdot 2 \cdot 5 \cdot 2 \cdot 7 \cdot \dots \cdot (2n-3) \cdot 2$$
$$= 2^{n-1} \cdot (3 \cdot 5 \cdot 7 \cdot \dots \cdot (2n-3)). \tag{2.66}$$

and similarly for the infinite series function (2.55):

$$f^{(n)}(x) = n(n-1)(n-2) \cdot \dots \cdot 2 \cdot k_n + \dots$$

and its value at zero as

$$f^{(n)}(0) = n! \cdot k_n. \tag{2.67}$$

If we equate the right-hand sides of formulas we obtain

$$n! \cdot k_n = 2^{n-1}(3 \cdot 5 \cdot 7 \cdot \dots \cdot (2n-3))$$

Solving this for k_n we obtain the formula for the nth Catalan number:

$$k_n = \frac{2^{n-1} \cdot (3 \cdot 5 \cdot 7 \cdot \dots \cdot (2n-3))}{n!} \tag{2.68}$$

Using this formula we can find any Catalan number and check some of them with the numbers found by recursion. Earlier we found that $k_4 = 5$, $k_5 = 14$. Using formula for $n = 4$ we get

$$k_4 = \frac{2^3 \cdot (3 \cdot 5)}{4!} = 5.$$

Or the fifth and sixth terms of the Catalan sequence:

$$k_5 = \frac{2^4 \cdot (3 \cdot 5 \cdot 7)}{5!} = 14$$
$$k_6 = \frac{2^5 \cdot (3 \cdot 5 \cdot 7 \cdot 9)}{6!} = 42.$$

Although the nth Catalan number is represented by a fraction, we understand that it is a natural number, so the numerator of the fraction must be a multiple of the denominator. Two consecutive Catalan numbers are connected by the relationship:

$$k_{n+1} = \frac{4n-2}{n+1} \cdot k_n.$$

The Catalan numbers can be evaluated using combinatoric formula as follows:

$$k_n = \frac{1}{n} \cdot C_{2n-2}^{n-1}, \quad C_n^k = \frac{n!}{k!(n-k)!} \tag{2.69}$$

Please prove this yourself and apply this formula by solving a triangulation problem as a homework exercise. You will see more combinatoric formulas in Chapter 3.

Chapter 3
Diophantine Equations and More

The ancient world left a legacy of several mathematical works that provide the foundation that more recent mathematicians have built upon. Even now, we continue to be amazed by the works of Pythagoras, Euclid, Archimedes, and Diophantus. Diophantus wrote *Arithmetica*, a compilation of thirteen books, in the 3rd century. Unfortunately, only six of the thirteen books remain today. After the fire in the Alexandria library, *Arithmetica* was lost for more than a millennium. But in 1464, the German scientist Regiomontanus found a copy of six of the thirteen books of Diophantus, and sparked the beginning of *Arithmetica's* reemergence in Europe. The mysterious *Arithmetica* was first translated to Latin in 1575. Later, Claude Gaspard Bachet de Muziriac prepared another translation, bringing *Arithmetica* to a wider audience. In his book, Diophantus freely operated with negative and rational numbers, possessed a letter notation for equations, and most importantly was able to find solutions in integer and rational numbers of linear, quadratic, and cubic equations and systems with two or more unknowns with integer coefficients. The great 17th-century mathematician, Fermat, was impressed with *Arithmetica* and studied the solutions of many such equations in integers under the influence of the works of Diophantus. In fact, it was in a translation of *Arithmetica* that Fermat wrote his famous margin note claiming to have a beautiful proof showing that

$$x^n + y^n = z^n$$

does not have solutions in integers for $n > 2$ but that he did not have enough room in the margin to write it. The statement now famously known as Fermat's Last Theorem was finally proven in 1995 by British mathematician Andrew Wiles. Wiles' proof took 130 pages and was based on modern methods particularly properties of elliptical curves. Some mathematicians still believe that Fermat's own shorter proof of his last theorem existed but this question remains a mystery.

Fermat was a genius and his contribution to number theory is priceless. In this chapter, we will discuss the so-called Fermat-Pell's equation. Fermat stated that $x^2 - Ny^2 = 1$ has infinitely many solutions for any natural number N that is not a perfect square. Interestingly, this problem is reducible to equations that were proposed

© Springer International Publishing AG, part of Springer Nature 2018
E. Grigorieva, *Methods of Solving Number Theory Problems*,
https://doi.org/10.1007/978-3-319-90915-8_3

in the work of Archimedes. In his letter to Eratosthenes, Archimedes described the "cattle problem", which included eight variables corresponding to different animals connected by seven equations. The equation that can be obtained as a result of solving this system is Pell's equation with a huge coefficient in front of the square of the second variable,

$$x^2 - 4,729,494y^2 = 1.$$

Fermat did not believe that Archimedes could solve this equation with a geometric approach. However, the fact that such equations were generally obtained in the works of ancient Greeks suggests that they could solve at least an easier version of it. Diophantine equations have since been at the center of attention of mathematicians and the methods of solving such equations are well developed. Still, many versions of Diophantine equations appear every year at different math contests. In this chapter, we will discuss methods of solving Diophantine equations and demonstrate how some of them are connected with many areas of modern mathematics. We will also solve other nonstandard equations in integers such as exponential equations, factorial equations, and many more.

3.1 Linear Equations in Two and More Variables

Starting this chapter I want to ask you to solve a problem. Find all ordered pairs (x, y) satisfying to the equation $2x + y = 7$.

What is your answer? Did you find the unique solution? Did you come up with many solutions? Most of my undergraduate students who are unfamiliar with number theory would think like this: The given equation can be written as $y = 7 - 2x$ that is a linear function of x and consequently has infinitely many solutions such as $(x, 7 - 2x)$ on the line. You just choose any number for x and then evaluate $y = 7 - 2x$.

Thus, points $(1, 5)$, $(0.5, 6)$, $(1.3, 4.4)$, $(-3.1, 13.2)$, and infinitely many others will be possible solutions of the problem. Yes, you are right! Now please think of how your answer would change if the same problem was to solve it over the set of natural numbers x and y. The listed last three ordered pairs would not satisfy the new condition. You could just take 1, 2, 3, and so on for x and evaluate y by the rule $y = 7 - 2x$ until y stays positive. This gives us just 3 possible solutions,

$$x = 1, y = 5$$
$$x = 2, y = 3$$
$$x = 3, y = 1.$$

It is sometimes difficult to find solutions by random checking, so we have to find some general approach for solving equations in integers. Equations $ax + by = c$ to be solved in integers are called linear Diophantine nonhomogeneous equations. Different methods of solving such equations are discussed in this book.

In general, an equation

$$a_1x_1 + a_2x_2 + ...a_nx_n = c, \quad a_i, x_i \in \mathbb{Z}, \, i \in \mathbb{N} \tag{3.1}$$

is called linear Diophantine equation. The theory of solution of such equations is very well developed and particular problems were solved in 4000 BC by Ancient Babylonians. The ancient Greek mathematician Diophantus in his book *Arithmetica* described general methods of solving such equations and methods of solving some nonlinear equations. His methods are very useful in our time because of his ingenious ideas. Many of Diophantus' ideas will be demonstrated in this chapter.

3.1.1 Homogeneous Linear Equations

A simple diophantine equation is a homogeneous linear equation in two variables to be solved in integers,

$$ax + by = 0$$

Many word problems can be reduced to solving such equations. For example, let us solve the following one.

Problem 120

A cycling relay race was held on a circular track; start and finish were in the same place, the length of the track is 55 km, and the length of each stage is 25 km. If the cyclists only go in one direction, how many relay trades were there? The starting point is also considered as a trade (switching) point. What is the distance between neighboring trade points?

Solution. The change of the bicyclists happens every 25 km on the circular track. Because start and finish coincide, and since $2 \cdot 25 = 50 < 55$, then during each race there is more than two switching points between bicyclists of the same team. During the relay, bicyclists of each team pass the path equals $25 \cdot k$ km, where k is the number of the switching points and it is a natural number. This path must be equal to the whole number of full circles of $55 \cdot n$ km of length. We need to solve the following equation:

$$25k = 55n$$

dividing both sides by 5, we obtain

$$5k = 11n$$

From which we understand that $n = 5$ and $k = 11$. Therefore, the number of switching points is 11, and the distance between switching points is $55/11 = 5$ km.
Answer. 11 and 5.

Remark. In general, when solving

$$5k = 11n$$

we would have a solution $k = 11m$, $n = 5m$, $m \in \mathbb{Z}$. Why in the problem above was our answer unique? This is because of the condition of the problem. First, we have to use only natural values of m and from common sense the minimal solution was obtained.

Theorem 15

Every diophantine equation $ax + by = 0$, where a and b are relatively prime, $((a,b) = 1)$ has infinitely many solutions that can be described by formulas,

$$x_n = b \cdot n,\ y_n = -a \cdot n,\ n \in \mathbb{Z}$$

Here n is a position of a solution in the sequence of all solutions.

Problem 121

The tram runs along a 5-km-long ring road. The beginning and end of the route are not the same. How many stops are on the tram route if the distance between stops is three kilometers? Find also the length of the tram's route.

Solution. The solution of the problem is analogous to the solution of the preceding problem; the number of stops on the route is five and the length of the route is 15 km.

3.2 Nonhomogeneous Linear Equations in Integers

In this section, we will solve some nonhomogeneous diophantine equations and the problems that can be reduced to solving such equations. Let us start from a word problem, "Mike wants to go camping and needs to buy salmon and tuna cans at 6 and 3 dollars each, respectively. He has 100 dollars. How many of each should he buy if he wants to spend all hundred bucks?"

The problem is reduced to the following equation:

$$6x + 3y = 100$$

and does not have solutions in integers because the left side is divisible by $3 = (6,3)$ but the right side is not. It is time for us to state the following theorems.

Theorem 16

If the greatest common divisor, d, of the coefficients a and b of the equation

$$ax + by = c$$

is greater than 1 and does not divide c, then the equation has no integer solutions.

Sometimes, you can use the fact that an even number cannot equal an odd number.

Example. We can see that the equation

$$4x + 6y = 9$$

has no solution in integers because the left side of it is always even but the right-hand side (9) is odd.

Theorem 17 (Bezout's Identity)

If the greatest common divisor of a and b is d, i.e., $(a,b) = d$, then an equation $ax + by = d$ has a solution for some integers x and y.

The proof of this theorem is omitted but can be done easily by working the Euclidean algorithm backward. Please do the proof on your own or after reading section 3.2.1. of this chapter.

Problem 122

Solve $10z + 18t = 14$ in integers.

Solution. By Theorem 17, $(10, 18) = 2$, then we can find x and y, such that $10x + 18y = 2$, for example, $x = 2$ and $y = -1$. Multiplying both sides of the equation by 7 we obtain,

$$10 \cdot 7x + 18 \cdot 7y = 14$$
$$z = 7x = 14, \quad t = 7y = -7.$$

Hence, $z = 14$ and $t = -7$ is one of the infinitely many other solutions.

In order to find all of the solutions, we will need to state the following theorem.

Theorem 18

If (x_0, y_0) is a solution of a linear equation

$$ax + by = c \tag{3.2}$$

then all other solutions can be found as follows:

$$x = x_0 + \frac{b}{d} \cdot u$$
$$y = y_0 - \frac{a}{d} \cdot u,$$

where $d = (a, b)$ and $u \in \mathbb{Z}$, i.e., $u = 0, \pm 1, \pm 2, \ldots$.

Proof. The proof of this theorem is simple. Substituting new formulas for x and y into (3.2), after collecting like terms and replacing $ax_0 + by_0 = c$, we obtain

$$a\left(x_0 + \frac{b}{d} \cdot u\right) + b\left(y_0 - \frac{a}{d} \cdot u\right) = ax_0 + by_0 + \frac{ab}{d}u - \frac{ab}{d}u = c.$$

The proof is complete.

3.2.1 Using Euclidean Algorithm to Solve Linear Equations

One of the earliest methods offered for such equations was probably Euclidean algorithm. Euclidean algorithm method is very useful when coefficients a and b are big numbers and finding any solution to the equations $ax + by = c$ is not easy. Though, this method is rarely used to solve actual linear equations, let us practice Euclidean algorithm.

Problem 123

Solve $314x + 159y = 1$ in integers.

Solution. First, we will find some solutions using Euclidean algorithm and underlining all remainders:

$$(314, 159) = ? = d = 1$$
$$314 = 1 \cdot 159 + \underline{155}$$
$$159 = 1 \cdot 155 + \underline{4}$$
$$155 = 38 \cdot 4 + \underline{3}$$
$$4 = 1 \cdot 3 + \underline{1}$$
$$3 = 3 \cdot 1 + \underline{0}.$$

Now we will find solution (x_0, y_0) working backward (using Euclidean algorithm in the opposite direction). Thus,

$$1 = 4 - 3 = (159 - 155) - (155 - 38 \cdot 4)$$
$$= 159 - 155 - 155 + 38(159 - 155) \text{ (we need 159 not 155)}$$
$$= 1 \cdot 159 - 2(314 - 159) + 38 \cdot (314 - 159)$$
$$= 39 \cdot 159 - 2 \cdot 314 + 2 \cdot 159 - 38 \cdot 314 + 38 \cdot 159$$
$$= 79 \cdot 159 - 40 \cdot 314 = 1 \text{ or}$$
$$314 \cdot (-40) + 159 \cdot 79 = 1.$$

It means that $(x_0, y_0) = (-40, 79)$.

In our case $a = 314$, $b = 159$, $d = 1$, $x_0 = -40$, $y_0 = 79$. Therefore,

$$x = -40 + 159 \cdot u,$$
$$y = 79 - 314 \cdot u, \ u \in \mathbb{Z}, \text{ or } u = 0, \pm 1, \pm 2, \ldots.$$

Giving u different values, we can find as many integer solutions to this equation as we want. Some of them are in the table:

u	$x = -40 + 159u$	$y = 79 - 314u$
0	-40	79
1	119	-235
-1	-199	393
-5	-835	1649
5	755	-1491
10	1550	-3061

Table 3.1 Solutions to $314x + 159y = 1$.

Are you ready to solve a word problem? Here it is a problem about honest Paul.

Problem 124 (Falin, MGU, 2005 Entrance Exam)

Paul made a few small purchases in the supermarket using cash. The cashier mistakenly switched the values of dollars and cents when giving him the change from $100 bill. Soon after that Paul bought a bandage in the pharmacy for $1.40 and could tell the supermarket cashier made a mistake. Paul found out that the amount the grocer gave him was 3 times the amount of what he should have received. What was the cost of all Paul's purchases?

Solution. Before we start, assume that correct change is supposed to be $72.64 but the cashier by mistake gave $64.72. This can be written in cents as 7264 and 6472, respectively, or as follows:

$$\$72.64 = 7264 = 100 \cdot 72 + 64$$
$$\$64.72 = 6472 = 100 \cdot 64 + 72$$

Let n be the number of dollars and m of cents in the correct change, then Paul's correct and wrong change can be written as

$$100 \cdot n + m$$

and

$$100m + n,$$

respectively. After Paul purchased a bandage in the pharmacy for 140 cents, he had only $100m + n - 140$ in his wallet. By the condition of the problem,

$$100m + n - 140 = 3 \cdot (100n + m)$$

This equation can be simplified as

$$97x - 299y = 140, \ \ x = m \le 99, \ \ y = n \ge 1. \tag{3.3}$$

The restrictions on the variables indicate that any possible change from $100 cannot be greater than $100. We obtained a linear Diophantine equation in two variables. Because it is difficult to find any solution right away, we will apply the Euclidean algorithm,

$$299 = 3 \cdot 97 + \underline{8}$$
$$97 = 12 \cdot 8 + \underline{1}$$

We obtain that

$$\boxed{1} = 97 - 12 \cdot 8 = 97 - 12 \cdot (299 - 3 \cdot 97) = \boxed{37 \cdot 97 - 12 \cdot 299}$$

Multiplying both sides of it by 140, we obtain

$$5180 \cdot 97 - 1680 \cdot 299 = 140$$
$$x_0 = 5180, \ \ y_0 = 1680$$

Applying Theorem 18, the general solution to (3.3) can be written as

$$\begin{cases} x = m = 5180 + 299k \le 99, \ k \in Z \\ y = n = 1680 + 97k \ge 1 \end{cases} \tag{3.4}$$

In order to find unique answers to the problem, we needed to impose the restriction on the variables. By solving together inequalities (3.4), we obtain that

$$-\frac{1679}{97} \le k \le -\frac{5089}{299}$$
$$k = -17$$

Substituting this value of a parameter k into (3.4), we obtain the values of our variables

$$x = m = 97$$
$$y = n = 31$$

Therefore, Paul's correct change in the supermarket would be \$31.97. The cashier gave him instead \$97.31. Now we can calculate Paul's total expenses,

$$\$100 - \$31.97 + \$1.40 = \$69.43.$$

We can see that Paul was very honest and came back to get correct change.
Answer. \$69.43

Sometimes we can find one of the solutions easily, then applying Theorem 18 all the other integer solutions will be found.

Problem 125

Find integers x and y that satisfy $85x + 34y = 51$.

Solution. First, we will simplify the equation as $5x + 2y = 3$, then $x_0 = 1$, $y_0 = -1$. Therefore,

$$x = x_0 + bu = 1 + 2u$$
$$y = y_0 - au = -1 - 5u, \quad u \in \mathbb{Z}.$$

Answer. $(x, y) = (1 + 2u, -1 - 5u)$, $u = 0, \pm 1, \pm 2, \dots$.

3.2.2 Extracting an Integer Portion of a Quotient

Let us rewrite the problem that started this section, $2x + y = 7$, differently by solving it for x in terms of y, i.e., find all natural numbers (x, y) that satisfy the equation

$$x = \frac{7 - y}{2}. \tag{3.5}$$

Because x can only be a positive integer, this can happen if and only if the numerator of fraction (3.5), $(7 - y)$ is a multiple of 2. Rewriting 7 as $(6 + 1)$ (6 is a multiple of 2) we obtain

$$x = \frac{6 + 1 - y}{2} = \frac{6}{2} + \frac{1 - y}{2} = 3 + \frac{1 - y}{2}. \tag{3.6}$$

In (3.6) we extracted the integer part (a number of 3) of the quotient (3.5).

In order for $\frac{1-y}{2}$ to be an integer, its numerator $(1 - y)$ must be an even number. This will be so for any odd integer y. However, in order to find only positive integer solutions (natural) we should restrain $x > 0$. Thus,

$$3 + \frac{1-y}{2} > 0 \Rightarrow \frac{1-y}{2} > -3 \Rightarrow y < 7.$$

There are only three odd natural numbers less than 7. They are $y = 1$, $y = 3$, and $y = 5$. Plugging them into (3.6) we will get the corresponding x: 3, 2, and 1.

Solving a relatively simple problem by a nontraditional method, we have demonstrated another method of solving many interesting problems in integers. In many problems involving integers we have to decide for what integer numbers n the given fraction $\frac{a(n)}{b(n)}$ is an integer. Solving such problems, we are extracting the integer portion of the quotient rewriting the given quotient as

$$\frac{a(n)}{b(n)} = c(n) + \frac{d}{b(n)}, \tag{3.7}$$

where $a(n)$, $b(n)$, $c(n)$ are expressions in n, and d is an integer. Then we have to find such n, at which $b(n)$ divides d. These values of n are needed.

Problem 126

For what integer numbers n is the expression $N = \frac{3n+2}{n-1}$ an integer?

Solution. Extracting the integer part in the quotient we obtain

$$\frac{3n+2}{n-1} = \frac{3n-3+5}{n-1} = 3 + \frac{5}{n-1}. \tag{3.8}$$

From (3.8) we notice that N will be an integer if and only if $(n-1)$ divides 5. This happens when

$$n-1 = \pm 1, \pm 5 \text{ or } n = 2, \; n = 0, \; n = 6, \; n = -4. \tag{3.9}$$

Evaluating N for each possible n from (3.9), we get

$$\begin{aligned}
n &= 2, & N &= 8 \\
n &= 0, & N &= -2 \\
n &= 6, & N &= 4 \\
n &= -4, & N &= 2.
\end{aligned}$$

Answer. $n = -4$, $n = 0$, $n = 2$, and $n = 6$.

Problem 127

Find all integer numbers n, for which $N = \frac{n^2+1}{n+2}$ is an integer.

Solution. Extracting the integer part in the quotient we obtain

$$\frac{n^2+1}{n+2} = \frac{(n+2)^2 - 4\cdot(n+2)+5}{n+2} = n+2-4+\frac{5}{n+2} = n-2+\frac{5}{n+2}.$$

Similarly to the previous problem, we can state that N will be an integer if and only if $(n+2)$ divides 5. This happens when

$$n+2 = \pm 1, \pm 5 \text{ or } n = -1, \, n = -3, \, n = 3, \, n = -7.$$

Answer. $n = -1, n = -3, n = 3$, and $n = -7$.

Problem 128 (Texas State Mathematics League, January 5, 1999)

What are both values of x for which all three of the expressions $\frac{x}{x-2}$, $\frac{x}{x-4}$, and $\frac{x}{x-6}$ have integral values?

Solution. Let us extract the integer part of the first expression.

$$\frac{x}{x-2} = \frac{x-2+2}{x-2} = 1+\frac{2}{x-2}.$$

The second quotient is an integer if and only if $x-2 = \pm 1$ or $x-2 = \pm 2$.

1. If $x-2 = 1$, $x = 3$, then $\frac{x}{x-2} = 3$, $\frac{x}{x-4} = -3$, and $\frac{x}{x-6} = -1$ are integers.
2. If $x-2 = -1$, $x = 1$, then $\frac{x}{x-2} = \frac{1}{1-2} = -1$ is an integer, but $\frac{x}{x-4} = -\frac{1}{3}$ is not an integer.
3. If $x-2 = 2$, $x = 4$, then $\frac{x}{x-2} = 2$, but $\frac{x}{x-4}$ is undefined for $x = 4$.
4. If $x-2 = -2$, then $x = 0$, and all quotients equal 0, hence we have an integer.

Answer. $x = 0$ and $x = 3$.

Problem 129

Find all integers n, such that the following expression is also integer $\frac{19n+17}{7n+11}$.

Solution.
Method 1. Extracting the integer part of the fraction

$$\frac{19n+17}{7n+11} = 3 - \frac{2n+16}{7n+11}$$

If $n = -8$, then the second faction is zero and our expression equals 3. If the second fraction is not zero, then $(2n+16)$ is divisible by $(7n+11)$, which can happen if

$$2n+16 = k(7n+11)$$

and at least the following must be true:

$$|2n+16| \geq |7n+11|$$
$$(2n+16)^2 \geq (7n+11)^2$$
$$(5n-5)(9n+27) \leq 0 \quad .$$
$$-3 \leq n \leq 1$$

However, within this interval only $n = -3, n = -2$ and $n = 1$ give us an integral values of the second fraction.

Remark. Note that above we used the following fact that if

$$|a| \geq |b|,$$
$$a^2 \geq b^2, \ b^2 \leq a^2,$$
$$(b-a)\cdot(b+a) \leq 0.$$

Method 2. Denote the given fraction by y:

$$y = \frac{19n+17}{7n+11} = 3 - \frac{2n+16}{7n+11}$$

Let us multiply both sides by 7 and continue extraction an integer part of the denominator:

$$7y = 21 - \frac{7(2n+16)}{7n+11} = 21 - \frac{2(7n+11)+90}{7n+11}$$
$$= 19 - \frac{90}{7n+11}$$

From here we know that $(7n+11)$ must divide 90. The following cases are valid:

$$7n+11 = -3, \quad n = -2, \quad y = 7$$
$$7n+11 = -10, \quad n = -3, \quad y = 4$$
$$7n+11 = 18, \quad n = 1, \quad y = 2$$
$$7n+11 = -45, \quad n = -8, \quad y = 3$$

Method 3. We can also find GCD of the numerator and denominator of the fraction $\frac{2n+16}{7n+11}$:

$$(7n+11, 2n+16) = (5n-5, 2n+16) = (3n-21, 2n+16)$$
$$= (2n+16, n-37) = (90, n-37)$$

Answer. $n = -8, -3, -2, 1$

1. If $n = -8$, then $\frac{19n+17}{7n+11} = 3$
2. If $n = 1$, then $\frac{19n+17}{7n+11} = 2$
3. If $n = -2$, then. $\frac{19n+17}{7n+11} = \frac{-38+17}{-14+11} = 7$
4. If $n = -3$, then $\frac{19n+17}{7n+11} = \frac{-57+17}{-21+11} = 4$

Sometimes the methods above are useful in solving nonlinear equations in integers. Let us see how it can be implemented.

Problem 130 (MGU, entrance exam)

Integers x, y, z are in geometric progression and $5x - 4, y^2, 3z + 2$ are in arithmetic progression. Find x, y, and z.

Solution. Please recall that in an arithmetic progression, a middle term is an arithmetic mean of its left and right neighboring terms, and in a geometric progression the middle term is a geometric mean of its left and right neighbors. For example, for an arithmetic progression with consecutive terms a, b, c, the following is true:

$$b = \frac{a+c}{2}.$$

On the other hand, for a geometric progression with consecutive terms m, n, k, the following is valid:

$$n^2 = m \cdot k.$$

(See more about progressions in my book *Methods of Solving Sequence and Series Problems*.) Using relationship for geometric and arithmetic means, we have the following system to solve:

$$\begin{cases} y^2 = xz \\ 2y^2 = 5x - 4 + 3z + 2 \end{cases} \Leftrightarrow \begin{cases} y^2 = xz \\ 2y^2 = 5x + 3z - 2 \end{cases}$$

Substituting the first equation of the system into the second, we obtain the following equation in two variables:

$$2xz = 5x + 3z - 2$$
$$z \cdot (2x - 3) = 5x - 2$$
$$z = \frac{5x-2}{2x-3}$$

Because z must be integer number, let us extract the integer portion of the fraction on the right of the last formula,

$$z = \frac{4x - 6 + x + 4}{2x - 3} = 2 + \frac{x+4}{2x-3}$$

It looks like we have no other move; however, multiplying both sides by 2, we can continue our extraction operation,

$$2z = 4 + \frac{2(x+4)}{2x-3}$$
$$2z = 4 + \frac{(2x-3)+11}{2x-3}$$
$$2z = 5 + \frac{11}{2x-3}$$

Now we want the fraction to be an integer number so that the entire expression on the right to be a multiple of 2. Additionally, the product of x and z must be a perfect square of y.

The first conditions can be written as

$$\begin{bmatrix} 2x - 3 = \pm 11 \\ 2x - 3 = \pm 1 \end{bmatrix} \Rightarrow \begin{cases} x = 7, \ z = 3 \\ x = -4, \ z = 2 \\ x = 2, \ z = 8 \\ x = 1, \ z = -3 \end{cases}$$

Applying the second condition, we choose only the third choice of the ordered pairs of x and z and evaluate the corresponding y,

$$x = 2, \ z = 8, \ y = \pm 4$$

Answer. $(x, y, z) = \{(2, 4, 8), \ (2, -4, 8)\}$

I want to offer you the following problem here.

Problem 131

Find all numbers that are simultaneously the terms of the both following arithmetic sequences 3, 7, 11, ..., 407 and 2, 9, 16, ..., 709.

Solution. The nth term of the first sequence is

$$a_n = 3 + 4(n - 1). \tag{3.10}$$

The kth term of the second sequence can be written as

$$b_k = 2 + 7(k - 1). \tag{3.11}$$

Therefore, we have to find such numbers n and k that $a_n = b_k$, $1 \leq n \leq 102$, $1 \leq k \leq 102$. Equating (3.10) and (3.11) we obtain

$$4n + 4 = 7k. \tag{3.12}$$

Equation (3.12) has integer solutions if and only if $k = 4s$. It is clear that s can be 1, 2, ..., 25, because $k = 1, 2, \ldots, 102$.

$$4(n + 1) = 7 \cdot 4s$$
$$n + 1 = 7s$$
$$n = 7s - 1.$$

Since $1 \leq n \leq 102$, then $1 \leq s \leq 14$.

Therefore, there are exactly 14 numbers that are terms of both arithmetic sequences. We can find all of them either from formula (3.10) using substitution $n = 7s - 1$ or from formula (3.11) using $k = 4s$, $s = 1, 2, 3, \ldots, 14$.

Answer. $23, 51, 79, \ldots, 387$.

The following problem was offered at the Hungarian Math Olympiad in 1894.

Problem 132 (Hungarian Math Olympiad)

Prove that expressions $2x+3y$ and $9x+5y$ are divisible by 17 for the same integer values of x and y. Find these (x,y).

Solution. Since both expressions are multiples of 17 at the same values of x and y, consider the system:

$$\begin{cases} 2x+3y = 17k \\ 9x+5y = 17m \end{cases} \tag{3.13}$$

Multiplying the first equation by 9 and the second by 2 and then subtracting them, using Gaussian elimination, we obtain

$$18x+27y = 17 \cdot 9k$$
$$18x+10y = 17 \cdot 2m$$
$$17y = 17(9k-2m),$$

and hence

$$y = n = 9k - 2m. \tag{3.14}$$

In order to get a general solution, we will replace y using formula (3.14) and obtain for x,

$$2x+3(9k-2m) = 17k$$
$$2x+10k = 6m$$

Then,

$$x = 3m - 5k. \tag{3.15}$$

Next, formulas (3.14) and (3.15) form a general solution for the problem. Both expressions are multiples of 17 at the same values of x and y, such that

$$(x,y) = (3m-5k, 9k-2m), \ k \in \mathbb{Z}, \ m \in \mathbb{Z}. \tag{3.16}$$

Note that in (3.16) parameters m and k change independently. For example, let $m = 5$ and $k = -4$, then $x = 35$ and $y = -46$. Moreover, $2x+3y = 70 - 138 = -68$ and $9x+5y = 85$ (both are multiples of 17).

Let us see the general proof of this problem. Denote

$$u = 2x+3y$$
$$v = 9x+5y,$$

or

$$3v = 27x + 15y$$
$$5u = 10x + 15y$$

and after subtraction we have

$$3v - 5u = 17x. \tag{3.17}$$

From (3.17) we obtain that
$$3v = 17x + 5u, \tag{3.18}$$

and
$$5u = 3v - 17x. \tag{3.19}$$

From equation (3.18) we can state that since v is a multiple of 17, then u must be a multiple of 17. From equation (3.19) we can state that since u is a multiple of 17, then v must be a multiple of 17. Therefore, the same pair of integers (x, y) will satisfy both given expressions. This pair can be written as (3.16).

Remark. The problem is equivalent to the statement that if $2x + 3y$ is divisible by 17, then $9x + 5y$ is also divisible by 17.

Proof. Let us extract $2x + 3y$ into $9x + 5y$ by rewriting it as

$$9x + 5y = 17 \cdot (x + y) - 4 \cdot (2x + 3y) = 17 \cdot k.$$

In the expression above, the first term is divisible by 17 so as the second term (given). This completes the proof.

3.3 Solving Linear Equations and Systems Using Congruence

We solved a few of the Diophantine equations earlier in this book. Now we will use a different approach, congruences. Let us find solutions to the following linear equation:
$$2x - 19y = 1, \tag{3.20}$$

or $2x = 19y + 1$, which can be read as "$2x$ divided by 19 gives a remainder of 1."
Using congruence notation we write it as

$$2x \equiv 1 \pmod{19}. \tag{3.21}$$

For example, $20 = 1 \cdot 19 + 1$. Let us show that (3.21) can also be written as

$$2x \equiv 20 \pmod{19}. \tag{3.22}$$

In general any equation $a \equiv r \pmod{b}$ can be written as $a \equiv (r + k \cdot b) \pmod{b}$ (by adding as many multiples of b as we wish). Thus, (3.21) is equivalent to (3.22) and to

$$2x = 19n + 20 = 19n + 19 + 1 = 19(n+1) + 1 = 19k + 1.$$

However, (3.22) can be solved for x by dividing both sides by 2.

$$x \equiv 10 \pmod{19},$$

which gives us solution to (3.20) right away. We have

$$x = 19m + 10, \quad m \in \mathbb{Z}. \tag{3.23}$$

Substituting (3.23) into (3.20) we obtain

$$2(19m + 10) = 19y + 1 \Rightarrow 2 \cdot 19m + 19 + 1 = 19y + 1 \Rightarrow 19(2m+1) = 19y,$$

which give us

$$y = 2m + 1, \quad m \in \mathbb{Z}. \tag{3.24}$$

Finally, (3.23) and (3.24) form the solution to the equation (3.20).

Remark. The congruence is a good tool for solving linear Diophantine equations. Thus, $ax + by = c$ can be written as $ax \equiv c \pmod{b}$ and $by \equiv c \pmod{a}$.

Theorem 19

Consider congruence equation $ax \equiv b \pmod{m}$. Three cases are valid.

1. If the greatest common divider of a and m, (a, m) does not divide b, then such an equation has no solutions.
2. If $(a, m) = 1$, then $ax \equiv b \pmod{m}$ has exactly one solution.
3. If $(a, m) = d$, and b is a multiple of d, then $ax \equiv b \pmod{m}$ has exactly d solutions.

We will consider an example for each case and learn how to solve Diophantine equations using congruence without making a mistake.

Problem 133

Prove that $6x + 8y = 7$ has no solutions in the integers.

Solution. The left side of the equation is even for any x and y, the right-hand side is odd (7). This equation does not have a solution for any integers x and y.

Now, consider an equivalent congruence equation: $6x \equiv 7 \pmod{8}$ and compare it with $ax \equiv b \pmod{m}$. So $a = 6$, $b = 7$, and $m = 8$. Since $(a, m) = (6, 8) = 2$, and 7 is not a multiple of 2, then this problem represents Case 1 of Theorem 19. Therefore there are no solutions.

Problem 134

Solve the equation $4x + 15y = 1$.

Solution. Let us rewrite this in the congruence form $4x \equiv 1 \pmod{15}$ and compare it with $ax \equiv b \pmod m$. Since $(4, 15) = 1$, and of course any number is divisible by 1 including 1, then this represents Case 2 of Theorem 19, and the congruence has exactly one solution.

Rewriting the congruence equations in a different form using the fact that $1 + 15 = 16$ is a multiple of 4:

$$
\begin{aligned}
4x &\equiv 1 \quad (\mathrm{mod}\ 15) \\
4x &\equiv 16 \quad (\mathrm{mod}\ 15) \Rightarrow x \equiv 4 \quad (\mathrm{mod}\ 15) \Rightarrow x = 4 + 15k.
\end{aligned}
\tag{3.25}
$$

Substituting the expression for x into the original equation we will obtain that $y = -4k - 1$.

Please check it yourself! There are an infinite number of ordered pairs of (x, y) for each integer k: $(4 + 15k, -4k - 1)$ such as $(4, -1)$ for $k = 0$, $(19, -5)$ for $k = 1$, $(34, -9)$ for $k = 2$, etc.

Answer. $(x, y) = (15k + 4, -4k - 1)$, $k \in \mathbb{Z}$.

Problem 135

Solve the following in integers: $6x \equiv 15 \pmod{33}$.

Solution. This problem represents Case 3 of Theorem 19: $(6, 33) = 3$, which is a multiple of 15. Therefore the congruence must have exactly 3 solutions.

$$
\begin{aligned}
6x &\equiv 15 \quad (\mathrm{mod}\ 33) \Rightarrow 2x \equiv 5 \quad (\mathrm{mod}\ 11) \\
2x &\equiv 16 \quad (\mathrm{mod}\ 11) \Rightarrow x \equiv 8 \quad (\mathrm{mod}\ 11)
\end{aligned}
$$

Hence, there are exactly three solutions to the last congruence that is less than 33: $x \equiv 8; 19; 30 \pmod{33}$.

Answer. $x \in \{8; 19; 30\}$.

Remark. If we think of solving the corresponding linear equation, then we will have the following general result:

$$
\begin{aligned}
6x + 33y &= 15 & \Rightarrow 2x + 11y &= 5 \\
x &= 11k + 8 & \Rightarrow 2(11k + 8) + 11y &= 5 \\
2 \cdot 11k + 16 + 11y &= 5 \Rightarrow 2 \cdot 11k + 11 + 11y = 0 \Rightarrow y = -2k - 1, \ k \in \mathbb{Z}.
\end{aligned}
$$

Answer. $(x, y) = (11k + 8, -2k - 1)$, $k \in \mathbb{Z}$.

3.3.1 Solving Linear Congruence Using Continued Fractions

The following statement is valid.

> **Lemma 12**
>
> Suppose numbers m and a are relatively prime, i.e., $(m,a) = 1$. If we have the following sequence of continued fractions:
>
> $$\frac{P_0}{Q_0}, \quad \frac{P_1}{Q_1}, \quad \frac{P_2}{Q_2}, \quad \cdots, \\ \frac{P_{s-1}}{Q_{s-1}}, \quad \frac{P_s}{Q_s} = \frac{m}{a},$$
>
> then the solution of the equation
>
> $$ax \equiv b \pmod{m} \tag{3.26}$$
>
> is $x \equiv (-1)^s b P_{s-1} \pmod{m}$.

Proof. We have $(a,m) = 1$ and

$$(P_s, Q_s) = 1 \Rightarrow \frac{P_s}{Q_s} = \frac{m}{a} \Rightarrow P_s = m, \ Q_s = a. \tag{3.27}$$

By the property of convergent fractions (see Section 2.1.2),

$$P_{s-1}Q_s - P_s Q_{s-1} = (-1)^s. \tag{3.28}$$

Substituting (3.27) into (3.28) we obtain $aP_{s-1} - mQ_{s-1} = (-1)^s$, or

$$aP_{s-1} = (-1)^s \pmod{m}. \tag{3.29}$$

Multiplying both sides of (3.29) by $(-1)^s b$ we obtain

$$(-1)^s b a P_{s-1} \equiv (-1)^{2s} b \pmod{m},$$

or

$$(-1)^s b a P_{s-1} \equiv b \pmod{m}. \tag{3.30}$$

Comparing (3.26) with (3.30) we obtain that

$$x \equiv (-1)^s b P_{s-1} \pmod{m}. \tag{3.31}$$

The proof is complete.

Problem 136

Solve the congruence $55x \equiv 7 \pmod{87}$.

Solution. Here $a = 55$, $b = 7$, and $m = 87$. Since $(55,87) = 1$, then the congruence has exactly one solution. This solution can be found using formula (3.31) from Lemma 12. It is necessary to find period s and the numerator of continued fraction P_{s-1}. $87/55$ was written as continued fraction in Section 2.1.3. As we remember

$$\frac{P_6}{Q_6} = \frac{87}{55} \ (s = 6),$$

and

$$\frac{P_5}{Q_5} = \frac{19}{12}.$$

Hence, $x \equiv ((-1)^6 \cdot 7 \cdot 19) \pmod{87}$, or

$$x \equiv 133 \pmod{87} \Rightarrow x \equiv 46 \pmod{87}.$$

Answer. $x \equiv 46 (mod\,87)$.

Problem 137

Solve the congruence $111x \equiv 75 \pmod{321}$.

Solution. Comparing this congruence with $ax \equiv b \pmod{m}$ of Section 2.1, we have $a = 111$, $b = 75$ and $m = 321$. Since $(111,321) = 3$ this congruence has exactly three solutions.

The given problem can be written as $37x \equiv 25 \pmod{107}$. $107/37$ was written as continued fraction in Section 2.1.3. It was found that $s = 3$, $s - 1 = 2$, and $P_{s-1} = P_2 = 26$. Then, $x \equiv ((-1)^3 25 \cdot 26) \pmod{107}$, or $x \equiv -650 \pmod{107}$. Hence, $x \equiv -8 \pmod{107}$. It means that $x \equiv 99; 206; 313 \pmod{321}$.

Answer. $x = 99 + 321n$, $x = 206 + 321k$, $x = 313 + 321m$.

Problem 138

Solve in integers x and y the following system:

$$\begin{cases} (x+2y) \equiv 3 \pmod{6} \\ (3x+y) \equiv 2 \pmod{6} \end{cases}$$

Solution. In order to solve this system we will apply Lemma 5.

1. We eliminate one of the variables:

$$(-3x - 6y) \equiv -9 \quad (\text{mod } 6)$$
$$(3x + y) \equiv 2 \quad (\text{mod } 6) \Rightarrow$$
$$(-5y) \equiv -7 \quad (\text{mod } 6) \Rightarrow 5y \equiv 7 \quad (\text{mod } 6)$$
$$7 + 3 \cdot 6 = 25 \Rightarrow 5y \equiv 25 \quad (\text{mod } 6).$$

Hence, $y \equiv 5 \ (\text{mod } 6)$ or

$$y = 6k + 5. \tag{3.32}$$

2. In order to find x we will substitute y from (3.32) into the first equation of the system:

$$(x + 2 \cdot 5 \quad (\text{mod } 6)) \equiv 3 \quad (\text{mod } 6) \Rightarrow x \equiv -7 \quad (\text{mod } 6).$$

Since $5 = -7 + 2 \cdot 6$ the relationship above can be written as $x \equiv 5 \ (\text{mod } 6)$ or $x = 6l + 5$.
Answer. $x = 6l + 5$, $y = 6k + 5$.

Problem 139

Find a multiple of 7 that leaves remainder of 1 when divided by 2, 3, 4, 5, or 6.

Solution. Since $(2,3) = 1$, $(2,5) = 1$ and $(3,5) = 1$ we can apply Theorem 4. From this theorem and Lemma 7 ($LCM = 60$), we obtain

$$7x \equiv 1 \quad (\text{mod } m), \ m = 2, 3, 4, 5, 6$$
$$7x \equiv 1 \quad (\text{mod } 2 \cdot 3 \cdot 2 \cdot 5)$$
$$7x \equiv 1 \quad (\text{mod } 60) \ (301 = 60 \cdot 5 + 1)$$
$$7x \equiv 301 \quad (\text{mod } 60) \Rightarrow x \equiv 43 \quad (\text{mod } 60) \Rightarrow x = 43, \ 7x = 301.$$

Answer. 301.

Problem 140

Solve $9x \equiv 4 \ (\text{mod } 2401)$ without a calculator.

Solution. Consider the decimal representation of 2401 and let us extract in it all multiples of 9. Then, $2401k + 4 = 9x$. Therefore,

$$(2 \cdot 10^3 + 4 \cdot 10^2 + 1) \cdot k + 4 = [2(999 + 1) + 4(99 + 1) + 1] \cdot k + 4 = 9x$$
$$\Rightarrow (2 \cdot 999 + 4 \cdot 99)k + 7k + 4 = 9x \Rightarrow k = 2.$$

Or by Lemma 8

$$4 + (2+4+0+1) \cdot k = 9m$$
$$4 + 7k = 9m$$
$$k = 2,$$

and $4 + 7k = 4 + 7 \cdot 2 = 18$ is divisible by 9, then

$$9x \equiv (4 + 2401 \cdot 2) \pmod{2401}$$
$$\Rightarrow 9x \equiv 4806 \pmod{2401}$$
$$\Rightarrow x \equiv 534 \pmod{2401}.$$

Answer. $x \equiv 534 \pmod{2401}$.

3.3.2 Using Euler's Formula to Solve Linear Congruence

Euler's Formula can be useful for solving linear congruence. Actually,

$$a^{\phi(m)} \equiv 1 \pmod{m}$$
$$ax \equiv b \pmod{m}. \tag{3.33}$$

Multiplying the first equation by b we obtain

$$(a^{\phi(m)}b) \equiv b \pmod{m},$$

and comparing this with the second equation of (3.33) we obtain that

$$ax \equiv (a^{\phi(m)}b) \pmod{m},$$

or solution can be written as

$$x_0 \equiv (a^{\phi(m)-1}b) \pmod{m}. \tag{3.34}$$

Problem 141

Solve the following congruence $9x \equiv 8 \pmod{34}$.

Solution.
Method 1. We have $(9,34) = 1$, and $a = 9$, $b = 8$, $m = 34$. Then,

$$34 = 2 \cdot 17 \Rightarrow \phi(34) = (17-1) \cdot (2-1) = 16.$$

By (3.34),

$$x_0 \equiv (9^{16-1} \cdot 8) \equiv (3^{30} \cdot 8) \equiv (8 \cdot (2187)^2) \quad (\mathrm{mod}\ 34)$$
$$2187 \equiv 11 \quad (\mathrm{mod}\ 34)$$
$$x_0 \equiv (8 \cdot 11^2) \quad (\mathrm{mod}\ 34)$$
$$x \equiv 16 \quad (\mathrm{mod}\ 34).$$

Method 2. Notice that $8 + 34 \cdot 4 = 144 = 9 \cdot 16$.

$$9x \equiv 8 \quad (\mathrm{mod}\ 34)$$
$$9x \equiv 144 \quad (\mathrm{mod}\ 34)$$
$$x \equiv 16 \quad (\mathrm{mod}\ 34)$$

Answer. $x \equiv 16(\ \mathrm{mod}\ 34)$.

Theorem 20 (Chinese Remainder Theorem. General Case)

Let $m_i,\ 1 \le i \le k$ be relatively prime and $M = m_1 \cdot m_2 \cdot \ldots \cdot m_k$ and let $a_i,\ 0 \le a_i \le m_i$ be integer numbers. Introduce numbers $M_i = \frac{M}{m_i}$ and let each number N_i satisfies the congruence

$$M_i N_i \equiv 1\,(mod\,m_i),\ i = 1...k., \tag{3.35}$$

then the system of congruences on $[0, M-1]$ $\begin{cases} x \equiv a_1\,(mod\,m_1) \\ x \equiv a_2\,(mod\,m_2) \\ ... \\ x \equiv a_k\,(mod\,m_k) \end{cases}$ has a

unique solution

$$x \equiv (a_1 N_1 M_1 + ... + a_k N_k M_k)\,(mod\,M) \tag{3.36}$$

Let us apply this theorem by solving the following problem.

Problem 142

Solve the system $\begin{cases} x \equiv a_1\,(mod\,4) \\ x \equiv a_2\,(mod\,5) \\ x \equiv a_3\,(mod\,7) \end{cases}$.

Solution. Define $m_1 = 4,\ m_2 = 5,\ m_3 = 7,\ M = 4 \cdot 5 \cdot 7 = 140$.

$$M_1 = \tfrac{M}{m_1} = \tfrac{140}{4} = 35$$
$$M_2 = \tfrac{M}{m_2} = \tfrac{140}{5} = 28$$
$$M_3 = \tfrac{M}{m_3} = \tfrac{140}{7} = 20$$

Next, we need to calculate $N_i,\ i = 1,2,3$ by solving the following congruences:

$$\begin{cases} M_1 N_1 \equiv 1 \,(mod\,4) \\ M_2 N_2 \equiv 1 \,(mod\,5) \\ M_3 N_3 \equiv 1 \,(mod\,7) \end{cases} \Leftrightarrow \begin{cases} 35 N_1 \equiv 1 \,(mod\,4) \\ 28 N_2 \equiv 1 \,(mod\,5) \\ 20 N_3 \equiv 1 \,(mod\,7) \end{cases} \tag{3.37}$$

Each congruence can be solved using Euler's Theorem. Consider the first equation of the system:

$$35 N_1 \equiv 1 \,(mod\,4). \tag{3.38}$$

it follows from Euler's formula that the following is true: $35^{\varphi(4)} \equiv 1 \,(mod\,4)$, $\varphi(4) = \varphi(2^2) = 2^2 - 2^1 = 2$, then

$$35^2 \equiv 1 \,(mod\,4) \tag{3.39}$$

Solving together (3.38) and (3.39) we can find that $N_1 \equiv 35 \,(mod\,4)$ or after simplification

$$N_1 \equiv 3 \,(mod\,4), \; N_1 = 3.$$

Consider now the second equation of the system and corresponding to it equation obtained by using Euler's formula and the fact that $\varphi(5) = 5 - 1 = 4$:

$$\begin{cases} 28 N_2 \equiv 1 \,(mod\,5) \\ 28^4 \equiv 1 \,(mod\,5) \end{cases} \Rightarrow N_2 \equiv 28^3 \,(mod\,5) \equiv 112 \,(mod\,5) \equiv 2 \,(mod\,5).$$

$$N_2 = 2.$$

Similarly, we can obtain the corresponding equations for the third equation of the system.

$$\begin{cases} 20 N_3 \equiv 1 \,(mod\,7) \\ 20^6 \equiv 1 \,(mod\,7) \end{cases} \Rightarrow N_3 \equiv 20^5 \,(mod\,7) \equiv 20 \,(mod\,7) \equiv 6 \,(mod\,7)$$

$$N_3 = 6.$$

Finally, substituting these values into (3.36) we have the answer,

$$\begin{aligned} x &\equiv (a_1 N_1 M_1 + \dots + a_k N_k M_k) \,(mod\,M) \\ &\equiv (35 \cdot 3 a_1 + 28 \cdot 2 a_2 + 20 \cdot 6 a_3) \,(mod\,140) \\ &\equiv (105 a_1 + 56 a_2 + 120 a_3) \,(mod\,140). \end{aligned}$$

Answer. $x \equiv (105 a_1 + 56 a_2 + 120 a_3) \,(mod\,140)$.

Remark. Of course, each of these equations can be solved without Euler's formula. For example, let us demonstrate it here for the third equation,

$$20 N_3 \equiv 1 \,(mod\,7)$$

We need to multiply 7 by such natural number that after adding one it would become a multiple of 20. Such a number can be found mentally as $7 \cdot 17 + 1 = 120$, then this equation can be written and easily solved as

$$20N_3 \equiv 120 \, (mod7)$$
$$N_3 \equiv 6 \, (mod7)$$

In general, we can always use Euler's formula to solve equations (3.35) as follows:

$$\begin{cases} M_i N_i \equiv 1 \, (mod \, m_i) \\ M_i^{\varphi(m_i)} \equiv 1 \, (mod \, m_i) \end{cases} \Leftrightarrow N_i \equiv M_i^{\varphi(m_i)-1} \, (mod \, m_i)$$

3.4 Nonlinear Equations: Applications of Factoring

Many nonlinear equations in two, three, or more variables that are subject to solution over the set of all integers can be essentially simplified if we apply factoring. In order to refresh your factoring skills I want to give you the most important formulas called special products.

- A Difference of Squares can be factored as $u^2 - v^2 = (u-v)(u+v)$, whatever we have for u and v.

  Example. $(x-1)^2 - 9 = (x-1)^2 - 3^2 = (x-1-3)(x-1+3) = (x-4)(x+2)$. Here $u = x - 1$ and $v = 3$.

- A Difference of Cubes can be factored as $u^3 - v^3 = (u-v)(u^2 + uv + v^2)$ for any u and v.
- A Sum of Cubes can be factored as $u^3 + v^3 = (u+v)(u^2 - uv + v^2)$ for any u and v.
- Difference of nth powers

$$u^n - v^n = (u-v)(u^{n-1} + u^{n-2}v + u^{n-3}v^2 + \cdots + u^2v^{n-3} + uv^{n-2} + v^{n-1}).$$

If $n = 2k$ (an even power) then

$$u^n - v^n = u^{2k} - v^{2k} = \left(u^2\right)^k - \left(v^2\right)^k$$
$$= (u^2 - v^2) \cdot (\text{polynomial of } (2k-2) \text{ degree})$$
$$= (u-v)(u+v) \cdot (\text{polynomial of } (2k-2) \text{ degree}).$$

Some number theory problems contain exponential expressions, so we have to list two of the most useful properties of exponents.

- Factoring of an exponent

$$a^{n+m} = a^n \cdot a^m$$
$$(a^n)^m = a^{n \cdot m} \Rightarrow a^{n \cdot m} - 1 = (a^n)^m - 1^m.$$

Example. Thus using the last formula, we can factor as follows:

$$4^{2x} - 1 = (4^x)^2 - 1^2 = (4^x - 1)(4^x + 1).$$

Above we used properties of exponents and again a difference of squares!

- Factoring the sum of squares,

$$(ac + bd)^2 + (ad - bc)^2 = (a^2 + b^2)(c^2 + d^2). \tag{3.40}$$

This formula can be easily proven. Please do it yourself. This formula is very important when you have to rewrite a number as sum of squares of other numbers. Though this formula is true for any a, b, c, and d, not every integer can be written as a sum of two squares. We will prove it and discuss it more in Chapter 4 of the book.

By doing problems in this chapter, you will see that all these formulae play a very important role in solving problems on integers and divisibility.

3.4.1 Newton's Binomial Theorem

You are probably familiar with the so-called Pascal's Triangle that allows us to raise $(x + y)$ or $(x - y)$ to an integer power and find appropriate coefficients. Thus, after making a triangle as follows and selecting, for example, row 5, we can evaluate the fifth power of $(x + y)$ (See Table 3.2): Hence, we have

0:									1							
1:								1		1							
2:							1		2		1						
3:						1		3		3		1					
4:					1		4		6		4		1				
5:				1		5		10		10		5		1			
6:			1		6		15		20		15		6		1		
7:		1		7		21		35		35		21		7		1	
8:	1		8		28		56		70		56		28		8		1

Table 3.2 Pascal's Triangle.

$$(x + y)^5 = 1 \cdot x^5 + 5 \cdot x^4 \cdot y + 10 \cdot x^3 \cdot y^2 + 10 \cdot x^2 \cdot y^3 + 5 \cdot x \cdot y^4 + 1 \cdot y^5.$$

This way you could find coefficients for any other power of $(x + y)$. Let us solve the following problem.

Problem 143

Prove that the equation

$$(x-y)^3 + (y-z)^3 + (z-x)^3 = 30$$

has no solutions in integer numbers.

Proof. Raising each term to the third power,

$$(x-y)^3 = x^3 - 3x^2y + 3xy^2 - y^3$$
$$(y-z)^3 = y^3 - 3y^2z + 3yz^2 - z^3$$
$$(z-x)^3 = z^3 - 3z^2x + 3zx^2 - x^3$$

and then adding the left and right sides of each equation, we obtain that the original equation can be written as

$$(x-y)^3 + (y-z)^3 + (z-x)^3 = 3xy^2 - 3x^2y + 3yz^2 - 3y^2z + 3zx^2 - 3z^2x = 30$$

We need to solve,

$$xy^2 - x^2y + yz^2 - y^2z + zx^2 - z^2x = 10$$

We can factor the left side of it step by step as follows:

$$y^2(x-z) + zx(x-z) + y(z^2 - x^2) = 10$$
$$y^2(x-z) + zx(x-z) + y(x-z)(x+z) = 10$$
$$(x-z)(y^2 + zx - y(x+z)) = 10$$
$$(x-z)(y^2 + zx - yx - yz) = 10$$
$$(x-z)(x(z-y) - y(z-y)) = 10$$
$$(x-z)(z-y)(x-y) = 10$$

The last equation can be rewritten as

$$(x-z)(y-x)(z-y) = 10.$$

In words, find such three factors of 10, the sum of which equals zero. We know that $\pm1; \pm2; \pm5; \pm10$ are all the factors of natural number 10, no combination of any three of these factors would ever give us 0. Therefore, the given equation has no solutions in natural numbers.

Using Pascal's triangle for small powers works fine. However, it is not very efficient way if the power is greater than 10. In order to expand any power of $(x+y)$ we will introduce the Newton's Binomial Theorem.

Theorem 21 (Newton's Binomial Theorem)

The following identity is valid:

$$(u+v)^n = \sum_{k=0}^{n} \frac{n!}{k!(n-k)!} u^{n-k} v^k$$
$$= u^n + nu^{n-1}v + \frac{n(n-1)}{2} u^{n-2}v^2 + \cdots + nuv^{n-1} + v^n. \qquad (3.41)$$

This identity is known as Newton's Binomial Theorem since it was discovered by Isaac Newton, and $(x+y)$ is a binomial, i.e., an expression with two terms. The formula $\frac{n!}{k!(n-k)!}$ is called the binomial coefficient and is often denoted by symbol

$$C_n^k = \frac{n!}{k!(n-k)!}.$$

Let us evaluate some of the binomial coefficients,

$$
\begin{aligned}
C_n^1 &= \frac{n!}{1!(n-1)!} = \frac{(n-1)!n}{(n-1)!} = n \\
C_n^2 &= \frac{n!}{2!(n-2)!} = \frac{(n-2)!(n-1)n}{2(n-2)!} = \frac{(n-1)n}{2} \\
C_n^{n-1} &= \frac{n!}{(n-1)!(n-(n-1))!} = \frac{(n-1)!n}{(n-1)!\cdot 1} = n.
\end{aligned}
$$

Each time we tried to factor a factorial on the top so that one of the factors would match with a factorial on the bottom. Since

$$\frac{n!}{k!(n-k)!} = \frac{n!}{(n-k)!(n-(n-k))!},$$

we also can state the following property of the binomial coefficients:

$$C_n^k = C_n^{n-k}.$$

Other properties of binomial coefficients can be found in my book *Methods of Solving Nonstandard Problems*. If you have taken a probability course or introductory statistics, then you are very familiar with binomial coefficients and their properties. In the course of statistics someone always selects a simple random sample of size n from a population of size N. I want to remind you that this as well can be done using binomial coefficients formula, in

$$C_N^n = \frac{N!}{n!(N-n)!}$$

different ways.

Problem 144

Find the power of 2 in the product

$$(n+1)(n+2)(n+3)\cdot...\cdot(2n-1)(2n).$$

Solution. The given expression can be also seen as

$$\frac{(2n)!}{n!} = \frac{1\cdot2\cdot3\cdot...\cdot(2n-1)\cdot(2n)}{1\cdot2\cdot3\cdot...\cdot(n-1)\cdot(n)}$$
$$= 1\cdot3\cdot5\cdot...\cdot(2n-1)\cdot\frac{2\cdot4\cdot6\cdot...\cdot2n}{1\cdot2\cdot3\cdot...\cdot n}$$
$$= 1\cdot3\cdot5\cdot...\cdot(2n-1)\cdot\frac{2}{1}\cdot\frac{4}{2}\cdot\frac{6}{3}\cdot...\cdot\frac{2n}{n}$$
$$= 1\cdot3\cdot5\cdot...\cdot(2n-1)\cdot2^n.$$

which can be also rewritten as $2^n\cdot\prod_{k=1}^{n}(2k-1)$.
Answer. 2 is raised to the power of n.

Formula (3.41) can be easily rewritten for the nth power of the difference of u and v, by placing a minus sign before each odd power of v. For example, if $v = 1$ and $n = 100$ we obtain,

$$(u-1)^{100} = u^{100} - 100u^{99} + \frac{100\cdot99}{2}u^{98} + \cdots - 100u + 1.$$

We can consider a Trinomial Square as

$$u^2 + 2uv + v^2 = (u+v)^2$$
$$u^2 - 2uv + v^2 = (u-v)^2,$$

and have the example: $x^2 + 1 - 2x = x^2 - 2x\cdot1 + 1^2 = (x-1)^2$.

Any quadratic equation $ax^2 + bx + c = 0$ with zeros x_1 and x_2 can be factored as

$$ax^2 + bx + c = a(x-x_1)(x-x_2) = 0,$$

where

$$x_{1,2} = \frac{-b\pm\sqrt{b^2-4ac}}{2a}. \tag{3.42}$$

Here $D = b^2 - 4ac$ is a **discriminant** of a quadratic equation. If $D \geq 0$ a quadratic equation has real roots, and if $D < 0$ no real roots. If the coefficient b of the linear term of a quadratic equation is an even, then it is better to use so-called $D/4$ formula. Dividing all terms of (3.42) by 2 we obtain

$$x_{1,2} = \frac{-\frac{b}{2}\pm\sqrt{\frac{b^2-4ac}{4}}}{2\cdot\frac{a}{2}}$$
$$= \frac{-\frac{b}{2}\pm\sqrt{\left(\frac{b}{2}\right)^2-ac}}{a}.$$

Moreover,

$$\frac{D}{4} = \left(\frac{b}{2}\right)^2 - ac. \tag{3.43}$$

The advantage of using this formula can be demonstrated as follows:

$$7x^2 - 4x - 1 = 0$$
$$\frac{D}{4} = 2^2 + 7 = 11$$
$$x_{1,2} = \frac{2 \pm \sqrt{11}}{7}.$$

The answer appears in the reduced form.

Problem 145 (MGU, entrance exam, 2003)

Find all integer solutions to

$$x^2 + 5y^2 + 34z^2 + 2xy - 10xz - 22yz = 0$$

Solution. Consider this equation as quadratic in one variable x by rewriting it as:

$$x^2 - 2(5z - y) \cdot x + (5y^2 + 34z^2 - 22yz) = 0$$

Since the coefficient of the linear term is even, we will evaluate its discriminant divided by 4, $D/4$:

$$\frac{D}{4} = (5z - y)^2 - (5y^2 + 34z^2 - 22yz)$$
$$= -9z^2 + 12yz - 4y^2 = -(3z - 2y)^2 \le 0.$$

We know that a quadratic equation will not have real solutions if its discriminant is negative, then the given equation can have only one zero and the following condition must be true:

$$\frac{D}{4} = 0, \; x = 5z - y$$

which implies that

$$3z - 2y = 0$$

The later equation is recognized as homogeneous linear Diophantine equation and it has infinitely many solutions. Solving two equations together, we obtain

$$\begin{cases} x = 5z - y \\ 3z = 2y \end{cases} \Leftrightarrow \begin{cases} x = 7k, \; k \in Z \\ z = 2k, \; y = 3k \end{cases}$$

Therefore, the given nonlinear equation has infinitely many solutions, each described by a triple:

$$(x, y, z) = (7k, 3k, 2k), \; k \in \mathbb{Z}.$$

Answer. $(x, y, z) = (7k, 3k, 2k), \; k \in \mathbb{Z}$

3.4.2 Difference of Squares and Vieta's Theorem

Difference of two squares formula is very important. For example, we know that the difference of any of two squares can be factored as

$$n^2 - m^2 = (n+m)(n-m). \tag{3.44}$$

Pierre Fermat used this formula in order to factor big odd numbers that do not pass divisibility tests right away. In order to understand Fermat's ideas, I will give a simple example. Assume that we want to find prime factorization of 899. Notice that

$$899 = 900 - 1.$$

Since both numbers on the right-hand side are perfect squares, we can factor it as

$$899 = 30^2 - 1^2 = (30-1) \cdot (30+1) = 29 \cdot 31,$$

which give us prime factorization of 899. Similarly you can factor 851 as

$$851 = 900 - 49 = 30^2 - 7^2 = (30-7) \cdot (30+7) = 23 \cdot 37.$$

However, 899 and 851 are "small" numbers and obviously it was easy to factor them. You can create your own many examples, such as

$$9991 = 10000 - 9 = 100^2 - 3^2 = (100-3)(100+3) = 97 \cdot 103.$$

The well-known Fermat's factorization method is based on evaluation squares by modulo n for integers x a little bigger than $\sqrt{n}$ with the goal to obtain a perfect square of y, y^2. This method works very fast if $n = p \cdot q$ and the numbers p, q and close to each other. Let's summarize the idea.

Suppose that we need to factor a big number n. If we can find such two numbers x and y that

$$x^2 - y^2 = n,$$

then

$$n = (x+y)(x-y)$$

The numbers $(x+y)$ and $(x-y)$ are factors and possibly trivial factors when the smallest factor equals unit. These two numbers, x and y, satisfying $x^2 - y^2 = n$ will be found if there exists an integer number x, such that $x^2 - n$ is a perfect square. Then $x^2 - (x^2 - n)$ is a difference of squares, equals n. Consider again a simple case of factorization of 11413:

$$107^2 - 11413 = 36 = 6^2,$$

then $11413 = 107^2 - 6^2 = 101 \cdot 113$.

Thus, Fermat started his search from $x = [\sqrt{n}] + 1$, the least number for which the difference $x^2 - n$ is positive. Then he increased x by one and evaluated $x^2 - n$ and repeated this procedure until $x^2 - n$ would become a perfect square. If this occurred, Fermat tried to decompose n as follows:

$$\left(x - \sqrt{x^2 - n}\right) \cdot \left(x + \sqrt{x^2 - n}\right)$$

and if this representation is trivial, he continued increasing x.

Let us apply Fermat's factorization method by solving the following problem. since

Problem 146

Factor $n = 364,729$ using Fermat's difference of squares approach.

Solution. We can find that the given number is between squares of 603 and 604:

$$603^2 = 363,609 < 364,729 < 364,816 = 604^2$$

which means that $[\sqrt{n}] + 1 = 604$. Evaluate

$$604^2 - 364729 = 87,$$

Because 87 is not a perfect square, we will increase the number by one:

$$605^2 - 364729 = 1296 = 36^2,$$

and we are done! The final step is

$$364,729 = 605^2 - 36^2 = (605 - 36)(605 + 36) = 569 \cdot 641.$$

Answer. $364,729 = 569 \cdot 641$.

Fermat found factorization of many big numbers. His method was much better than the way of trying to divide a big number n by all possible prime factors less than $\sqrt{n}$. However, this method did not guarantee that for a big n, after taking several consecutive differences of $x^2 - n$, any of them would be a perfect square soon. What is the solution? Suppose that we evaluated several values of $x^2 - n$, and neither one is a perfect square, but the product of them is a perfect square, for example, y^2. Let X be the product $X = \prod x_i$, for which

$$X = \prod \left(x_i^2 - n\right) = y^2,$$

then $X^2 - y^2 = k \cdot n$ is divisible by n. Applying again the difference of squares to the left side of it, we obtain

$$(X - y) \cdot (X + y) = k \cdot n$$

and hence, at least one of the factors of this bigger number will be divisible by one of the factors of the original number n. Next, we can find this factor by using the Euclidean algorithm: we can find the greatest common divisor of each of the two factors and number n. Unfortunately, we can again find only trivial factors of n and then we have to choose other factors the product of which is a perfect square. Let us implement the idea above by solving another problem.

Problem 147

Factor 2189 if it is known that $579^2 - 18^2 = 2189 \cdot k$.

Solution. How can we use the given information? Ok. Let us factor the left side of the given expression using difference of squares,

$$(579 - 18) \cdot (579 + 18) = 561 \cdot 597.$$

One of these factors must have common factor with 2189. Using Euclidean algorithm, we obtain

$$
\begin{aligned}
2189 &= 3 \cdot 561 + 506 \\
561 &= 1 \cdot 506 + 55 \\
506 &= 9 \cdot 55 + 11 \\
55 &= 5 \cdot 11.
\end{aligned}
$$

Therefore, 2189 is divisible by 11 and finally can be factored as

$$2189 = 11 \cdot 199.$$

Of course, this factorization could be found much easier, if we noticed that the given number is divisible by 11.

Problem 148

Using Fermat's method find factorization of 12499.

Solution. Notice that $\left\lceil \sqrt{12499} \right\rceil = 111$, and $112^2 = 12544 > 12499$. Starting considering $x^2 - n$ for different $x \geq 112$, we would not have luck right away. However, we can notice that

$$113^2 - 12499 = 270 = 2 \cdot 5 \cdot 3^3$$

and that

$$127^2 - 12499 = 3630 = 2 \cdot 3 \cdot 5 \cdot 11^2$$

Multiplying the left and right sides of the two equations, we will get a perfect square on the right-hand side!

$$\left(113^2 - 12499\right) \cdot \left(127^2 - 12499\right) = \left(2 \cdot 3^2 \cdot 5 \cdot 11\right)^2$$

this can be written as

$$(127 \cdot 113)^2 - n \cdot 113^2 - n \cdot 127^2 + n^2 = \left(2 \cdot 3^2 \cdot 5 \cdot 11\right)^2$$

or as

$$(127 \cdot 113)^2 - (990)^2 = k \cdot 12499$$

This can be simplified using the notation already introduced,

$$X = \prod x_1 = 113 \cdot 127, \ y = 990, \ X - y = 14351 - 990 = 13361, \ X + y = 15341$$

Next, we need to find *gcd* of numbers $X - y$ and n and maybe $X + y$ and n:

$$(X - y, n) = (14351, 12499) = (1852, 12499) = (862, 12499) = 431$$

By dividing 12,499 by its nontrivial factor, 431, we complete factorization as

$$12499 = 431 \cdot 29$$

Answer. $12,499 = 431 \cdot 29$.

Fermat's method of factorization in its modification with help of computer algo-rithms is used by different agencies in encryption of data. I hope that you will remember this wonderful formula and use it in solving problems in numbers.

When demonstrating Fermat's factorization method, we tried to factor odd num-bers only, because any even number is divisible by 2 and eventually becomes odd. Consider again the difference of squares of some natural numbers. For example, $289 - 144 = 17^2 - 12^2 = (17 + 12)(17 - 12) = 29 \cdot 5 = 145$ (the difference of odd and even perfect numbers) $144 - 100 = 12^2 - 10^2 = (12 + 10)(12 - 10) = 22 \cdot 2 = 44$ (the difference of two even perfect squares) $225 - 25 = 15^2 - 5^2 = (15 + 5)(15 - 5) = 20 \cdot 10 = 200$ (the difference of two odd perfect squares). $2^2 - 1^2 = 3, \ 5^2 - 4^2 = 9, \ 3^2 - 2^2 = 5, \ 4^2 - 3^2 = 7$ (the difference of two consecu-tive squares). Hence, if we have two perfect squares, we always can subtract them and rewrite applying the formula. On the other hand, it would be nice to know what numbers a priory can be written as difference of two squares.

Problem 149

Prove that 2 cannot be written as difference of two squares.

Proof. Assume contradiction and that $n^2 - m^2 = (n + m)(n - m) = 2$. Since n and m are both natural numbers and we must have $n > m$, we see that $n + m > n - m$ and $n - m$ and $n + m$ are both natural numbers. Since 2 is prime, it follows that $n + m = 2$ and $n - m = 1$. Adding these two equations, we get $2n = 3$, which means n is not a natural number. We obtain contradiction.

Therefore, 2 cannot be written as the difference of two squares.

However, we know that some even numbers, such as 44 or 100 mentioned above or other even numbers, can be written as difference of two squares:

$$8 = 3^2 - 1^2, \ 16 = 5^2 - 3^2, \ 12 = 4^2 - 2^2$$

can be written as difference of two squares.

Problem 150

Prove that any even number that can be written as a difference of two integer squares must be divisible by 4.

Proof. Suppose that $N = n^2 - m^2 = 2l$ is an even number. N can be rewritten as

$$N = (n+m) \cdot (n-m) = 2l.$$

Because $n+m-(n-m) = 2m$, both factors differ by an even number, then they are either both odd or both even. Since N is even, both quantities must be even.

Let $n - m = 2k$, then $n + m = 2k + 2m = 2p$. Since both $n+m$ and $n-m$ are even, then l is even and then the difference of squares is divisible by 4:

$$N = (n+m) \cdot (n-m) = 2k \cdot 2p = 4 \cdot t.$$

This completes the proof.

Problem 151

In how many ways 400 can be written by the difference of two squares?

Solution. Any multiple of 4 can be written as follows:

$$n^2 - m^2 = (n-m)(n+m) = 4j$$
$$st = 4j$$

Since the right side is divisible by 4, we may choose two factors, $s = n+m$ and $t = n-m$, of $4j$ so that both s and t are even.

$$\begin{cases} s = n+m \\ t = n-m \end{cases}$$

$$n = \frac{s+t}{2}, \ m = \frac{s-t}{2}$$

Next, we can practice this algorithm.
1. Choose a number divisible by 4.
2. Factor the number is all ways as product of two even factors, s, t
3. Solve equations above and find corresponding m, n
4. Rewrite given number as difference of two squares.

400 can be written as the product of two even factors only in three ways $400 = 2 \cdot 200 = 4 \cdot 100 = 8 \cdot 50$

Hence, it can be written by the difference of squares exactly in three different ways.

$$(s,t): \ (2,200), (4,100), (8,50)$$
$$(n,m): \ (101,99), (52,48), (29,21)$$
$$400 = 101^2 - 99^2 = 52^2 - 48^2 = 29^2 - 21^2$$

Answer. In three different ways.

The problem on representation of an odd number by the difference of squares is offered in the Homework section. The next problems again emphasize the importance of the difference of squares formula.

Problem 152

Solve the equation

$$y^2 - x(x+1)(x+2)(x+3) = 1$$

in natural numbers $x, y \in \mathbb{N}$

Solution. By multiplying separately middle and outer factors in the second term, and then moving terms around, we can use the difference of squares formula:

$$(y-1)(y+1) = (x^2 + 3x)(x^2 + 3x + 2)$$

The numbers must be natural; moreover, on the left and on the right, we have a product of two quantities that differ by 2, then solution to the problem can be obtained by solving the following system:

$$\begin{cases} y - 1 = x^2 + 3x \\ y + 1 = x^2 + 3x + 2 \end{cases} \Leftrightarrow y = x^2 + 3x + 1, \ x \in N$$

Answer. $(x,y) = (x, \ x^2 + 3x + 1), \ x \in \mathbb{N}$.

Problem 153

Can a quadratic equation $ax^2 + bx + c = 0$ with integer coefficients have a discriminant equals 23?

Solution. Assume that it is possible and that

$$b^2 - 4ac = 23.$$

Adding 2 to both sides and moving $4ac$ to the right-hand side, we can rewrite it as

$$b^2 - 25 = 4ac - 2$$

Applying the difference of squares to the left-hand side and factoring the right side, we have

$$(b-5)(b+5) = 2(2ac - 1)$$

We can see that the two factors on the left differ by an even number 10, then they are either both even or both odd. If the factors are odd then this equation has no solutions, because the right-hand side is an even. If both factors on the left are even, then the left side is divisible by 4, while the right side $2(2ac - 1)$ is divisible by 2 not by 4. Hence this equation has no solutions in integers and therefore the discriminant of the quadratic equation cannot be 23.

We are used to solving quadratic equations but it is interesting to know that quadratic equations appeared as a tool to solve different applications, for example, geometric problems stated by Ancient Babylonians. In this section we will talk about theorem that has the name of French mathematician Vieta's who lived in 17th century, but based on the tablets dug up by archeologists was probably first stated 1000 BC by Babylonians, and was used to solve the following problem: Find all rectangles with the unit area and semi-perimeter of a. Ancient Babylonians recorded it as

$$xy = 1, \quad x + y = a.$$

If a is semi-perimeter, then this geometric problem can be rephrased as follows.

Find a quadratic equation, the roots of which satisfy the system above. By solving the system we obtain the equation,

$$x + \frac{1}{x} = a$$
$$x^2 - ax + 1 = 0.$$

Zeros of a quadratic equation can be found using the Vieta's Theorem.

Theorem 22 (Vieta's Theorem)

Let the quadratic equation $ax^2 + bx + c = 0$ have the roots x_1 and x_2. Then,

$$x_1 \cdot x_2 = \frac{c}{a}$$
$$x_1 + x_2 = -\frac{b}{a}.$$

The reversed Theorem is also true.

Theorem 23 (Reversed Vieta's Theorem)

If the numbers x_1 and x_2 satisfy the system of equations:

$$\begin{aligned} x_1 \cdot x_2 &= \frac{c}{a} \\ x_1 + x_2 &= -\frac{b}{a}. \end{aligned}$$

then they are the solutions to the quadratic equation $ax^2 + bx + c = 0$.

This formula is very useful, especially when you can guess one of the zeros or that zero is given.

Example. You can see that $x = 1$ is the root of $17x^2 + 4x - 21 = 0$. Since $a = 17$, $b = 4$, $c = -21$, and $x_1 = 1$, then from the first formula of Theorem 22 we obtain right away that the second root is $x_2 = \frac{-21}{17}$. If $a = 1$ factoring looks easier

$$x^2 + bx + c = (x - x_1)(x - x_2),$$

where

$$\begin{aligned} x_1 \cdot x_2 &= c \\ x_1 + x_2 &= -b. \end{aligned} \tag{3.45}$$

Therefore we are looking for such numbers x_1 and x_2, the product of which equals a constant term c, and the sum of which adds up to a negative coefficient of b.

Example. We have: $x^2 - 7x + 6 = (x - 1)(x - 6)$ because

$$\begin{aligned} 1 \cdot 6 &= 6 \\ 1 + 6 &= -(-7) = 7. \end{aligned}$$

Since the sum of the roots is a negative coefficient of x (linear term) and the product of the roots is a constant term of the quadratic equation, the reversed Vieta's Theorem is useful in finding a quadratic equation to which the given roots satisfy.

Example. If it is given that the sum of the roots of a quadratic equation is -7 and their product is 2, then the equation is

$$x^2 + 7x + 2 = 0.$$

Some students can factor quadratic trinomial mentally using so-called FOIL. If you have such skills go ahead and factor, but do not forget to check your factorization multiplying parentheses. Of course, every technique has an explanation. Let us expand the expression

$$(x+\alpha)(x+\beta) = x^2 + (\alpha+\beta)x + \alpha\beta.$$

Working backward this formula is FOIL.

Example. Factor $x^4 - 2x^2 - 8$. Let $u = x^2$, then the given expression becomes quadratic with respect to u and

$$u^2 - 2u - 8 = (u-4)(u+2) = (x^2-4)(x^2+2) = (x-2)(x+2)(x^2+2).$$

To complete factoring we applied the formula of the difference of squares as well.

You never can overestimate the importance of Vieta's Theorem. There will be several problems in this book, including one below.

Problem 154

Prove that the equation $x^2 - 2019x + 2018a - 1 = 0$ has no integer solutions for any integer value of a.

Proof. Assuming that integer solutions $x_1, x_2 \in Z$ exist, and applying Vieta's Theorem, we obtain the system of two equations in two variables:

$$\begin{aligned} x_1 \cdot x_2 &= 2018a - 1 \\ x_1 + x_2 &= 2019. \end{aligned}$$

Since $2018a - 1$ is an odd number, it follows from the first equation that the both roots of the quadratic equation must be also odd. However, such case is impossible because then from the second equation, the sum of these two odd numbers must be an odd number 2019. Therefore we obtain the contradiction. Our assumption was wrong and hence, the given equation never has integer roots, whatever the value of a parameter a is.

3.4.3 Homogeneous Polynomials

If a polynomial $f(x,y)$ possesses the property $f(tx,ty) = t^n f(x,y)$ for some natural number n, then $f(x,y)$ is said to be a homogeneous polynomial of degree n in two variables x, y. For example, $f(x,y) = x^3 + 2x^2y + y^3$ is a homogeneous polynomial of degree 3 because

$$f(tx,ty) = (tx)^3 + 2(tx)^2(ty) + (ty)^3 = t^3(x^3 + 2x^2y + y^3) = t^3 f(x,y).$$

There exists a general approach to factoring homogeneous polynomials as

$$f(x,y) = y^n f(1,u),\qquad(3.46)$$

where $u = \frac{x}{y}$.

Let us consider a homogeneous polynomial of the 4th degree:

$$f(x,y) = 6x^4 + 25x^3y + 12x^2y^2 - 25xy^3 + 6y^4.\qquad(3.47)$$

Next, we apply (3.46) to it. For this we factor out y^4:

$$f(x,y) = y^4 \left(6 \left(\tfrac{x}{y}\right)^4 + 25 \left(\tfrac{x}{y}\right)^3 + 12 \left(\tfrac{x}{y}\right)^2 - 25(\tfrac{x}{y}) + 6 \right)$$
$$= y^4 \left(6u^4 + 25u^3 + 12u^2 - 25u + 6 \right)\qquad(3.48)$$

If polynomial (3.47) can be factored, then the polynomial within brackets must have rational zeros. Using the rational zeros test we find $u = -2$, $u = -3$, $u = 1/2$, and $u = 1/3$. Now the polynomial (3.48) can be factored as

$$f(x,y) = y^4[(u+3)(u+2)(3u-1)(2u-1)]$$
$$= [y(u+3)][y(u+2)][y(3u-1)][y(2u-1)].\qquad(3.49)$$

Substituting $u = x/y$ into (3.49), we obtain that

$$f(x,y) = (x+3y)(x+2y)(3x-y)(2x-y).$$

Remark. I have to mention here that some homogeneous polynomials of second degree can be factored mentally using FOIL as

$$ax^2 + bxy + cy^2 = (mx+py)(nx+sy),$$

if we can find m, n, p, and s such that $mn = a$, $ps = c$, and $ms + pn = b$. Thus, $15x^2 - 11xy + 2y^2 = (5x-2y)(3x-y)$, because $5 \cdot 3 = 15$, $(-2)(-1) = 2$, and $5 \cdot (-1) + (-2) \cdot 3 = -11$.

Problem 155

Prove that the equation

$$x^5 + 3x^4y - 5x^3y^2 - 15x^2y^3 + 4xy^4 + 12y^5 = 33$$

has no solution in integers.

Proof. The left side is a homogeneous polynomial of fifth degree. Dividing the left side and multiplying it by y^5 we obtain the following:

$$y^5 \cdot (u^5 + 3u^4 - 5u^3 - 15u^2 + 4u + 12), \quad u = \frac{x}{y}.$$

Since a polynomial inside parentheses has unit leading coefficient, then all its rational solutions (if exist) are integers and the factors of the constant term, 12. Trying $u = 1$ makes the polynomial zero, so as $u = -1, 2, -2, -3$, then $(u-1)$, $(u+1)$, $(u-2)$ and $(u+2)$, and $(u+3)$ are its factors. All factors can be also obtained by applying consecutive synthetic division. We have

$$y^5(u-1)(u+1)(u-2)(u+2)(u+3) = (\tfrac{x}{y} - 1)y(\tfrac{x}{y} + 1)y(\tfrac{x}{y} - 2)y(\tfrac{x}{y} + 2)y(\tfrac{x}{y} + 3)y$$
$$= (x-y)(x+y)(x-2y)(x+2y)(x+3y).$$

Now our problem is equivalent to solving the equation,

$$(x-y)(x+y)(x-2y)(x+2y)(x+3y) = 33.$$

Obviously $x \neq \pm y, \pm 2y, \neq -3y$.

Also if $y = 0$, we obtain that $x^5 = 33$ that does not have integer solutions. In all other cases, we need to factor 33 into five distinct factors, which is impossible, because $33 = 3 \cdot 11 \cdot 1 \cdot (-1)^2$, so at least two factors would be the same, which is not the case. Therefore the given equation does not have any integer solutions.

Special Property of Homogeneous Equations of Even Degree. For a homogeneous equation of second, fourth, sixth, and so on (any even degree) if (a,b) is a solution, then $(-a,-b)$ will be a solution as well. From formula (3.46) we find $f(-a,-b) = (-1)^n f(a,b)$. But if $n = 2k$ (even degree), then $f(-a,-b) = f(a,b)$. Therefore, if (a,b) is a solution, then $(-a,-b)$ is another solution.

Problem 156

Find all whole numbers x and y satisfying the equation

$$15x^2 - 11xy + 2y^2 = 7.$$

Solution. Introducing a new variable $u = y/x$ and replacing each y by ux, we can rewrite the left side as

$$15x^2 - 11xy + 2y^2 = 15x^2 - 11x \cdot ux + 2u^2x^2$$
$$= x^2(2u^2 - 11u + 15). \tag{3.50}$$

Quadratic trinomial $2u^2 - 11u + 15$ can be factored as $(2u-5)(u-3)$, then the left side of (3.50) can be represented as

$$15x^2 - 11xy + 2y^2 = x^2(2u-5)(u-3)$$
$$= (2ux - 5x)(ux - 3x) = (5x - 2y)(3x - y). \tag{3.51}$$

On the other hand, the right side of the given equation equals a number of 7, that is a prime, and can be factored as

$$7 = 1 \cdot 7 = 7 \cdot 1 = (-1)(-7) = (-7)(-1). \tag{3.52}$$

Combining (3.51) and (3.52) we can obtain four systems in x and y. However, because we deal with the homogeneous polynomial of second (even) order, we can find all solutions by solving only two systems:

1.

$$\begin{cases} 5x - 2y = 1 \\ 3x - y = 7 \end{cases} \Leftrightarrow x = 13, \ y = 32.$$

2.

$$\begin{cases} 5x - 2y = 7 \\ 3x - y = 1 \end{cases} \Leftrightarrow x = -5, \ y = -16.$$

Answer. $(13, 32)$, $(-13, -32)$, $(5, 16)$, and $(-5, -16)$.

Summary. Every time you have a nonlinear equation in integers first try to factor it. For this you have to be able to recognize a special product or rewrite the given expression in such a way that you can use a special product. The problems below will help you to mastering your skills.

Problem 157 (Sophie Germain Problem)

Prove that $n^4 + 4$ cannot be prime, if $n > 1$.

Proof. Let us show that this number can be factored. Thus,

$$n^4 + 4 = n^4 + 4 - 4n^2 + 4n^2 = n^4 + 4n^2 + 4 - (2n)^2 = (n^2 + 2)^2 - (2n)^2$$
$$= (n^2 + 2n + 2)(n^2 - 2n + 2) = ((n+1)^2 + 1)((n-1)^2 + 1).$$

Therefore, this number as a product of two other numbers, of course, cannot be prime.

Problem 158 (Dudley)

DeBouvelles (1509) claimed that one or both of $(6n + 1)$ and $(6n - 1)$ are primes for all positive integers. Show that there are infinitely many n such that $(6n - 1)$ and $(6n + 1)$ are composite.

Solution. Since sum and difference of cubes can be factored, denote $n = 36m^3$, and

$$6n + 1 = 6 \cdot 36m^3 + 1 = (6m)^3 + 1^3 = (6m + 1)(36m^2 - 6m + 1)$$
$$6n - 1 = 6 \cdot 36m^3 - 1 = (6m)^3 - 1^3 = (6m - 1)(36m^2 + 6m + 1)$$

There are other possible values for n that allow us to factor $(6n+1)$ and $(6n-1)$. For example, if $n = 6k^2$, then $6n - 1$ can be factored applying the difference of squares formula:

$$6n - 1 = 6 \cdot 6k^2 - 1 = 36k^2 - 1 = (6k+1)(6k-1).$$

Problem 159 (American Mathematical Monthly 1977)

The difference of two consecutive cubes is a perfect square of some number. Prove that this number can be represented as the sum of two consecutive squares.

Proof. Let us consider the difference of two consecutive cubes:

$$(x+1)^3 - x^3 = 3x^2 + 3x + 1 = y^2.$$

Multiply both sides by 4,

$$4(3x^2 + 3x + 1) = 4y^2$$
$$12x^2 + 12x + 4 = 4y^2.$$

Completing the square on the left and moving one to the right-hand side:

$$3(4x^2 + 4x + 1) + 1 = 4y^2$$
$$3(2x+1)^2 \qquad = 4y^2 - 1$$
$$3(2x+1)^2 \qquad = (2y-1)(2y+1).$$

Since $(2y-1, 2y+1)$ are relatively prime, we have two possible cases:
Case 1.

$$2y - 1 = 3m^2, \ (n,m) = 1$$
$$2y + 1 = n^2, \quad n^2 - 3m^2 = 2,$$

Case 2.

$$2y - 1 = m^2, \ (n,m) = 1$$
$$2y + 1 = 3n^2, \ m^2 - 3n^2 = -2.$$

Of course m and n are odd integers.

Case 1 is not possible because it leads to the equation

$$n^2 - 3m^2 = 2 \ \Leftrightarrow \ n^2 \equiv 2 \pmod 3$$

that does not have solution in integers (square of any number divided by 3 leaves remainder of 0 or 1, not 2).

From Case 2 we have

$$2y - 1 = m^2, \ (n,m) = 1$$
$$2y \quad = m^2 + 1$$
$$2y \quad = (2k+1)^2 + 1$$
$$2y \quad = 4k^2 + 4k + 2 = 2(k^2 + 2k + 1 + k^2)$$
$$2y \quad = 2[(k+1)^2 + k^2]$$
$$y \quad = (k+1)^2 + k^2.$$

The last formula represents y as the sum of two consecutive squares.

Note. Many problems on representation of a number of two or more squares can be found in Chapter 4.

Problem 160

Find all integers x and y that satisfy the equation $xy = x + y$.

Solution. Let us rewrite the equation in the form

$$xy - x - y + 1 = 1.$$

Factoring by grouping the left-hand side we obtain

$$x(y-1) - (y-1) = 1$$
$$(y-1)(x-1) = 1$$

We obtain very interesting situation. Since a number on the right side is one, then we have only two opportunities.

1. $y - 1 = 1$ and $x - 1 = 1$, then $y = 2$ and $x = 2$.
2. $y - 1 = -1$ and $x - 1 = -1$, then $x = 0$ and $y = 0$.

Answer. $(0,0)$ and $(2,2)$.

Problem 161

Find all natural x and y that satisfy the equation: $2x^3 + xy - 7 = 0$.

Solution. This problem may look unusual, because it has one equation and two variables x and y. Of course, you noticed that 7 is a prime number. Using ideas from the previous problem, we can see an advantage of moving 7 to the right-hand side of the equation:

$$2x^3 + xy = 7. \tag{3.53}$$

The left side of (3.53) can be factored as

$$x(2x^2 + y) = 7. \tag{3.54}$$

The right side (number 7) in turn can be factored as

$$\begin{aligned}
7 &= 1 \cdot 7 \\
7 &= 7 \cdot 1 \\
7 &= (-1) \cdot (-7) \\
7 &= (-7) \cdot (-1).
\end{aligned} \tag{3.55}$$

From relations (3.54) and (3.55) we have four possible situations for numbers x and y that are reducible to four systems:

1.

$$\begin{cases} x = 1 \\ 2x^2 + y = 7 \end{cases}$$

We obtain $x = 1$, $y = 5$.

2.

$$\begin{cases} x = 7 \\ 2x^2 + y = 1 \end{cases}$$

We find $x = 7$, $y = -93$.

3.

$$\begin{cases} x = -1 \\ 2x^2 + y = -7 \end{cases}$$

We have $x = -1$, $y = -9$.

4.

$$\begin{cases} x = -7 \\ 2x^2 + y = -1 \end{cases}$$

We obtain $x = -7$, $y = -95$.

We find four ordered pairs of (x, y) that could satisfy the given equation, but among them only pair $(1, 5)$, i.e., $x = 1$ and $y = 5$ satisfies the condition of the problem.

Answer. $x = 1$ and $y = 5$.

> **Problem 162**
>
> Find natural solutions to $xy - 2011(x + y) = 0$.

Solution. Let us add to both sides of the given equation 2011^2 and then factor it as

$$\begin{aligned}
xy - 2011x - 2011y + 2011^2 &= 2011^2 \\
(x - 2011)(y - 2011) &= 2011^2
\end{aligned}$$

Because 2011 is prime, then solutions to this equation will be obtained by solving the systems of equations:

1.

$$\begin{cases} x - 2011 = 1 \\ y - 2011 = 2011^2 \end{cases}$$

We obtain $x = 2012$ and $y = 4046132$

2.

$$\begin{cases} y - 2011 = 1 \\ x - 2011 = 2011^2 \end{cases}$$

We obtain $y = 2012$ and $x = 4046132$

3.

$$\begin{cases} x - 2011 = 2011 \\ y - 2011 = 2011 \end{cases}$$

We have $x = y = 4022$

Since our original equation does not change if we exchange x and y variables, we could reduce the number of cases from three to two. Hence, if $x = a, y = b$ is a solution, then $x = b, y = a$, will be another solution to the given equation.

Answer. $(2012, 4046132), (4046132, 2012), (4022, 4022)$.

> **Problem 163**
>
> Find all integers n and m such that $nm \geq 9$ and satisfying the equation
>
> $$2mn + n = 14.$$

Solution. Factoring the given equation gives us $(2m + 1)n = 14$. Since $(2m + 1)$ is an odd number, n must be even. This can be written as

1.

$$\begin{cases} 2m + 1 = 1 \\ n = 14 \end{cases}$$

We obtain $m = 0$ and $n = 2$, then $nm = 0$, and it is not a solution.

2.

$$\begin{cases} 2m + 1 = 7 \\ n = 2 \end{cases}$$

We find $m = 3$ and $n = 2$, but $mn = 6 < 9$, and it is not a solution as well.

3.

$$\begin{cases} 2m + 1 = -7 \\ n = -2 \end{cases}$$

We have $m = -4$ and $n = -2$, but $nm = 8 < 9$, and again it is not a solution.

4.

$$\begin{cases} 2m + 1 = -1 \\ n = -14 \end{cases}$$

We obtain $m = -1$ and $n = -14$, where $nm = 14 > 9$ true.

Answer. $m = -1$ and $n = -14$.

Problem 164

Find any prime number k greater than 2, such that $(17k+1)$ is a perfect square.

Solution. Let us translate the problem into a mathematical language. Since any perfect square can be written in the form n^2 we are looking for solutions of the equation:

$$17k+1 = n^2 \qquad (3.56)$$

such that k is a prime integer and n is an integer. Now we rewrite (3.56) in the form,

$$17k = n^2 - 1. \qquad (3.57)$$

Applying the difference of squares formula to the right side of equation (3.57) we obtain,

$$17k = (n-1)(n+1). \qquad (3.58)$$

This gives us an interesting situation when the left-hand side of (3.58) is divisible by 17 and the right-hand side is represented as a product of two integers that differ by 2; these two numbers can be either both even or both odd. However, since k is prime greater than 2, then numbers $(n-1)$ and $(n+1)$ must be both odd, such as 5 and 7.

In order for equation (3.58) to have any solution, the right side of it, $(n-1)(n+1)$, must be divisible by 17 as well. Hence either $(n-1) = 17$, then $(n+1) = k = 19$, or $(n+1) = 17$, then $(n-1) = k = 15$ but this k is not prime.

Now, please, look at equation (3.58) again. Given the condition that k is a prime, we have only one unique solution, $n = 18$ and $k = 19$.

Check: $17 \cdot 19 + 1 = 18^2$ true.

Answer. $k = 19$.

Let us see how factoring will help us to solve the following problem.

Problem 165

Find natural solutions of the equation $x^2 - xy = 2x - 3y + 11$.

Solution. We want to factor the given equation. Let us go through the steps:

$$x^2 - xy - 2x + 3y = 11$$
$$x^2 - 3x + 3y - xy + x - 3 = 11 - 3$$
$$x(x-3) - y(x-3) + (x-3) = 8.$$

From which we obtain a beautiful equation in the form,

$$(x-3)(x-y+1) = 8 = 1 \cdot 8 = (\pm 1) \cdot (\mp 8) = (\pm 8) \cdot (\mp 1) = (\pm 2) \cdot (\mp 4) = (\pm 4) \cdot (\mp 2).$$

This equation can be solved by considering ten systems of equations (for all factors of 8. Only five systems will give us natural solutions:

$$\begin{cases} x-3 = 8, \ x = 11 \\ x-y+1 = 1, \ y = 11. \end{cases}$$

or

$$\begin{cases} x-3 = 2, \ x = 5 \\ x-y+1 = 4, \ y = 2. \end{cases}$$

or

$$\begin{cases} x-3 = -1, \ x = 2 \\ x-y+1 = -8, \ y = 11. \end{cases}$$

or

$$\begin{cases} x-3 = 4, \ x = 7 \\ x-y+1 = 2, \ y = 6. \end{cases}$$

or

$$\begin{cases} x-3 = -2, \ x = 1 \\ x-y+1 = -4, \ y = 6. \end{cases}$$

Answer. $(x,y) = (1,6),(2,11),(5,2),(7,6),(11,11).$

Problem 166 (Chirsky)

Solve the equation $3^{2x} - 2^y = 1$ over the set of natural numbers. (Find all positive integers x and y satisfying the equation.)

Solution. Let us rewrite the equation in a different form and apply the property of exponents to 3^{2x}:

$$3^{2x} - 1 = 2^y$$
$$(3^x)^2 - 1^2 = 2^y. \tag{3.59}$$

The left-hand side of (3.59) can be factored as a difference of squares:

$$(3^x - 1)(3^x + 1) = 2^y. \tag{3.60}$$

Using another property of exponents we can rewrite

$$2^y = 2^{y_1} \cdot 2^{y_2}, \tag{3.61}$$

where $y_1 + y_2 = y$. Assuming that $y_1 < y_2$, then $2^{y_1} < 2^{y_2}$.

If equation (3.59) has integer solutions, then, by (3.60) and (3.61), they must satisfy the following system:

$$\begin{cases} 3^x - 1 = 2^{y_1} \ (a) \\ 3^x + 1 = 2^{y_2} \ (b) \end{cases} \tag{3.62}$$

Subtracting equation (a) of system (3.62) from (b) yields

$$2^{y_2} - 2^{y_1} = 2$$
$$2^{y_1}(2^{y_2-y_1} - 1) = 2 \cdot 1. \tag{3.63}$$

Since 2 is a prime, then the relation (3.63) gives us

$$2^{y_1} = 2 \Leftrightarrow y_1 = 1$$

$$2^{y_2-y_1} - 1 = 1 \Leftrightarrow 2^{y_2-y_1} = 2 \Rightarrow y_2 - y_1 = 1 \Rightarrow y_2 = 2. \tag{3.64}$$

On the other hand, $y = y_1 + y_2$. Combining this with (3.64) we obtain that

$$y = 1 + 2 = 3.$$

The value of x can be obtained by adding equations (a) and (b) of system (3.62):

$$3^x = \frac{2^{y_1} + 2^{y_2}}{2} = \frac{2^1 + 2^2}{2} = 3 \Rightarrow x = 1.$$

Check: $3^{2\cdot1} - 2^3 = 9 - 8 = 1$ true.
Answer. $x = 1$ and $y = 3$.

Remark. In the problem above, we used unusual way of factoring:

$$2^a - 2^b = 2^{b+a-b} - 2^b = 2^b(2^{a-b} - 1).$$

Let us try to apply the same technique to the more complicated problem in integers.

Problem 167

Solve the equation $n^x + n^y = n^z$ for all natural numbers n, x, y, and z.

Solution. Looking at the given equation

$$n^x + n^y = n^z \tag{3.65}$$

we can notice that it does not have a solution if $n = 1$ because

$$1^x + 1^y = 2 \neq 1^z.$$

However, we may be able to find some solutions for $n > 1$. If $n = 2$ we have a new equation,

$$2^x + 2^y = 2^z.$$

This equation has an obvious solution, a triple $(t,t,t+1)$ or $x = t$, $y = t$, and $z = t+1$ because $2^x + 2^x = 2 \cdot 2^x = 2^{x+1}$ is true.

Let us prove that equation (3.65) does not have any other solutions. Using proof by contradiction we will consider the following cases:

- **Case 1.** If $x \leq y$, then (3.65) yields $x \leq y < z$. Isolating n^x on the left-hand side of (3.65) we obtain

$$n^x = n^z - n^y. \tag{3.66}$$

Using the factoring technique mentioned above we will get

$$n^x = n^{x+z-x} - n^{y+x-x}$$
$$n^x = n^x(n^{z-x} - n^{y-x}).$$

From the last equation we can conclude that

$$n^{z-x} - n^{y-x} = 1. \tag{3.67}$$

If $y > x$ and $z > x$, introducing $y - x = k$ and $z - x = m$, we have

$$n^m - n^k = 1,$$

then $n^k(n^{m-k} - 1) = 1$ implies that $n = 1$. However, above we assumed that $n > 1$. We obtained a contradiction.

Therefore, if $x \leq y$, equation (3.65) does not have other solutions than $n = 2$, $x = t$, $y = t$, and $z = t+1$ for any natural t.

- **Case 2.** If $x = y$, then from (3.67) we obtain

$$n^{z-x} - 1 = 1$$
$$n^{z-x} = 2. \tag{3.68}$$

The last equation in (3.68) has solutions only for $n = 2$, then $z - x = 1$ and $z = x + 1$, $y = x$. However, this is the same answer as it was obtained earlier.

Hence, the given equation has solutions only for $n = 2$. Solutions can be written in the form $(2,t,t,t+1)$ or $n = 2$, $x = t$, $y = t$, and $z = t+1$, $t \in \mathbb{N}$. For example, $(2,1,1,2)$, $(2,4,4,5)$, and $(2,71,71,72)$ are particular solutions of the given equation.

Check: $n^x + n^y = n^z$, for $n = 2$, $x = 4$, $y = 4$, $z = 5$. We have

$$2^4 + 2^4 = 2^5$$
$$16 + 16 = 32$$

is true. Now you can check other cases by yourself!

Answer. $(x,y,z) = (t,t,t+1)$, $n = 2$, $t \in \mathbb{N}$.

Problem 168 (Hungarian Math Olympiad 1947)

Prove that if n is odd, then $46^n + 296 \cdot 13^n$ is divisible by 1947.

Proof. We will prove this statement directly by factoring the given expression:

$$p = 46^n + 296 \cdot 13^n = 46 \cdot 46^{n-1} - 46 \cdot 13^{n-1} + 46 \cdot 13^{n-1} + 296 \cdot 13^n$$
$$= 46(46^{n-1} - 13^{n-1}) + (46 + 296 \cdot 13)13^{n-1}$$
$$= 46(46^{n-1} - 13^{n-1}) + 2 \cdot 1947 \cdot 13^{n-1}.$$

Let us consider the first term of the expression (the second is obviously multiple of 1947). Since n is odd, then let $n = 2k + 1$, then $n - 1 = 2k$ (even power). As we know the difference of even powers of u and v always can be factored with one of the factors as difference of squares $u^2 - v^2 = (u - v)(u + v)$. Therefore,

$$p = 46(46^{n-1} - 13^{n-1}) + 2 \cdot 1947 \cdot 13^{n-1} = 46(46 - 13)(46 + 13)m + 2 \cdot 1947 \cdot 13^{n-1},$$

and $p = 1947m$.

Let us see how Newton's Binomial Theorem can help us to solve some problems involving power of a number or the last digits of such a number.

Problem 169

Find all remainders that an integer raised to the 200th power can leave when it is divided by 125.

Solution. If an integer is divisible by 5 (5, 10, 15, 20, ..., 5465, etc.), then obviously, its 200th power will be divisible by 125 and the remainder will be zero. If an integer is not divisible by 5, then it can be represented as $(5k + 1)$, $(5k + 2)$, $(5k + 3)$, $(5k + 4)$. We can also rewrite the cases in the following form: $5k \pm 1$ or $5k \pm 2$. You can check that $(5k - 1)$ will represent all numbers that leave the remainder 4 when divided by 5 and $(5k - 2)$ will represent all numbers that would leave the remainder 3 when divided by 5.

According to Theorem 21,

$$(5k \pm 1)^{200}$$
$$= (5k)^{200} \pm \tfrac{200 \cdot 199}{2}(5k)^{199} \pm \cdots + \tfrac{200 \cdot 199}{2} \cdot (5k)^2 \pm 200(5k) + 1.$$

We can see that every term except the last one is a multiple of 125, then such numbers will always leave the remainder 1 when they are divided by 125. Next, for all numbers if the form $(5k + 2)$ or $(5k - 2)$ we have

$$(5k \pm 2)^{200} = (5k)^{200} \pm \tfrac{200 \cdot 199}{2} \cdot (5k)^{199} \cdot 2 \pm \ldots$$
$$+ \tfrac{200 \cdot 199}{2} \cdot (5k)^2 \cdot (2)^{198} \pm 200 \cdot (5k) \cdot 2^{199} + 2^{200}.$$

Again, these numbers will leave the same remainder when divided by 125 as the last term. How can we find it? We obviously have

$$2^{200} = (2^2)^{100} = (5-1)^{100} = 125n + 1.$$

Therefore, the 200th power of any number divided by 125 leaves a remainder 0 or 1.
Answer. 0 or 1.

Remark. Solving the problem above we additionally found that $2^{200} = 125n + 1$. This means that the 200th power of 2 leaves the remainder 1 when it is divided by $5, 25$, and 125. This is very interesting result because now we can find the last three digits of 2^{200}.

Let us solve the following problem.

Problem 170

Find the last three digits of 2^{200}.

Solution. We have

$$2^{200} = 125n + 1 = 25m + 1 = 5l + 1.$$

Starting from the end of the formula above we find that the last digit can be only 6 (because 2^{200} is an even number).

Next, we can list the candidates for the last two digits of 2^{200}: 26, 76. Out of two possibilities we will keep only 76 since 2^{200} is a multiple of 4, so, as we proved earlier in this book, the number written by its last two digits. In fact, $76 = 19 \cdot 4$. Finally, the candidate for the last three digits of 2^{200} is only $376 = 125 \cdot 3 + 1$ because only 376 is a multiple of 8 so as 2^{200}.
Answer. The last three digits of 2^{200} are 376.

Next, I want to give you the following problem. It can be solved using a different method, for example, Indices, but I believe that all other ways are good if they lead us to correct solution and the answer.

Problem 171

It is necessary to solve $3^{4x} \equiv 11^{x+3} \pmod{13}$.

Solution. Let us rewrite the given equation in a different form,

$$\left(3^4\right)^x \equiv \left(11^3 \cdot 11^x\right) \pmod{13}$$
$$81^x \equiv \left(11^3 \cdot 11^x\right) \pmod{13}$$
$$81^x \equiv \left(1331 \cdot 11^x\right) \pmod{13}.$$

Extracting a multiple of 13 from 81 and 1331 we obtain

$$(78+3)^x \equiv (1326+5) \cdot 11^x \pmod{13}.$$

Using Theorem 21 we can rewrite it as

$$3^x \equiv \left(5 \cdot 11^x\right) \pmod{13}.$$

Multiplying both sides by 6^x we get

$$\left(6^x \cdot 3^x\right) \equiv \left(6^x \cdot 5 \cdot 11^x\right) \pmod{13}$$
$$18^x \equiv \left(5 \cdot (66)^x\right) \pmod{13}.$$

Extracting a multiple of 13 again we obtain

$$(13+5)^x \equiv \left(5 \cdot (65+1)^x\right) \pmod{13}$$
$$5^x \equiv 5 \pmod{13}$$
$$5^{x-1} \equiv 1 \pmod{13}.$$

In order to find x, let us consider some powers of 5 and use Theorem 13,

$$625 = 5^4 \equiv 1 \pmod{13}$$
$$5^{12} \equiv 1 \pmod{13}$$
$$5^{4k} \equiv 1 \pmod{13}.$$

Therefore, $x-1 = 4k$, or $x = 4k+1$.
Answer. $x = 4k+1$, $k \in \mathbb{N}$, i.e., $x = 1, 5, 9, 13, \ldots$.

Higher Order Polynomial Equations in Integers.
In the following problem, we can use completing a square method.

Problem 172

Solve the equation in integers $5x^4 - 40x^2 + 2y^6 - 32y^3 = -208$.

Solution. Because $40 = 2 \cdot 5 \cdot 4$, $16 = 2 \cdot 1 \cdot 8$, we can try to complete a square

$$5x^4 - 40x^2 + 80 + 2y^6 - 32y^3 + 128 = -208 + 80 + 128$$
$$5\left(x^4 - 8x^2 + 16\right) + 2\left(y^6 - 16y^3 + 64\right) = 0$$
$$5\left(x^2 - 4\right)^2 + 2\left(y^3 - 8\right)^2 = 0$$

The sum of two squares equals zero if and only if each quantity on the left is zero. We obtain a system to solve

$$\begin{cases} (x-2)(x+2) = 0 \\ y^3 - 8 = 0 \end{cases} \Leftrightarrow x = \pm 2, \ y = 2.$$

Answer. $(x,y) = \{(2,2), \ (-2,2)\}$

Problem 173

Find all natural solutions to the equation $x^4 + x^3 + x^2 + x + 1 = y^2$.

Solution. Let us move one to the right-hand side and factor by grouping terms on the left side,

$$(x^4 + x^2) + (x^3 + x) = y^2 - 1$$
$$x^2(x^2 + 1) + x(x^2 + 1) = (y-1)(y+1)$$

Considering now this equation,

$$(x^2 + 1) \cdot x \cdot (x+1) = (y-1)(y+1), \tag{3.69}$$

we can see that since the left side of it contains a product of two consecutive integers, then it is divisible by 2. On the right-hand side we have a product of two factors differed by 2, which means that they are either both odd (when y is even) or both even (when y is odd). When both factors on the right are odd, our equation would not have any solution.

Assume that both factors on the right-hand side are even, then the following is true:

$$y = 2k+1, \ y-1 = 2k, \ y+1 = 2k+2$$
$$(y-1)(y+1) = 2k \cdot (2k+2) = 4k(k+1) = 8n$$

In this case, the right side of this equation is always divisible by 8. Can the left side of (3.69) be divisible by 8?

All possible values of x when divided by 8 are: $x = 8m$, $x = 8m+1$, $x = 8m+2$, $x = 8m+3$, $x = 8m+4$, $x = 8m+5$, $x = 8m+6$, $x = 8m+7$. But the left side of (3.69) is divisible by 8 only if

$$x = 8m, \ x = 8m+3, \ x = 8m+7,$$

and then only these three cases will be investigated.

First, we separately consider each case for $m = 0$:

1. $x = 8m+3$. If $m = 0$, then $x = 3$. Then the equation (3.69) will be rewritten as:

$$3 \cdot 4 \cdot 10 = y^2 - 1,$$

  or $y^2 = 121$. From here we find $y = 11$. We have an ordered pair $x = 3, y = 11$.
2. $x = 8m + 7$. If $m = 0$, then $x = 7$. The (3.69) will be rewritten as:

$$7 \cdot 8 \cdot 50 = y^2 - 1,$$

or $y^2 = 2801$. It has no natural solutions for y.

Next, by considering three cases for $m \geq 1$, it can be shown that other solutions do not exist. Here we will demonstrate our proof for $x = 8m + 3$. The proofs for other cases are similar. Let us rewrite the equation (3.69) for $x = 8m + 3$, $m \geq 1$ and $y = 2k + 1$, $k \geq 1$ as

$$(8m + 3)(2m + 1)\left(64m^2 + 48m + 10\right) = k(k + 1),$$

or after multiplying the first and the third factors it can be represented as,

$$(2m + 1)\left(2(256m^3 + 288m^2 + 112m + 14) + 2\right) = k(k + 1).$$

The right-hand side of it is the product of two consecutive natural numbers and the left side is the product of an odd number $2m + 1$ and an even number $2(256m^3 + 288m^2 + 112m + 14) + 2$, while the second factor is greater than the first. Due to the right-hand side, the equality is possible if the following is true

$$2m = 2(256m^3 + 288m^2 + 112m + 14),$$

or

$$256m^3 + 288m^2 + 111m + 14 = 0.$$

This equation does not have natural solutions. Hence, the equation (3.69) does not have a solution at $x = 8m + 3$, $m \geq 1$.

Therefore, the given equation has only one solution: $x = 3, y = 11$.

Answer. $x = 3, y = 11$.

Finally, here is another problem on proof.

Problem 174 (MGU, oral exam, 2007)

Prove that the sum of three consecutive cubes is always divisible by 9.

Proof. We will prove this directly

$$\begin{aligned}
(n - 1)^3 + n^3 + (n + 1)^3 &= n^3 - 3n^2 + 3n - 1 + n^3 + n^3 + 3n^2 + 3n + 1 \\
&= 3n^3 + 6n = 3n(n^2 + 2) \\
&= 3n(n^2 - 1 + 3) \\
&= 9n + 3(n - 1)n(n + 1) = 9k
\end{aligned}$$

By extracting $(n^2 - 1)$ inside parentheses, after applying distributive law, we obtained a multiple of 9 and a number 3 multiplied by the product of three consecutive integers. Therefore the entire right-hand side is always divisible by 9.

3.5 Second-Order Diophantine Equations in Two or Three Variables

3.5.1 Methods and Equations

For the following problems, you will have to deal with quadratic expressions in two or three variables. Solving such problems in integers usually can be done in two different ways:

1. Consider the expression as a quadratic with respect to one variable, and solve it as a quadratic, using the fact that its discriminant must be a perfect square. You might use divisibility properties as well.
2. Try to factor the given quadratic expression, and then use divisibility.

In order to demonstrate the first way I want to offer you the following problems.

Problem 175 (USAMO 2015)

Given $x^2 + xy + y^2 = (\frac{x+y}{3} + 1)^3$. Solve it in integers.

Solution. From the condition of the problem we can see that the numerator of the fraction on the right-hand side must be divisible by 3, $x + y = 3n$, then

$$\frac{x+y}{3} + 1 = n + 1,$$
$$y = 3n - x.$$
$$(3.70)$$

Thus the original equation becomes

$$(3n)^2 - x(3n - x) = (n + 1)^3$$

which can be simplified to a quadratic equation in variable x:

$$x^2 - 3n \cdot x - (n^3 - 6n^2 + 3n + 1) = 0$$

Using quadratic formula, we obtain formula for x,

$$x_{1,2} = \frac{3n \pm \sqrt{(3n)^2 + 4(n^3 - 6n^2 + 3n + 1)}}{2}$$
$$(3.71)$$

Consider the expression under the square root, the discriminant of the quadratic equation is:

$$D = 4n^3 - 15n^2 + 12n + 4$$

We can see that $n = 2$ is zero of the discriminant, then $(n - 2)$ is a factor. Using Horner synthetic division (Grigorieva, *Methods of Solving Nonstandard Problems*), the discriminant can be written in the factorized form as

$$D = (n - 2)^2 \cdot (4n + 1)$$

In order for the discriminant to have integer solutions and because $(4n + 1)$ is an odd, then it must be a perfect square of an odd number. This can be written as

$$4n + 1 = (2m + 1)^2$$
$$4n + 1 = 4m^2 + 4m + 1$$
$$4n = 4m^2 + 4m$$
$$n = m^2 + m$$

which also can be written as

$$n = m \cdot (m + 1) \tag{3.72}$$

Substituting this n into the quadratic formula, we obtain,

$$x_{1,2} = \frac{3n \pm \sqrt{(n-2)^2 \cdot (2m+1)^2}}{2}$$
$$x_{1,2} = \frac{3(m^2+m) \pm (m^2+m-2) \cdot (2m+1)}{2}$$
$$x_1 = 3m^2 + m^3 - 1$$
$$x_2 = 3m - m^3 + 1.$$

Substituting each x into formula for y, we have

$$y_1 = 3m - m^3 + 1$$
$$y_2 = 3m^2 + m^3 - 1.$$

We can see that $x_1 = y_2$ and that $x_2 = y_1$. This is not strange, it follows from the symmetry of the original equation, so variables x and y are interchangeable. The answer can be written as an ordered pair:

$$(x, y) = (3m^2 + m^3 - 1, 3m - m^3 + 1), \quad m \in Z.$$

We can check our answer by substituting this pair into the original equation and confirm that the left side equals the right side.

$$(3m^2 + m^3 - 1)^2 + (3m^2 + m^3 - 1)(3m - m^3 + 1) + (3m - m^3 + 1)^2$$
$$= m^6 + 3m^5 + 6m^4 + 7m^3 + 6m^2 + 3m + 1 = ((m^2 + m) + 1)^3 = (n + 1)^3.$$

Answer. $(x, y) = (3m^2 + m^3 - 1, 3m - m^3 + 1), \quad m \in Z.$

Problem 176 (AIME 2000)

Find the sum of all positive integers, n, for which $n^2 - 19n + 99$ is a perfect square.

Solution. If the perfect square is represented by x^2, then the equation is $n^2 - 19n + 99 - x^2 = 0$. The quadratic formula yields

$$n = \frac{19 \pm \sqrt{361 - 4(99 - x^2)}}{2} = \frac{19 \pm \sqrt{4x^2 - 35}}{2}.$$

In order for this to be an integer, the discriminant must also be a perfect square, so $4x^2 - 35 = y^2$ for some nonnegative integer y.

This factors to $(2x + y)(2x - y) = 35$. Number 35 has two pairs of positive factors: $(1, 35)$ and $(5, 7)$. Respectively, these yield 9 and 3 for x, which results in $n = 1, 9, 10$, and 18. The sum is therefore 38.

Answer. 38.

Problem 177

Solve the equation $x^2 - 3xy + 2y^2 = 3$ for integers, x and y.

Solution. We suggest two methods to solve this problem.

Method 1. Considering the given equation as a quadratic with respect to x we obtain:

$$x^2 - 3xy + 2y^2 - 3 = 0 \Rightarrow x_{1,2} = \frac{3y \pm \sqrt{y^2 + 12}}{2}. \tag{3.73}$$

Because $x \in \mathbb{Z}$ (a whole number), we understand from (3.73) the discriminant must be a perfect square. This yields

$$y^2 + 12 = k^2, \tag{3.74}$$

where $k \in \mathbb{Z}$.

Since (3.74) is symmetric then if a pair (y, k) is a solution then $(-y, -k)$ is also solution. Moreover, equation (3.74) can be written differently and factored as

$$12 = k^2 - y^2 \Leftrightarrow 12 = (k - y)(k + y). \tag{3.75}$$

From (3.75) we see that since $(k - y)$ and $(k + y)$ must divide 12 then they both are even. By checking we can find pairs (x, y) satisfying (3.73) and (3.75) simultaneously, obtaining

$$(x, y) : (5, 2), (1, 2), (-1, -2), (-5, -2)$$

This problem can be solved differently using factoring of homogeneous polynomials.

Method 2. The left side of the equation $x^2 - 3xy + 2y^2 = 3$ above is a homogeneous polynomial. Any polynomial $ax^2 + bxy + cy^2$, $a \neq 0$ in two variables x and y we can consider as a quadratic trinomial in one variable $u = \frac{x}{y}$. This can be factored as

$$ax^2 + bxy + cy^2 = y^2(au^2 + bu + c) = ay^2(u - u_1)(u - u_2) = a(x - u_1 y)(x - u_2 y).$$

Let us apply this to the equation $x^2 - 3xy + 2y^2 = 3$. Factoring the left side of the equation we can rewrite it as

$$(x - y)(x - 2y) = 3. \tag{3.76}$$

Because the right side of (3.76) is a prime number 3, then if the equation has any integer solutions (x, y), they must satisfy the four possible systems:

1.

$$\begin{cases} x - y = 3 \\ x - 2y = 1 \end{cases} \Leftrightarrow y = 2, \, x = 5.$$

2.

$$\begin{cases} x - y = 1 \\ x - 2y = 3 \end{cases} \Leftrightarrow y = -2, \, x = -1.$$

3.

$$\begin{cases} x - y = -3 \\ x - 2y = -1 \end{cases} \Leftrightarrow y = -2, \, x = -5.$$

4.

$$\begin{cases} x - y = -1 \\ x - 2y = -3 \end{cases} \Leftrightarrow y = 2, \, x = 1.$$

We have the same answer.

Answer. $\{(x, y) : (5, 2), (1, 2), (-1, -2), (-5, -2)\}$.

Problem 178

Prove that the equation $x^2 - 2y^2 + 8z = 3$ does not have solutions over the set of integers.

Proof. We will prove it by contradiction. Assuming that there exist some integers x, y, and z that are solutions of the given equation, we rewrite it in a different form as

$$x^2 = 2y^2 - 8z + 3. \tag{3.77}$$

Then, we will consider two possible cases:

1. Let y be an even number or $y = 2k$, then from (3.77) we find,

$$x^2 = 2 \cdot 4k^2 - 8z + 3 = 8k^2 - 8z + 3.$$

This means that x^2 divided by 8 must leave a remainder of 3. Nevertheless, by the property proven above, it is impossible for an even number, x, or an odd. (Squares of all even numbers are either multiples of 8 or leave a remainder of 4. Squares of all odd numbers leave a remainder of 1. See Problem 8.)

2. Let y be an odd number, then $y = 2k + 1$, then

$$x^2 = 2(2k+1)^2 - 8z + 3$$
$$= 2(4k^2 + 4k + 1) - 8z + 3 = 8(k^2 + k - z) + 5.$$

A square of no integers x (an even or an odd) divided by 8 leaves a remainder of 5.

Therefore, we have proven that the given equation has no solutions over the set of integers.

Problem 179

Solve in integers $2x^2 - 2xy + 9x + y = 2$.

Solution. Rewrite it as

$$2x^2 - 2xy + 10x - x + y = 2$$
$$2x(x - y + 5) - x + y - 5 = 2 - 5$$
$$(x - y + 5)(2x - 1) = -3$$
$$(1 - 2x)(x - y + 5) = 3$$

Because 3 is prime, then we have to solve four systems:

$$\begin{cases} 1 - 2x = 1 \\ x - y + 5 = 3 \end{cases} \Rightarrow x = 0, \ y = 2$$

$$\begin{cases} 1 - 2x = 3 \\ x - y + 5 = 1 \end{cases} \Rightarrow x = -1, \ y = 3$$

$$\begin{cases} 1 - 2x = -1 \\ x - y + 5 = -3 \end{cases} \Rightarrow x = 1, \ y = 9$$

$$\begin{cases} 1 - 2x = -3 \\ x - y + 5 = -1 \end{cases} \Rightarrow x = 2, \ y = 8$$

Answer. $(x, y) = \{(0, 2), (-1, 3), (1, 9), (2, 8)\}$

Problem 180

Find integer solutions to $x^2 - 5y^2 = 3$.

Solution. Let us rewrite this equation as

$$x^2 = 5y^2 + 3$$

The right side of this equation consists of a multiple of 5 that can end in 0 or 5 and number 3, then for any integer y, $5y^2 + 3$ can end either in 3 or in 8. On the other hand, the left side is a square of an integer number and can end in $0, 1, 4, 5, 6, 9$. Therefore the last digits of left and right sides never match.

Answer. The given equation has no solutions.

3.5.2 Problems Leading to Pell's Equation

In Chapter 1 we showed that each square number can be written as a sum of two consecutive triangular numbers. For example, $4 = 1 + 3$, $9 = 3 + 6$, $16 = 6 + 10$, etc.. Ancient Greeks also geometrically proved that the sum of two consecutive triangular numbers is a square number, i.e., $T(n-1) + T(n) = n^2$.

If you list at least ten consecutive triangular numbers, such as

$$T(1) = 1, \; T(2) = 3, \; T(3) = 6, \; T(4) = 10,$$
$$T(5) = 15, \; T(6) = 21, \; T(7) = 28, \; T(8) = 36, ...$$

you can notice that some of the numbers (such as 1 and 36) are perfect squares. Can we find all such numbers? Let us try to find the general rule for this by solving the following problem.

Problem 181

What triangular numbers are also square numbers?

Solution. Hence we need to solve the equation,

$$T(n) = m^2$$
$$\frac{n \cdot (n+1)}{2} = m^2$$
$$n^2 + n = 2m^2$$

At this point the way to solve the last equation can vary. For example, we can keep it in the factorized form as $n(n+1) = 2m^2$ and to use the fact of the right-hand side be even and that the left side is the product of two consecutive natural numbers. On the other hand, we can work with the last equation of the set and multiply both sides of it by 4 in order to be able to complete the square on the left-hand side of it. We proceed as

$$4n^2 + 4n = 8m^2$$
$$4n^2 + 4n + 1 = 8m^2 + 1$$
$$(2n+1)^2 - 8m^2 = 1.$$

We can see that we have to solve the second-order Diophantine equation of the following type:

$$x^2 - 8y^2 = 1,$$
$$x = 2n+1, \ y = m. \tag{3.78}$$

In the formula above, n is associated with the nth triangular number, $T(n)$. We are going to use the known information in order to solve the equation. Thus, the following is true.

$$T(1) = 1 = 1^2, \ n = 1, \ m = 1$$
$$(2 \cdot 1 + 1)^2 - 8 \cdot 1^2 \tag{3.79}$$
$$x = 3, \ y = 1.$$

Now for $T(8) = 36$.

$$T(8) = 36 = 6^2, \ n = 8, \ m = 6$$
$$(2 \cdot 8 + 1)^2 - 8 \cdot 6^2$$
$$17^2 - 8 \cdot 6^2 = 289 - 288 = 1 \tag{3.80}$$
$$x = 17, \ y = 6.$$

Because (3.78) models the case of all triangular numbers that are also squares, it is clear that there are infinitely many of them. This equation is the so-called Pell's equation and in the following section we will explain some known methods of finding a general solution for them. For example, one can check that as soon we know at least one nontrivial solution, all the other solutions can be obtained as a result of **consecutive iterations**

$$(x, y) \rightarrow (3x + 8y, x + 3y) \tag{3.81}$$

Please check yourself that replacement of variable x by $(3x + 8y)$ and y by $(x + 3y)$ and their then substitution into (3.78) will keep it in the original form, since,

$$(3x + 8y)^2 - 8 \cdot (x + 3y)^2 = x^2 - 8y^2 = 1.$$

Therefore, the first several ordered pairs are

$$(1, 0) \rightarrow (3, 1) \rightarrow (17, 6) \rightarrow (99, 35) \rightarrow (577, 204) \rightarrow (3363, 1189)$$

The first coordinate is the value of x, the corresponding triangular number $T(n)$ and its position n can be easily found from solving

$$n = \frac{x-1}{2}$$

We can now list the first triangular numbers satisfying this rule,

$x = 3, n = 1, T(1) = 1, m = 1$
$x = 17, n = 8, T(8) = 36, m = 6$
$x = 99, n = 49, T(49) = \frac{49 \cdot 50}{2} = 1225 = 35^2, m = 35$
$x = 577, n = 288, T(288) = \frac{288 \cdot 289}{2} = 41616 = 204^2, m = 204$
$x = 3363, n = 1681, T(1681) = \frac{1681 \cdot 1682}{2} = 1413721 = 1189^2, m = 1189$
$\dots$

or as a quadruple,

$(x, n, T(n), m) : (3, 1, 1, 1), (17, 8, 36, 6), (99, 49, 1225, 35), (577, 288, 41616, 204)\dots$

Answer. $(1, 0) \rightarrow (3, 1) \rightarrow (17, 6) \rightarrow (99, 35) \rightarrow (577, 204) \rightarrow (3363, 1189), \dots$

Now let us find all triangular numbers similar to $T(4) = 10 = 3^2 + 1$ that are greater than a square by the unit. This can be formulated as the next problem.

> **Problem 182**
>
> Find all triangular numbers that are greater than a square plus 1.

Solution. We start as $T(n) = m^2 + 1$, which will lead us to the equation in two variables that can be simplified as follows:

$$\frac{n(n+1)}{2} = m^2 + 1$$
$$n(n+1) = 2m^2 + 2$$
$$4n(n+1) = 8m^2 + 8$$
$$4n^2 + 4n + 1 = 8m^2 + 9$$
$$(2n+1)^2 - 8m^2 = 9.$$

If we substitute into the last row $n = 4$ and $m = 3$, we obtain correct equality $(2 \cdot 4 + 1)^2 - 8 \cdot 3^2 = 9$. The equation in the last row can be also rewritten as

$$x^2 - 8y^2 = 9$$
$$x = 2n + 1, \; y = m. \tag{3.82}$$

How can we solve this equation? It looks very similar to (3.78) but it has 9 instead of 1 on the right-hand side, which is a perfect square of 3. Why do not we use this fact and divide both sides of (3.82) by 9? The following chain of equations is valid:

$$\left(\frac{x}{3}\right)^2 - 8 \cdot \left(\frac{y}{3}\right)^2 = 1. \tag{3.83}$$

Introducing new variables,

$$\tilde{x} = \frac{x}{3}, \ \tilde{y} = \frac{y}{3}$$

we can rewrite it as

$$(\tilde{x})^2 - 8(\tilde{y})^2 = 1 \tag{3.84}$$

Solution $(\tilde{x}, \tilde{y})$ is the same as one obtained for (3.78), and after returning to the original variables

$$x = 3 \cdot \tilde{x}, \ \ y = 3 \cdot \tilde{y},$$

we get

$$(x, y) = (3, 0) \rightarrow (9, 3) \rightarrow (51, 18) \rightarrow (297, 105) \leftrightarrow \cdots$$

Let us demonstrate that the third ordered pair is valid. If $x = 51$ and $y = m = 18$, then $n = 25$ then $T(25) = \frac{25 \cdot 26}{2} = 325 = 18^2 + 1$.

Answer. $T(1) = 1 = 0^2 + 1, T(4) = 10 = 3^2 + 1, T(25) = 325 = 18^2 + 1, T(148) = 11026 = 105^2 + 1, \ldots$

3.5.3 Fermat-Pell's Equation

Pell's equation is

$$x^2 - Ny^2 = 1 \tag{3.85}$$

It was named after mathematician, Pell, by Leonard Euler the Pell's Equation. Later it was established that Pell never saw the equation. However, Fermat did see the equation and was able to solve a portion of it. So, it would be valid to call this equation Fermat equation. However, historically it is known as Pell's so we will try keeping both names when we can in this text. The equation plays very important role among all second-order equations in two variables. Some information about Pell's equation was found in the work of Ancient Greeks and Indians. For example, in 1773, Lessing published the "cattle problem" formulated by Archimedes. Lessing believed that Archimedes proposed this problem to challenge Eratosthene. The problem had 8 unknowns (animals of different kinds) satisfying seven linear equations and the condition by which some of the variables are perfect squares. The problem could be reduced to

$$x^2 - 4729494y^2 = 1.$$

Its minimal nontrivial solution was found by Amthor in 1880 and had 41 digits.

Of course, ancient mathematicians could not solve such equations but they probably could solve some easier version of such equations! Fermat, also in his time did not believe that Ancient people could solve the equation because he did not see any geometric solution. He challenged mathematical world by presenting this equation. In the middle of 17th century Fermat formulated it as follows.

Theorem 24 (Pell-Fermat Equation)

Let N be a natural number, not a perfect square, then there are infinitely many such squares that this number times that square plus 1 is another perfect square,

$$N \cdot y^2 + 1 = x^2.$$

Irish mathematician William Brouncker (1620–1684) was the first who attempted to solve the equation. He actually attempted to decompose $\sqrt{N}$ as infinite fraction. Frenicle de Bessy created the tables of solutions of Pell's equations up to $N = 150$ and asked Brouncker to solve

$$x^2 - 313y^2 = 1.$$

There is an anecdotal story that Brouncker solved it in two hours. Fermat was the first who stated that (3.85) has infinitely many solutions. The first rigorous proof of this fact is in the work of Lagrange (1766). Lagrange's method will be discussed later. But, first, let us find some interesting properties of the equation.

Obviously, Pell's equation (3.85) has at least two solutions: $x = 1, y = 0$ and $x = -1, y = 0$. We can call those a trivial solution. If $x = a \in N$, then $x = -a$ is also a solution. Moreover, N cannot be a perfect square.

Problem 183

Prove that Fermat equation $x^2 - Ny^2 = 1$ does not have solutions for natural N that is a perfect square.

Proof. We will prove this statement by contradiction. Assume that $N = n^2$. Then Fermat equation can be written as

$$x^2 - n^2 y^2 = 1.$$

Factoring difference of squares, we obtain

$$(x - ny)(x + ny) = 1.$$

However, it is obvious that the equation above does not have a solution in integers. Moreover, a difference of two integer squares cannot equal one.

The following lemma is valid.

Lemma 13

If $N = n^2 + 1$, then a nontrivial solution to Fermat-Pell's equation always exists and is given by
$$x = 2n^2 + 1, \ y = 2n.$$

Proof. We will prove this directly: $(2n^2 + 1)^2 - (n^2 + 1)(2n)^2 = 1$. We have that
$$4n^4 + 4n^2 + 1 - 4n^2(n^2 + 1) = 1$$
true.

Let us solve the following problem.

Problem 184

Solve in integers, $x^2 - 10 \cdot y^2 = 1$.

Solution. Since $10 = 3^2 + 1$, then by Lemma 13, we obtain
$$x = 2 \cdot 9 + 1 = 19, \ y = 2 \cdot 3 = 6.$$

Answer. $x = 19, \ y = 6$

Question. We found one nontrivial solution
$$19^2 - 10 \cdot 6^2 = 1$$

How can we find all other solutions?

Problem 185

Find all integer solutions to the equation $x^2 - 2y^2 = 1$.

Solution. Assume that we found one solution to the given equation, (x, y). It can be shown that then
$$(3x + 4y, 2x + 3y)$$
is also a solution.

In order to prove this, we will substitute it in the Pell's equation,

$$(3x+4y)^2 - 2(2x+3y)^2 = 9x^2 + 24xy + 16y^2 - 2 \cdot (4x^2 + 12xy + 9y^2) = x^2 - 2y^2 = 1.$$

Hence, $(3x+4y, 2x+3y)$ is a solution.

Starting from a trivial solution $x_0 = 1, y_0 = 0$, then using the generating formula, we can generate as many solutions as we want,

$$(1,0) \rightarrow (3,2) \rightarrow (17,2) \rightarrow (99,70) \rightarrow (577,408) \rightarrow (3363,2378) \rightarrow \cdots$$

Are there general methods of solving Pell's equations?

When I had a similar problem at the district math Olympiad (problem on finding square triangular numbers), I noticed that both, $x = 1$, $y = 0$ and $x = 3$, $y = 1$, are solutions to the given equation. I found that if x is replaced by $(3x+8y)$ and y by $(x+3y)$, then we will have the same equation,

$$\begin{aligned} (3x+8y)^2 &- 8 \cdot (x+3y)^2 \\ &= 9x^2 + 48xy + 64y^2 - 8(x^2 + 6xy + 9y^2) \\ &= x^2 - 8y^2 = 1 \end{aligned}$$

Since $(1,0)$ is the first trivial solution, using this substitution we obtain the chain of other valid solutions,

$$(x,y) \rightarrow (3x+8y, x+3y): \ (1,0) \rightarrow (3,1) \rightarrow (17,6) \rightarrow (99,35) \rightarrow (577,122) \rightarrow \ldots$$

All of these pairs are consecutive solutions to the given equation. This equation is called Pell's equation and there are methods of solving such equations that have been developed about which I had no idea in age 17. Some of the methods are briefly explained here.

Matrix Iterative Approach. Because $x^2 - N \cdot y^2$ is a quadratic form, I usually make my students look for a matrix of linear transformation, B, that is responsible for linear changes of the variables that preserve a quadratic form, characterized by matrix A so that $B^T A B = A$. We will attempt to use the matrix approach to develop methods of finding all possible solutions. Let us rewrite the given equation in a matrix form by recognizing quadratic form in it.

$$x^2 - N \cdot y^2 = 1$$

$$Z^T A Z = [x \ y] \cdot \begin{pmatrix} 1 & 0 \\ 0 & -N \end{pmatrix} \cdot \begin{bmatrix} x \\ y \end{bmatrix} = 1, \quad Z^T = [x \ y] \qquad (3.86)$$

Assume that z_k is a solution to (3.86) then

$$z_k^T A z_k = 1$$

and assume that there exists a matrix of linear transformation $B = \begin{pmatrix} a & b \\ c & d \end{pmatrix}$, with integer entries such that produces a new solution, next to it,

$$Bz_k = z_{k+1}$$

Substituting it into the given equation of a quadratic form, we will obtain

$$(Bz_k, ABz_k) = (z_k, B^T ABz_k) = (z_k, Az_k)$$

From this, we obtain the following matrix equation:

$$(z_k, (B^T AB - A) z_k) = 0$$

There are now two possible cases:

Case 1.

$$\mathbf{B}^T \mathbf{AB} - \mathbf{A} = 0 \qquad (3.87)$$

Case 2. Vectors z_k, $(\mathbf{B}^T \mathbf{AB} - A) z_k$ are orthogonal, i.e., if $z_k = \begin{pmatrix} x_k \\ y_k \end{pmatrix}$, then the vector orthogonal to it would have the type,

$$z_k^{\perp} = \alpha \cdot \begin{pmatrix} 0 & 1 \\ -1 & 0 \end{pmatrix} \begin{pmatrix} x_k \\ y_k \end{pmatrix}$$

and the following matrix equation would be true:

$$(\mathbf{B}^T \mathbf{AB} - A) \cdot z_k = \alpha \begin{pmatrix} 0 & 1 \\ -1 & 0 \end{pmatrix} \cdot z_k, \ \alpha \neq 0, \ \alpha \in \mathbb{R}. \qquad (3.88)$$

Considering Case 1, let us find matrix **B**, by first assuming that

$$\mathbf{B} = \begin{pmatrix} a & b \\ c & d \end{pmatrix}$$

After substituting this matrix into (3.87), we obtain

$$\begin{pmatrix} a & c \\ b & d \end{pmatrix} \cdot \begin{pmatrix} 1 & 0 \\ 0 & -N \end{pmatrix} \cdot \begin{pmatrix} a & b \\ c & d \end{pmatrix} = \begin{pmatrix} 1 & 0 \\ 0 & -N \end{pmatrix}$$
$$\begin{pmatrix} a^2 - Nc^2 & ab - Ncd \\ ab - Ncd & b^2 - Nd^2 \end{pmatrix} = \begin{pmatrix} 1 & 0 \\ 0 & -N \end{pmatrix}$$

This will lead to the following system of nonlinear equations with integer coefficients:

$$\begin{cases} ab - Ncd = 0 \\ a^2 - Nc^2 = 1 \\ b^2 - Nd^2 = -N \end{cases}$$

From the first equation, we can express b as $b = \frac{Ncd}{a}$, then substitute it into the third equation we have

$$\frac{N^2 c^2 d^2}{a^2} - Nd^2 = -N$$
$$\frac{d^2 (Nc^2 - a^2)}{a^2} = -1$$

Using the second equation of the system, we finally have $\frac{d^2}{a^2} = 1$, from which we obtain that $d = a$ or $d = -a$ and that $b = Nc$ or $b = -Nc$, respectively.

What is the meaning of matrix $\mathbf{B}$? What does it do to the pair of coordinates (x,y) that are also a solution to the equation (3.86)? We can see this matrix of linear transformation, step by step moves an integer point on hyperbola $x^2 - Ny^2 = 1$ to another integer point of the same hyperbola, closest to the previous one. Matrices of orthogonal transformation have determinant of ± 1. Note that matrix with $b = -Nc$ and $d = -a$

$$\mathbf{B_1} = \begin{pmatrix} a & -Nc \\ c & -a \end{pmatrix}$$

is not suitable because its every even power is the identity matrix and so the matrix $\mathbf{B_1}$ moves the trivial solution $(1,0)$ to the solution (a,c) and then moves it back to the trivial without producing any new solutions.

Finally, matrix B has the following type:

$$\mathbf{B} = \begin{pmatrix} a & Nc \\ c & a \end{pmatrix} \tag{3.89}$$

Let us consider Case 2. We need to solve the following system of nonlinear equations:

$$\begin{cases} ab - Ncd = \alpha \\ a^2 - Nc^2 = 1 \\ ab - Ncd = -\alpha \\ b^2 - Nd^2 = -N \end{cases}$$

This case leads us to a conclusion that $\alpha = 0$, which is not applicable regarding (3.88). Therefore we should proceed with Case 1.

Using the matrix (3.89), one can find any ordered pair solution to (3.86). Let us order all solutions to this equation. Matrix has only two parameters, a and c, the trivial solution is always $x = 1$ and $y = 0$, then in order to find matrix $\mathbf{B}$, we need to find only the first nontrivial solution. This least nontrivial solution (a,c) will be the numbers of the first column of matrix $\mathbf{B}$, and its second column will be (Nc,a). Moreover, if we know matrix of transformation, (3.89), the first solution and the nth solutions are $z_1 = Bz_0$, $z_n = B^n \cdot z_0$, respectively. Thus,

$$z_1 = \begin{pmatrix} a & Nc \\ c & a \end{pmatrix} \cdot \begin{pmatrix} 1 \\ 0 \end{pmatrix} = \begin{pmatrix} a \\ c \end{pmatrix}$$

Hence, as soon as one finds the first nontrivial solution, the matrix $\mathbf{B}$ and all other solutions are known.

For example, for (3.86), the 4th solution is $z_4 = \begin{pmatrix} 3 & 8 \\ 1 & 3 \end{pmatrix}^4 \cdot \begin{pmatrix} 1 \\ 0 \end{pmatrix} = \begin{pmatrix} 577 \\ 122 \end{pmatrix}$.

Can we find the matrix if we know any solution, not necessarily the very first one, different from the trivial? For example, we know any solution z_k, then we can multiply it from the left by the inverse of matrix **B** until we obtain the trivial solution $(1,0)$.

$$z_{k-1} = \mathbf{B}^{-1} \cdot z_k, \quad \mathbf{B}^{-1} = \begin{pmatrix} a & -Nc \\ -c & a \end{pmatrix} \tag{3.90}$$

We can apply our ideas right away for the following simple problem.

Problem 186

Find all natural solutions of $x^2 - 3y^2 = 1$.

Solution. We can find that $x = 2$ and $y = 1$ make this equation true. This is the least natural nontrivial solution. Then we can make matrix of linear transformation using (3.89)

$$\mathbf{B} = \begin{pmatrix} 2 & 3 \\ 1 & 2 \end{pmatrix}$$

Then all solutions are obtained by subsequent multiplication of this matrix by the trivial solution and so on.

$$(x,y) \rightarrow (2x + 3y, x + 2y)$$

$$(1,0) \rightarrow (2,1) \rightarrow (7,4) \rightarrow (26,15) \rightarrow (97,56) \rightarrow (362,209), \dots$$

Answer. $(x,y) : (1,0) \rightarrow (2,1) \rightarrow (7,4) \rightarrow (26,15) \rightarrow (97,56) \rightarrow (362,209), \dots$

Are there other methods of building all other solutions by knowing one nontrivial solution? Let us prove the following Lemma.

Lemma 14

For any set of real numbers x, y, z, t, N, the following formula is true:

$$(x^2 - Ny^2) \cdot (z^2 - Nt^2) = (xz + Nyt)^2 - N \cdot (xt + yz)^2 \tag{3.91}$$

Proof. Let us apply the difference of squares formula to each factor on the left of (3.91) and then multiply the left and right sides of both equations by additionally by interchanging the factors's order. Then apply the difference of squares formula again,

$$x^2 - Ny^2 = (x - \sqrt{N}y) \cdot (x + \sqrt{N}y)$$
$$z^2 - Nt^2 = (z - \sqrt{N}t) \cdot (z + \sqrt{N}t)$$
$$(x^2 - Ny^2) \cdot (z^2 - Nt^2)$$
$$= (x + \sqrt{N}y) \cdot (z + \sqrt{N}t)(x - \sqrt{N}y) \cdot (z - \sqrt{N}t)$$
$$= (xz + Nyt + \sqrt{N}(xt + yz)) \cdot (xz + Nyt - \sqrt{N}(xt + yz))$$
$$= (xz + Nyt)^2 - N \cdot (xt + yz)^2.$$

The proof is complete.

Let us see how we can use this formula to find additional integer solutions to Pell's equation, if all variables are integer or natural numbers and N is a natural number but it is not a perfect square. Now we can find other solutions to the equation:

$$x^2 - N \cdot y^2 = \pm M.$$

Theorem 25

If $x^2 - Ny^2 = a$ and $z^2 - Nt^2 = b$, then the pair

$$(X, Y) = (xz + Nyt, xt + yz)$$

satisfies another Pell's type equation

$$X^2 - NY^2 = ab.$$

The proof of this theorem follows from Lemma 14.

Problem 187

Find a nontrivial solution to the equation $x^2 - 5y^2 = 44$.

Solution. Consider two other equations, the nontrivial solutions of which can be easily found:

$$x^2 - 5y^2 = -11$$
$$3^2 - 5 \cdot 2^2 = -11$$
$$x = 3, \ y = 2$$

and

$$x^2 - 5y^2 = -4$$
$$1^2 - 5 \cdot 1^2 = -4$$
$$z = 1, \ t = 1$$

Then substituting

$$x + 5yt = 3 + 5 \cdot 2 = 13$$
$$xt + yz = 3 + 2 \cdot 1 = 5.$$

The following will be true:

$$13^2 - 5 \cdot 5^2 = (-11) \cdot (-4) = 44$$

This nontrivial solution will help generating all other solutions to the equation $x^2 - 5y^2 = 44$.

Answer. $x = 13$, $y = 5$.

The next lemma establishes how to construct all solutions to Pell's equation if we know just one, nontrivial solution to it.

Lemma 15

Let a be the least natural number, for which there exists another natural number c and that

$$a^2 - Nc^2 = 1,$$

then the Pell's equation $x^2 - Ny^2 = 1$ does not have any other solutions except those obtained from the trivial $(1,0)$ using the iteration rule:

$$(x,y) \rightarrow (ax + Ncy, cx + ay). \qquad (3.92)$$

Remark. You can see that this formula (3.92) is equivalent to formula (3.89) and they both produce absolutely identical chain of solutions to Pell's equation (3.85). The only problem now is how we can find this nontrivial solution? Is there a general method?

For some equations such as $x^2 - 61 \cdot y^2 = 1$ it is very hard to find the first, nontrivial solution. It is interesting that this equation was solved in 11th century by an Indian mathematician Bhaskara, but his solution was not available to Pierre Fermat who challenged the mathematics world by asking for its solutions in 1657, and so this equation was solved much later in 1732, again by Leonard Euler.

Method 1. Geometric Fitting Approach. Let us try to find the generating formula as we were able to find in solving the previous problem. We will consider Pell's equation $x^2 - Ny^2 = 1$ as a quadratic equation in two variables. The graph of this equation is a **hyperbola** with the determinant of a quadratic form $-N < 0$ and with positive discriminant $(D/4)$ of a quadratic expression $N > 0$.

As we know a quadratic form

$$ax^2 + 2bxy + cy^2$$

will describe a hyperbola as long as the determinant of the matrix of a quadratic form is negative or the discriminant of the quadratic expression is positive. Step by step we will start with $x^2 - Ny^2 = 1$ and then rewrite the equation in a new form

$g(x,y) = ax^2 + 2bxy + cy^2$ by replacing either $x \to x+y$ or $y \to x+y$. Obviously, we will obtain the following two quadratic expressions:

$$\begin{aligned} g(x+y,y) &= a(x+y)^2 + 2b(x+y)y + cy^2 \\ &= ax^2 + 2(a+b)xy + (a+2b+c)y^2 \end{aligned} \tag{3.93}$$

$$\begin{aligned} g(x,x+y) &= ax^2 + 2bx(x+y) + c(x+y)^2 \\ &= (a+2b+c)x^2 + 2(b+c)xy + cy^2 \end{aligned} \tag{3.94}$$

If we calculate determinants of their symmetric matrices or the discriminants for each of the transformed expressions, we can see that the replacement of variable x by $x+y$ will not qualitatively change anything and that $ax^2 + 2(a+b)xy + (a+2b+c)y^2$ is a hyperbola with the same discriminant:

$$\frac{D}{4} = (a+b)^2 - a(a+2b+c) = b^2 - ac.$$

as hyperbola $(a+2b+c)x^2 + 2(b+c)xy + cy^2$, obtained by replacement $y \to x+y$:

$$\frac{D}{4} = (b+c)^2 - (a+2b+c)c = b^2 - ac.$$

In Pell's equations, the coefficients of x^2 and y^2 must have opposite signs. Then if $a+2b+c < 0$, we will choose the first formula and replace $x \to x+y$, otherwise, if $a+2b+c > 0$, we will use the second formula and replace $y \to x+y$.

Considering the sign of the coefficients of the consecutive quadratic expressions, after finite number of iterations, we will obtain a generating formula for the solutions of Pell's equation. The following theorem is valid.

Theorem 26 (Pell's Solution Generator)

After several substitutions, such as $x \to x+y$ or $y \to x+y$ the following equality will be obtained:

$$(\alpha \cdot x + \beta \cdot y)^2 - n \cdot (\gamma \cdot x + \delta \cdot y)^2 = x^2 - n \cdot y^2, \tag{3.95}$$

where $\alpha, \beta, \gamma, \delta$ are natural numbers. If (x,y) is a solution to the Pell's equations, then

$$(\alpha \cdot x + \beta \cdot y, \gamma \cdot x + \delta \cdot y)$$

is also a solution.

Let us see how the solution to the previous problem can be obtained by implementing Method 1.

$$x^2 - 2y^2 = 1$$

Step by step we will make proper variable substitutions:

$x^2 - 2y^2, \ 1 - 2 < 0$

$x \to x + y$

$(x+y)^2 - 2y^2 = x^2 + 2xy - y^2, \ a = 1, \ b = 1; \ c = -1, \ a + 2b + c = 1 + 2 - 1 = 2 > 0$

$y \to x + y$

$x^2 + 2x(x+y) - (x+y)^2 = (x+x+y)^2 - 2(x+y)^2 = 2x^2 - y^2,$

$\quad a = 2, \ b = 0, \ c = -1, \ a + 2b + c = 2 - 1 = 1 > 0$

$y \to x + y$

$2x^2 - (x+y)^2 = (3x+y)^2 - 2(2x+y)^2 = x^2 - 2xy - y^2, \ a + 2b + c = 1 - 2 - 1 = -2 < 0$

$x \to x + y$

$(x+y)^2 - 2(x+y)y - y^2 = (3(x+y)+y)^2 - 2(2(x+y)+y)^2 = (3x+4y)^2 - 2(2x+3y)^2.$

Therefore we obtain the rule of iteration,

$$(x, y) \Rightarrow (3x + 4y, 2x + 3y).$$

Remark. Consider transformation $x^2 - 2y^2 \to x^2 + 2xy - y^2$. We can calculate $a = 1, \ b = 1, \ c = -1, \ \frac{D}{4} = b^2 - ac = 1^2 + 1 = 2 = d$.

Problem 188

Find all natural solutions of $x^2 - 11y^2 = 1$ using Method 1 of subsequent fitting steps.

Solution.

$x^2 - 11y^2, \ 1 - 11 < 0$

$x \to x + y$

$= (x+y)^2 - 11y^2 = x^2 + 2xy - 10y^2, \ 1 + 2 - 10 < 0$

$x \to x + y$

$(x+y)^2 + 2(x+y)y - 10y^2 = (x+2y)^2 - 11y^2 = x^2 + 4xy - 7y^2, \ 1 + 4 - 7 < 0$

$x \to x + y$

$(x+y)^2 + 4(x+y)y - 7y^2 \ = (x+3y)^2 - 11y^2 = x^2 + 6xy - 2y^2, \ 1 + 6 - 2 > 0$

$y \to x + y$

$x^2 + 6x(x+y) - 2(x+y)^2 \ = (4x+3y)^2 - 11(x+y)^2 = 5x^2 + 2xy - 2y^2, \ 5 + 2 - 2 > 0$

$y \to x + y$

$5x^2 + 2x(x+y) - 2(x+y)^2 = (7x+3y)^2 - 11(2x+y)^2 = 5x^2 - 2xy - 2y^2, \ 5 - 2 - 2 > 0$

$y \to x + y$

$5x^2 - 2x(x+y) - 2(x+y)^2 = (10x+3y)^2 - 11(3x+y)^2 = x^2 - 6xy - y^2, \ 1 - 6 - 1 < 0$

$x \to x + y$

$(x+y)^2 - 6(x+y)y - y^2 \ = (10x+13y)^2 - 11(3x+4y)^2 = x^2 - 4xy - 6y^2, \ 1 - 4 - 6 < 0$

$x \to x + y$

$(x+y)^2 - 4(x+y)y - 6y^2 \ = (10x+23y)^2 - 11(3x+7y)^2 = x^2 - 2xy - 9y^2, \ 1 - 2 - 9 < 0$

$x \to x + y$

$(x+y)^2 - 2(x+y)y - 9y^2 \ = (10x+33y)^2 - 11(3x+10y)^2 = x^2 - 11y^2.$

Therefore we obtain the chain of iterations,

$$(x,y) \Rightarrow (10x+33y, 3x+10y).$$

Again, starting with $x = 1, y = 0$, we obtain the following periodic solutions:

$$(1,0) \rightarrow (10,3) \rightarrow (199,60) \rightarrow (3970,1197) \rightarrow \ldots$$

$$3970^2 - 11 \cdot 1197^2 = 1.$$

Answer. $(1,0) \rightarrow (10,3) \rightarrow (199,60) \rightarrow (3970,1197), \ldots$

Obviously this method would not give an answer quickly for an equation such as

$$x^2 - 61y^2 = 1.$$

Method 2. Using Periodic Continued Fractions. Irrational numbers are real numbers that cannot be written as a quotient of two integers. However, it was shown in the earlier Section 1.7 that all irrational numbers can be written as infinite continued fractions and that all quadratic irrationalities, such as $a + \sqrt{N} \cdot b$ are represented as periodic continued fractions.

Since $x^2 - Ny^2 = 1$ can be factored as $(x - \sqrt{N}y)(x + \sqrt{N}y) = 1$, the general solution of the Fermat equation is related to the rational approximation of $\sqrt{N}$. This approximation is obtained by infinite convergent continued fractions.

Fermat's equation was the occasion for the introduction of this technique into number theory. Euler and then Lagrange in 1776 proved that Pell's equation has infinitely many solutions using quadratic forms and quadratic irrationalities. Lagrange also developed a method of solving any equation of the form:

$$ax^2 + 2bxy + cy^2 + 2dx + 2ey + f = 0$$

in integers.

Assume that k is the length of the period of decomposition of $\sqrt{N}$ into periodic continued fractions. Then all solutions of the Fermat equation in integers will be written as $x = P_{kn-1}, y = Q_{kn-1}$, where n is any natural number such that nk is even and P_i, Q_i are convergent continued fractions. Let us remember that $n \cdot k - 1$ must be odd.

Let us rewrite $\sqrt{2}$ by a continued fraction.

$$a_0 + \cfrac{1}{a_1 + \cfrac{1}{a_2 + \cfrac{1}{\ddots + \cfrac{1}{a_s + \cfrac{1}{\ddots}}}}} = [a_0; a_1, a_2, \ldots, a_s \ldots], \qquad (3.96)$$

$$\sqrt{2} = 1 + \left(\sqrt{2} - 1\right) = 1 + \frac{1}{1+\sqrt{2}}$$
$$= 1 + \frac{1}{2+(\sqrt{2}-1)} = 1 + \frac{1}{2+\frac{1}{1+\sqrt{2}}}$$
$$= 1 + \frac{1}{2+\frac{1}{2+\frac{1}{1+\sqrt{2}}}} = \dots 1 + \frac{1}{2+\frac{1}{2+\frac{1}{2+\dots}}}$$

Additionally $\sqrt{2}$ can be written as $\sqrt{2} = [1; 2, 2, 2\dots]$

By cutting infinite fraction, we obtain the following consecutive convergent fractions:

$$\frac{1}{1}; \; 1 + \frac{1}{2} = \frac{3}{2}; \; 1 + \frac{1}{2+\frac{1}{2}} = \frac{7}{5}; \; 1 + \frac{1}{2+\frac{1}{2+\frac{1}{2}}} = \frac{17}{12}$$

Let $\frac{x}{y}$ be a reduced convergent fraction of an irrational number $\sqrt{N}$, then only the numerator and denominator of such a fraction can be a solution for the equation $x^2 - Ny^2 = 1$. Some convergent fractions will lead to a solution and some will not. However, all possible solutions must be searched only among convergent fractions, approaching to $\sqrt{N}$.

Thus,

$$1^2 - 2 \cdot 1^2 = -1$$
$$\boxed{3^2 - 2 \cdot 2^2 = 1}$$
$$7^2 - 2 \cdot 5^2 = -1$$
$$\boxed{17^2 - 2 \cdot 12^2 = 1}$$

Consider

$$x^2 - Ny^2 = (x + y\sqrt{N})(x - y\sqrt{N}) = 1.$$

Irrational numbers in the form $x + y\sqrt{N}$ are closely connected with solutions of Pell's equation. Irrational number

$$\alpha = r + s\sqrt{N}$$

will generate all other solutions to the Pell's equation if and only if

$$r^2 - Ns^2 = 1.$$

Lemma 16

If $\alpha = r + s\sqrt{N}$ gives a solution to $x^2 - Ny^2 = 1$, then its conjugate $\overline{\alpha} = r - s\sqrt{N}$ will also generate a solution.

The proof of this lemma is obvious.

Lemma 17

If α and β give solutions to $x^2 - Ny^2 = 1$, then their product $\alpha \cdot \beta$ will also give a new solution.

Proof. Multiplying two quantities, we obtain

$$\alpha \cdot \beta = (a + b\sqrt{N})(c + d\sqrt{N}) = (ac + bdN) + (ad + bc)\sqrt{N}$$

If we now substitute $x = c + d\sqrt{N}$ and $y = ad + bc$ into Pell's equation, and apply Lemma 14, we obtain

$$(ac + Nbd)^2 - n(ad + bc)^2 = (a^2 - Nb^2)(c^2 - Nd^2) = 1,$$

that confirms that the new values of x and y are the solutions and completes the proof. $\blacksquare$

Let us apply this lemma and solve the following problem.

Problem 189

Given $x^2 - 8y^2 = 1$ and two of its solutions, $(x,y) = (3,1)$ and $(x,y) = (17,6)$. Find another solution to the equation.

Solution. Let $\alpha = 3 + 1 \cdot \sqrt{8}$ and $\beta = 17 + 6 \cdot \sqrt{8}$, then $\gamma = \alpha \cdot \beta$ will produce a new solution,

$$\begin{aligned} \gamma &= (3 + 1 \cdot \sqrt{8})(17 + 6 \cdot \sqrt{8}) = 51 + 18\sqrt{8} + 17\sqrt{8} + 48 \\ &= 99 + 35\sqrt{8} \end{aligned}$$

it is easy to check that $x = 99$, $y = 35$ is another solution:

$$99^2 - 8 \cdot 35^2 = 1.$$

Answer. $(99, 35)$.

Let us solve the following problem:

Problem 190

Find solutions for $x^2 - 7y^2 = 1$ using a continued fraction technique.

Solution. First, find a continued fraction representation for $\sqrt{7}$,

$$\sqrt{7} = 2 + \sqrt{7} - 2 = 2 + \frac{1}{\alpha_1}$$

where $\alpha_1 = \frac{1}{\sqrt{7}-2}$. Rationalize the denominator for α_1: $\frac{1}{\sqrt{7}-2}\left(\frac{\sqrt{7}+2}{\sqrt{7}+2}\right) = \frac{\sqrt{7}+2}{3} = 1 + \frac{1}{\alpha_2}$. Now, $\alpha_2 = \frac{1}{\frac{\sqrt{7}+2}{3}-1} = \frac{1}{\frac{\sqrt{7}-1}{3}} = \frac{3}{\sqrt{7}-1}$. Rationalize the denominator for α_2:

$\frac{3}{\sqrt{7}-1}\left(\frac{\sqrt{7}+1}{\sqrt{7}+1}\right) = \frac{3(\sqrt{7}+1)}{6} = \frac{\sqrt{7}+1}{2} = 1 + \frac{1}{\alpha_3}$. Now, $\alpha_3 = \frac{1}{\frac{\sqrt{7}+1}{2}-1} = \frac{1}{\frac{\sqrt{7}-1}{2}} = \frac{2}{\sqrt{7}-1}$.

Rationalize the denominator for α_3: $\frac{2}{\sqrt{7}-1}\left(\frac{\sqrt{7}+1}{\sqrt{7}+1}\right) = \frac{2(\sqrt{7}+1)}{6} = \frac{\sqrt{7}+1}{3} = 1 + \frac{1}{\alpha_4}$. Now,

$\alpha_4 = \frac{1}{\frac{\sqrt{7}+1}{3}-1} = \frac{1}{\frac{\sqrt{7}-2}{3}} = \frac{3}{\sqrt{7}-2}$. Rationalize the denominator for α_4: $\frac{3}{\sqrt{7}-2}\left(\frac{\sqrt{7}+2}{\sqrt{7}+2}\right) =$

$\frac{3(\sqrt{7}+2)}{3} = \sqrt{7}+2 = 4 + \frac{1}{\alpha_5}$. Now, $\alpha_5 = \frac{1}{\sqrt{7}+2-4} = \frac{1}{\sqrt{7}-2}$, which can be rationalized:

$\frac{1}{\sqrt{7}-2}\left(\frac{\sqrt{7}+2}{\sqrt{7}+2}\right) = \frac{\sqrt{7}+2}{3} = \alpha_1$.

The period has ended.

Now we can express $\sqrt{7}$ in continued fraction form:

$$\sqrt{7} = 2 + \cfrac{1}{1 + \cfrac{1}{1 + \cfrac{1}{1 + \cfrac{1}{4 + \cfrac{1}{\ddots}}}}}$$

or using a different notation:

$$\sqrt{7} = 2; \overline{1114}$$

Now using this continued fraction expansion of $\sqrt{7}$, we will look for a solution to the equation

$$x^2 - 7y^2 = 1.$$

If we go to the end of the period, we would have a good approximation for $\sqrt{7}$, but this does not give a solution to the equation,

$$\sqrt{7} = 2 + \cfrac{1}{1 + \cfrac{1}{1 + \boxed{1 + \cfrac{1}{4}}}} = 2 + \cfrac{1}{1 + \cfrac{1}{\boxed{1 + \cfrac{1}{\frac{5}{4}}}}} = 2 + \cfrac{1}{\boxed{1 + \cfrac{1}{\frac{9}{5}}}} = 2 + \boxed{\cfrac{1}{\frac{14}{9}}} = 2 + \frac{9}{14} = \frac{37}{14}$$

Trying $x = 37, y = 14$ in the equation, we discover that it is not a solution. So we will try determining a solution by stopping earlier in the continued fraction:

$$\sqrt{7} = 2 + \cfrac{1}{1 + \cfrac{1}{1 + \frac{1}{1+1}}} = 2 + \cfrac{1}{1 + \boxed{\cfrac{1}{1 + \frac{1}{2}}}} = 2 + \cfrac{1}{\boxed{1 + \cfrac{1}{1 + \frac{3}{2}}}} = 2 + \boxed{\cfrac{1}{\frac{5}{3}}} = 2 + \frac{3}{5} = \frac{13}{5}$$

Trying $x = 13, y = 5$ in the equation, we discover that it also is not a solution. So we will try determining a solution by stopping earlier in the continued fraction:

$$\sqrt{7} = 2 + \cfrac{1}{1 + \cfrac{1}{1 + \frac{1}{1+1}}} = 2 + \cfrac{1}{1 + \frac{1}{1 + \frac{1}{2}}} = 2 + \cfrac{1}{1 + \frac{3}{2}} = 2 + \frac{2}{3} = \frac{8}{3}$$

Trying $\boxed{x = 8, y = 3}$ in the equation, we discover that it is a solution.

This solution is also obtained with the use of the theory developed by Lagrange. (See also Section 2.2.)

Since the period of the infinite continued fraction is $k = 4$ (an even number), then the first solution will appear for $k - 1 = 3$, the third approximation of the square root as

$$x = P_{kn-1}, \quad y = Q_{kn-1}$$

The successive approximations for the period are $2, 3, \frac{5}{2}, \frac{8}{3}, \frac{13}{5}, \frac{37}{14}$. These alternate being less than or greater than the actual value for $\sqrt{7}$ (oscillate), but get closer and closer to the actual value, i.e., $2.645751311067...$
We have found that the first solution to the equation $x^2 - 7y^2 = 1$ is $x = P_3 = 8, y = Q_3 = 3$. This solution can be used with the $\sqrt{7}$ to make what is called a solution "generator":

$$\theta = 8 + 3\sqrt{7}$$

Using our generator (See, for example, Dudley), we can find additional solutions to the same equation using θ^k, where $k = 2, 3, 4, ...$ To find a second solution, $\theta^2 = (8 + 3\sqrt{7})^2 = (8 + 3\sqrt{7})(8 + 3\sqrt{7}) = 127 + 48\sqrt{7}$. This gives a solution of $x = 127, y = 48$.

To find a third solution, $\theta^3 = (8 + 3\sqrt{7})^3 = (8 + 3\sqrt{7})(127 + 48\sqrt{7}) = 2024 + 765\sqrt{7}$. This gives a solution of $x = 2024, y = 765$. In a similar way, the next solutions are found,

$$x = 32,257, y = 12,192$$

$$x = 514,088, y = 194,307$$

$$x = 8,193,151, y = 3,096,720$$

Problem 191

Find all natural solutions of $x^2 - 17y^2 = 1$.

Solution.
Method 1. Using Lemma 13, notice that

$$N = 17 = 16 + 1 = 4^2 + 1, \Rightarrow n = 4$$
$$x = 2n^2 + 1 = 2 \cdot 16 + 1 = 33$$
$$y = 2 \cdot n = 2 \cdot 4 = 8.$$

This was possible because 17 can be represented as a sum of perfect square and 1. For many other radicand values the method of infinite continued fractions will be applied.

Method 2. Let us rewrite $\sqrt{17}$ into continued fraction:

$$\sqrt{17} = 4 + \sqrt{17} - 4 = 4 + \frac{1}{\alpha_1}$$
$$\alpha_1 = \frac{1}{\sqrt{17}-4} = \sqrt{17} + 4 = 8 + \frac{1}{\alpha_2}$$
$$\alpha_2 = \frac{1}{\sqrt{17}+4-8} = \frac{1}{\sqrt{17}-4} = 8 + \frac{1}{\alpha_3} = \alpha_1$$
$$\sqrt{17} = [4; \overline{8}]$$

or

$$\sqrt{17} = 4 + \cfrac{1}{8 + \cfrac{1}{8 + \cfrac{1}{8 + \cfrac{1}{8 + \cfrac{1}{\ddots}}}}}$$

Because period $k = 1$. We will select the first approximation of it as $4 + 1/8 = 33/8$. Therefore, $x = 33$, $y = 8$. We obtain

$$33^2 - 17 \cdot 8^2 = 1089 - 1088 = 1.$$

Consider now $(33 + \sqrt{17} \cdot 8)^2 = 2177 + 528\sqrt{17}$, hence we obtain the second solution as $(x = 2177, y = 528)$. Multiplying this radical expression by $(33 + \sqrt{17})$, we obtain the following $143649 + 34840\sqrt{17}$, and a new solution $x = 143649, y = 34840$, etc.

Answer. $x = 33$, $y = 8$ is the least nontrivial solution.

Problem 192

Find the number represented by the expression:

$$4 + \cfrac{1}{8 + \cfrac{1}{8 + \cfrac{1}{8 + \dots}}}.$$

Solution. The number can be written as $N = 4 + x$. On the other hand,

$$x = \frac{1}{8+x}$$
$$x^2 + 8x - 1 = 0, \quad x > 0$$
$$x = -4 + \sqrt{17}.$$

Therefore, $N = 4 - 4 + \sqrt{17} = \sqrt{17}$.
Answer. $\sqrt{17}$.

Problem 193

Find all natural solutions of $x^2 - 11y^2 = 1$ using continued fractions technique.

Solution.

$$\sqrt{11} = 3 + \sqrt{11} - 3 = 3 + \frac{1}{\alpha_1}$$
$$\alpha_1 = \frac{1}{\sqrt{11}-3} = \frac{\sqrt{11}+3}{2} = 3 + \frac{1}{\alpha_2}$$
$$\alpha_2 = \frac{1}{\frac{\sqrt{11}+3}{2}-3} = \frac{2}{\sqrt{11}-3} = \sqrt{11} + 3 = 6 + \frac{1}{\alpha_3}$$
$$\alpha_3 = \frac{1}{\sqrt{11}+3-6} = \frac{1}{\sqrt{11}-3} = \alpha_1$$

$$\sqrt{11} = [3; \overline{363}] = 3 + \cfrac{1}{3 + \cfrac{1}{6 + \cfrac{1}{3+...}}}.$$

Since the period is $k = 3$, then if we approximate $\sqrt{11}$ by $\frac{P_2}{Q_2}$ as

$$\sqrt{11} \approx 3 + \frac{1}{3} = \frac{10}{3},$$

it must give us the least nontrivial solution to the given equation. Let us check $x = 10$, $y = 3$, as we remember by solving this problem in a different way, this pair must be the least solution:

$$10^2 - 11 \cdot 3^2 = 1.$$

Further, if we will approximate $\sqrt{11}$ as

$$\sqrt{11} \approx 3 + \cfrac{1}{3 + \cfrac{1}{6+\frac{1}{3}}} = \frac{199}{60},$$

then 199 and 60 are also solution, but not the least, which we need as further solutions generator and

$$199^2 - 11 \cdot 60^2 = 1.$$

Thus, $\alpha = 10 + 3 \cdot \sqrt{11}$ and its consecutive powers will give us all other solutions to the equation.
Answer. $x = 10$ and $y = 3$.

Problem 194

Find all natural solutions of $x^2 - 22y^2 = 1$.

Solution. Let us rewrite $\sqrt{22}$ into continued fraction,

$$\sqrt{22} = 4 + \frac{1}{\alpha_1}$$
$$\alpha_1 = \frac{1}{\sqrt{22}-4} = \frac{\sqrt{22}+4}{6} = 1 + \frac{1}{\alpha_2}$$
$$\alpha_2 = \frac{1}{\frac{\sqrt{22}+4}{6}-1} = \frac{6}{\sqrt{22}-2} = \frac{\sqrt{22}+2}{3} = 2 + \frac{1}{\alpha_3}$$
$$\alpha_3 = \frac{1}{\frac{\sqrt{22}+2}{3}-2} = \frac{3}{\sqrt{22}-4} = \frac{\sqrt{22}+4}{2} = 4 + \frac{1}{\alpha_4}$$
$$\alpha_4 = \frac{1}{\frac{\sqrt{22}+4}{2}-4} = \frac{\sqrt{22}+4}{3} = 2 + \frac{1}{\alpha_5}$$
$$\alpha_5 = \frac{1}{\frac{\sqrt{22}+4}{3}-2} = \frac{3}{\sqrt{22}-2} = \frac{\sqrt{22}+4}{6} = 1 + \frac{1}{\alpha_6}$$
$$\alpha_6 = \frac{1}{\frac{\sqrt{22}+2}{6}-1} = \frac{6}{\sqrt{22}-4} = \sqrt{22}+4 = 8 + \frac{1}{\alpha_7}$$
$$\alpha_7 = \frac{1}{\sqrt{22}+4-8} = \frac{\sqrt{22}+4}{6} = \alpha_1.$$

Therefore,

$$\sqrt{22} = [4; \overline{1,2,4,2,1,8}] = 4 + \cfrac{1}{1 + \cfrac{1}{2 + \cfrac{1}{4 + \cfrac{1}{2 + \cfrac{1}{1 + \cfrac{1}{8 + \cfrac{1}{1+\dots}}}}}}}.$$

The period contains six terms. Since $k = 6$, then $x_0 = P_5$, $y_0 = Q_5$ are minimal values of x and y. Since $P_5 = 197$, $Q_5 = 42$, then $x_0 = 197$, and $y_0 = 42$.
Answer. $x = 197$ and $y = 42$.

Remark.

All other solutions to this equation can be obtained from

$$\alpha = 197 + 42 \cdot \sqrt{22}$$

by raising it to natural powers. Thus, if after simplification,

$$\alpha^n = a + b \cdot \sqrt{22},$$

then

$$a = x_n, \ b = y_n$$

and

$$a^2 - 22b^2 = 1.$$

You can check that the following three ordered pairs are the second, third, and fourth solutions to the given equation:

$$x_2 = 77617, \ y_2 = 16548$$
$$x_3 = 30580901, \ y_3 = 6519870$$
$$x_4 = 1204879738, \ y_4 = 2568812232.$$

If you like matrices or if you have a calculator or a computer, then the easy way to get all other solutions, after you obtained the first, nontrivial, $(x = 197, y = 42)$ is to multiply it from the left by the matrix of linear transformation

$$\mathbf{B} = \begin{pmatrix} 197 & 924 \\ 42 & 197 \end{pmatrix}$$

or by any of its integer power.

Remark. Ok. We can see that rewriting N as continued periodic fraction allowed us to find the **least nontrivial solution** even for strange values of N when it was difficult to do it mentally at first glance. Another efficient way to find the solution to $x^2 - Ny^2 = 1$ is by trial, by making a table of the values $1 + Ny^2$ for $y = 1, 2, 3, \ldots$ and inspect the result for a full perfect square. The least value of y producing a perfect square of x will generate for us all other solutions to the equation! Let $y = b$ produces a perfect square of $x = a$, and makes the following true:

$$a^2 = 1 + N \cdot b^2.$$

Then irrational number

$$\alpha = a + b \cdot \sqrt{N}. \tag{3.97}$$

is our solution generator. Remember how we found the least natural nontrivial solution to equation $x^2 - 17y^2 = 1$? It was $y = b = 8$, $x = a = 33$, then $\alpha = 33 + 8 \cdot \sqrt{17}$ is the solution generator. Raising this irrational number to the second, third, fourth, and other powers we will obtain another irrational number and the corresponding new pair of the solutions to the given equation.

$$(\alpha)^n = \beta = r + s\sqrt{N}, \ r^2 - Ns^2 = 1.$$

For example, the second and third solutions to $x^2 - 17y^2 = 1$ will be obtained from squaring and cubing of α, respectively,

$$(33 + 8 \cdot \sqrt{17})^2 = 2177 + 528\sqrt{17}, \Rightarrow x = 2177, \ y = 528$$
$$(33 + 8 \cdot \sqrt{17})^3 = 143649 + 34840\sqrt{17}, \Rightarrow x = x = 143649, \ y = 34840$$

On the other hand, if we know the least nontrivial solution, (a, c), then we can create a matrix **B**

$$\mathbf{B} = \begin{pmatrix} a & Nc \\ c & a \end{pmatrix}$$

and then using the trivial solution $z_0 = (x_0, y_0)^T = (1, 0)^T$, any nontrivial solution can be obtained as

$$\begin{pmatrix} x_k \\ y_k \end{pmatrix} = z_k = B^k \begin{pmatrix} 1 \\ 0 \end{pmatrix} = B z_{k-1}$$

3.5.4 Pell's Type Equations and Applications

Consider the equation

$$(x - \sqrt{2}y)(x + \sqrt{2}y) = x^2 - 2y^2 = \pm 1. \tag{3.98}$$

Earlier we found several consecutive solutions to this equation with one on the right-hand side. Let us start raising $(1 + \sqrt{2})$ to second, third, fourth, and all other powers.

$$(1 + \sqrt{2})^2 = 1 + 2\sqrt{2} + 2 = 3 + 2 \cdot \sqrt{2}, \ x = 3; y = 2$$

$$(1 + \sqrt{2})^3 = (1 + \sqrt{2})^2 \cdot (1 + \sqrt{2}) = 7 + 5\sqrt{2}, \ x = 7; y = 5$$

Notice that the obtained pairs $(3, 2)$, $(7, 5)$ are also solutions to our equation. Using Newton's Binomial Theorem, we can predict the nth power of $(1 + \sqrt{2})$,

$$(1 + \sqrt{2})^n = x_n + y_n\sqrt{2}, \ x_n, y_n \in N \tag{3.99}$$

Can we find the $(n + 1)$th power of it? Of course!

$$(1 + \sqrt{2})^{n+1} = (1 + \sqrt{2})^n \cdot (1 + \sqrt{2}) = (x_n + y_n\sqrt{2})(1 + \sqrt{2})$$
$$= (x_n + 2y_n) + (x_n + y_n) \cdot \sqrt{2}. \tag{3.100}$$

From (3.99) we can state the following recursive formulas:

$$x_{n+1} = x_n + 2y_n$$
$$y_{n+1} = x_n + y_n. \tag{3.101}$$

It is not hard to predict formulas for the conjugate of the quadratic irrationality similar to (3.99),

$$(1 - \sqrt{2})^n = x_n - y_n\sqrt{2}, \ x_n, y_n \in N. \tag{3.102}$$

By adding and subtracting formulas (3.99) and (3.102), we can easily obtain exact formulas for any solution to (3.98):

$$x_n = \frac{(1+\sqrt{2})^n + (1-\sqrt{2})^n}{2}$$
$$y_n = \frac{(1+\sqrt{2})^n - (1-\sqrt{2})^n}{2\sqrt{2}}. \tag{3.103}$$

Formulas (3.101) give solutions to equations with 1 and -1 on the right-hand side in alternation, starting from the least solution to $x^2 - 2y^2 = -1$ as $x_1 = 1$, $y_1 = 1$.

All solutions with even subscripts will belong to $x^2 - 2y^2 = 1$ (for example, $x_2 = 3$, $y_2 = 2$). On the other hand, formulas (3.103) for an even $n = 2k$ produce solutions to $x^2 - 2y^2 = 1$ and for odd $n = 2k - 1$ solutions for $x^2 - 2y^2 = -1$.

Let us solve a problem from geometry.

Problem 195

The most famous right triangle has sides 3, 4, 5, and its leg's lengths differ by one. Find all such right triangles.

Solution. Denote by y the length of hypothenuse and by z the length of the shorter leg, then the second leg will have length $(z+1)$. Using Pythagorean formula, we get the equation to solve,

$$z^2 + (z+1)^2 = y^2.$$

Expanding the square and after multiplying both sides by 2, we will complete the square on the left. All steps are shown below.

$$2z^2 + 2z + 1 = y^2$$
$$4z^2 + 4z + 2 = 2y^2$$
$$(2z+1)^2 - 2y^2 = -1.$$

Introducing a new variable

$$x = 2z + 1,$$

and after substitution, we obtain Pell's type equation, but not Pell's equation,

$$x^2 - 2 \cdot y^2 = -1. \tag{3.104}$$

How can we solve this equation? Can we use an approach developed for solving Pell's equations? Can we somewhere how find the solution from the previously found solutions to $x^2 - 2 \cdot y^2 = 1$? Let us think.

It follows from the theory developed above that if

$$x^2 - 2y^2 = \pm 1,$$

then the equation

$$X^2 - 2Y^2 = \mp 1, \quad X = x + 2y, \ Y = x + y$$

Applying this operation one more time, we obtain new, correct coordinates,

$$\widehat{X} = (x+2y) + 2(x+y) = 3x + 4y$$
$$\widehat{Y} = (x+2y) + (x+y) = 2x + 3y.$$

which I hope you recognized as a familiar formula

$$(x,y) \rightarrow (3x+4y, 2x+3y)$$

that also can be given by a matrix of linear transformation

$$\mathbf{B} = \begin{pmatrix} 3 & 4 \\ 2 & 3 \end{pmatrix}$$

applied to the least solution $x = 1, y = 1$ to obtain the first consecutive solutions:

$$(1,1) \rightarrow (7,5) \rightarrow (41,29) \rightarrow (239,169) \rightarrow (1393,985)...$$

Finally, since $z = \frac{x-1}{2}$, $y = y$, we can obtain the following Pythagorean triples with the required property,

$$(z, z+1, y): (3,4,5) \rightarrow (20,21,29) \rightarrow (119,120,169) \rightarrow (696,697,985)...$$

Answer. The first four right triangles, the integer legs of which differ by one, are $(3,4,5),(20,21,29),(119,120,169),(696,697,985)$.

Problem 196

Find all natural solutions to the equation $x^2 - 2y^2 = 7$.

Solution. We know that if we replace $x \rightarrow (3x+4y)$ and $y \rightarrow (2x+3y)$, then the left side of the equation will not change. For Pell's equation $x^2 - 2y^2 = 1$, we then generated all other possible solutions by simply using the trivial solution $(1,0)$. However, for this problem we cannot use the trivial solution, so let us find the least natural solution to it. It is $x = 3$, $y = 1$ and next to it looks like $x = 5$, $y = 3$. Applying linear transformation

$$(x,y) \rightarrow (3x+4y, 2x+3y)$$

using the first pair and we obtain the following chain

$$(3,1) \rightarrow (13,9) \rightarrow (75,53) \rightarrow (437,309) \rightarrow (2547,1801)...$$

Notice that another obvious solution $x = 5, y = 3$ is not in this chain! Applying the same matrix $\mathbf{B}$ to starting solution $(5,3)$, we obtain the second chain:

$$(5,3) \rightarrow (27,19) \rightarrow (157,111) \rightarrow (915,647) \rightarrow (5333,3771)...$$

Answer. $(3,1),(5,3),(13,9),(27,19),(75,53),(157,11),(437,309),(915,647)...$

The following problem is also interesting.

Problem 197

Find all such natural numbers n that the sum of all natural numbers from 1 to m $(1 < m < n)$ equals the sum of all numbers from $m+1$ to n.

Solution. The condition of the problem can be written as

$$1+2+3+\ldots+m = (m+1)+(m+2)+\ldots+n.$$

We can see that on the left side we have the sum of all natural numbers from 1 to m and on the right side the difference between all natural numbers from 1 to n and the sum of all numbers from 1 to m,

$$\frac{m(m+1)}{2} = \frac{n(n+1)}{2} - \frac{m(m+1)}{2}$$

which can be simplified as

$$m(m+1) = \tfrac{n(n+1)}{2}$$
$$2m^2 + 2m = n^2 + n.$$

Using the standard approach, we can multiply the both sides of the equation by 4 and then complete the squares on each side of it.

$$2(4m^2 + 4m + 1) - 2 = 4n^2 + 4n$$
$$2(2m+1)^2 - 1 = 4n^2 + 4n + 1 = (2n+1)^2$$

This can be rewritten as Pell's type equation,

$$x^2 - 2y^2 = -1, \quad x = 2n+1, \quad y = 2m+1 \tag{3.105}$$

After solving this equation, and after finding only its odd solutions, the values of n and m can be found as

$$n = \tfrac{x-1}{2}$$
$$m = \tfrac{y-1}{2}. \tag{3.106}$$

The trivial solution of (3.105) is $(x_0 = 1, y_0 = 1)$, then all other solutions can be obtained by the following recurrence formula:

$$x_{k+1} = 3x_k + 4y_k$$
$$y_{k+1} = 2x_k + 3y_k. \tag{3.107}$$

The first solution is $x_1 = 7$, $y_1 = 5$, then the corresponding $n_1 = 3, m_1 = 2$. Obviously, then in the sum

$$(1+2)+3$$

the sum of the first two terms $(1+2)$ equals the last term (3). Combining (3.107) and (3.106) we finally obtain the formulae for all consecutive solutions for n and m:

$$n_{k+1} = 3n_k + 4m_k + 3$$
$$m_{k+1} = 2n_k + 3m_k + 2. \qquad (3.108)$$

Applying (3.107) several times, we can find all pairs of n and m. The first four pairs with this property are

$$(n_1 = 3, m_1 = 2), (n_2 = 20, m_2 = 14), (n_3 = 119, m_3 = 84), (n_4 = 696, m_4 = 492)$$

Let us check, suppose $n = 20$, then based on our calculations, if we take the sum of the first 14 terms then it will be equal to the remaining sum from 15 to 20. Indeed,

$$1 + 2 + 3 + ... + 14 = \frac{14 \cdot 15}{2} = 105 = 15 + 16 + ... + 20 = \frac{15 + 20}{2} \cdot 6 = 105.$$

Answer. $(n, m) : (3, 2), (20, 14), (119, 84), (696, 492), ...$

Problem 198

Prove that the equation $x^3 + y^3 + z^2 = w^4$ has infinitely many solutions in natural numbers.

Proof. Consider again the sum of the first n cubes of natural numbers:

$$1^3 + 2^3 + 3^3 + ... + (n-2)^3 + (n-1)^3 + n^3 = \left(\frac{n(n+1)}{2} \right)^2 \qquad (3.109)$$

Since the first $(n-2)$ cubes again can be rewritten using the same formula, that is a square of a natural number,

$$1^3 + 2^3 + 3^3 + ... + (n-2)^3 = \left(\frac{(n-2)(n-1)}{2} \right)^2 \qquad (3.110)$$

Substituting (3.110) into (3.109) we will obtain

$$\left(\frac{(n-2)(n-1)}{2} \right)^2 + (n-1)^3 + n^3 = \left(\frac{n(n+1)}{2} \right)^2$$

then the solution to the given problem will be found as

$$x = n-1, \ y = n, \ z = \frac{(n-2)(n-1)}{2}, \ w = k \qquad (3.111)$$

as soon as we can find such numbers n that

$$\frac{n(n+1)}{2} = k^2$$

How can we solve this equation? After multiplying both sides by 2, we obtain

$$n^2 + n = 2k^2 \tag{3.112}$$

We can find right away a solution $n = 1, k = 1$. Are there other solutions? Let us multiply both sides of (3.112) by 4 and add unit to both sides:

$$4n^2 + 4n + 1 = 8k^2 + 1$$

Completing a square on the left-hand side and after rearranging terms, this equation can be rewritten as follows:

$$(2n+1)^2 - 2 \cdot (2k)^2 = 1.$$

This equation is Pell's equation and it has infinitely many solutions in integers. Therefore the given problem has infinitely many solutions.

The next problem of this section is also interesting.

Problem 199

Solve in natural numbers $z^2 - 5y^2 = \pm 4$.

Solution. Replacing $z = 2x - y$ and after substitution and dividing by 4, we can instead of the given equation to consider the auxiliary equation,

$$x^2 - xy - y^2 = \pm 1$$

x	0	1	1	2	3	5	8	13	21
y	1	0	1	1	2	3	5	8	13
$x^2 - xy - y^2$	-1	1	-1	1	-1	1	-1	1	-1

Table 3.3 Solutions to $x^2 - y^2 - xy = \pm 1$.

If we look at this table, we can see that each consecutive solution for x is obtained by adding the coordinates of x and y and that the second coordinate takes the value of the preceding x. The right-hand side takes the value of 1 or -1 in alternation. The more interesting is the fact that the numbers recorded in the first row are Fibonacci numbers! It looks like

$$(X, Y) \rightarrow (x + y, x) \Rightarrow X^2 - XY - Y^2 = \mp 1$$

which can be proven immediately by substitution,

$$(x+y)^2 - (x+y)\cdot x - x^2 = -x^2 + xy + y^2 = \mp 1$$

Next if $x = F_{n+1}$, $y = F_n$, then for an equation

$$z^2 - 5\cdot y^2 = \pm 4$$

and because $z = 2x - y$, we can obtain that

$$z = 2\cdot F_{n+1} - F_n = F_{n+1} + (F_{n+1} - F_n)$$
$$= F_{n+1} + F_{n-1}$$

Therefore, solution to the equation is obtained by the following ordered pairs:

$$(z,y) = (F_{n+1} + F_{n-1},\ F_n)$$

Answer. $(z,y) = (F_{n+1} + F_{n-1},\ F_n)$

Problem 200

Solve in natural numbers $x^2 - xy - y^2 = 1$.

Solution. Let us multiply both sides of the equation by 4 and complete the square on the left side.

$$4x^2 - 4xy - 4y^2 = 4$$
$$(2x - y)^2 - 5y^2 = 4.$$

We have Pell's type equation,

$$z^2 - 5y^2 = 4,\ z = 2x - y,\ y = y,\ x = \frac{z+y}{2} \tag{3.113}$$

I can see two methods to solve it.

1. Divide both sides by 4 (perfect square) and then solve the equation:

$$Z^2 - 5Y^2 = 1,\ Z = \frac{z}{2},\ Y = \frac{y}{2},\ z = 2Z,\ y = 2Y \tag{3.114}$$

2. Solve the equation $x^2 - 5y^2 = 1$ instead of $x^2 - 5y^2 = 4$, find matrix of linear transformation for Pell's equation, and then after finding nontrivial solutions to the equation, in which the right-hand side is 4, generate all other possible solutions.

We can demonstrate that both methods are equivalent.

Consider the equation

$$x^2 - 5y^2 = 1 \tag{3.115}$$

Because $N = 5 = 2^2 + 1$, $n = 2$, then its nontrivial solution can be obtained from Lemma 13. Its first nontrivial solution is $x_1 = 9, y_1 = 4$ because $9^2 - 5\cdot 4^2 = 1$. Then the solution generator is

$$(x,y) \rightarrow (9x + 20y, 4x + 9y)$$

which also can be described by matrix of linear transformation that preserves the quadratic form $x^2 - 5y^2$:

$$\mathbf{B} = \begin{pmatrix} 9 & 20 \\ 4 & 9 \end{pmatrix} \tag{3.116}$$

Multiplying $(9,4)$ by the inverse of matrix B, we obtain the trivial solution $(1,0)$. Multiplying $(9,4)$ from the left by powers of B, we obtain new solutions:

$$\begin{pmatrix} 9 & 20 \\ 4 & 9 \end{pmatrix}^{-1} \begin{pmatrix} 9 \\ 4 \end{pmatrix} = \begin{pmatrix} 1 \\ 0 \end{pmatrix}$$

$$\begin{pmatrix} 9 & 20 \\ 4 & 9 \end{pmatrix} \begin{pmatrix} 9 \\ 4 \end{pmatrix} = \begin{pmatrix} 161 \\ 72 \end{pmatrix}, \quad 161^2 - 5 \cdot 72^2 = 1$$

$$\begin{pmatrix} 9 & 20 \\ 4 & 9 \end{pmatrix}^2 \begin{pmatrix} 9 \\ 4 \end{pmatrix} = \begin{pmatrix} 2889 \\ 1292 \end{pmatrix}, \quad 2889^2 - 5 \cdot 1292^2 = 1$$

Solutions to the equation can be obtained by multiplying each coordinate by 2 or by multiplying a vector solution from the left by a double matrix B,

$$(z,y) \rightarrow (18z + 40y, 8z + 18y): \\ (18,8) \rightarrow (322, 144) \rightarrow (4414, 1962) \tag{3.117}$$

In order to find solutions to (3.113) using (3.116), we need to find nontrivial solutions of (3.113) and multiply then from the left by matrix (3.116). By looking at (3.113), we can see that the following is true,

$$3^2 - 5 \cdot 1^2 = 4, \quad 7^2 - 5 \cdot 3^2 = 4$$

Using the first solution $(3,1)$ and multiplying it by matrix B we obtain

$$\begin{pmatrix} 9 & 20 \\ 4 & 9 \end{pmatrix} \begin{pmatrix} 3 \\ 1 \end{pmatrix} = \begin{pmatrix} 47 \\ 21 \end{pmatrix}$$

$$\begin{pmatrix} 9 & 20 \\ 4 & 9 \end{pmatrix}^2 \begin{pmatrix} 3 \\ 1 \end{pmatrix} = \begin{pmatrix} 843 \\ 377 \end{pmatrix}$$

$$(z,y) \rightarrow (9z + 20y, 4z + 9y): \ (3,1) \rightarrow (47,21) \rightarrow (843,377) \tag{3.118}$$

Using the second solution $(7,3)$ and doing the same we get new solutions:

$$\begin{pmatrix} 9 & 20 \\ 4 & 9 \end{pmatrix} \begin{pmatrix} 7 \\ 3 \end{pmatrix} = \begin{pmatrix} 123 \\ 55 \end{pmatrix} \quad 123^2 - 5 \cdot 55^2 = 4$$

$$\begin{pmatrix} 9 & 20 \\ 4 & 9 \end{pmatrix}^2 \begin{pmatrix} 7 \\ 3 \end{pmatrix} = \begin{pmatrix} 2207 \\ 987 \end{pmatrix} \quad 2207^2 - 5 \cdot 987^2 = 4$$

$$(z,y) \rightarrow (9z+20y, 4z+9y) : (7,3) \rightarrow (123,55) \rightarrow (2207,987) \qquad (3.119)$$

Finally, because $x = \frac{z+y}{2}$ and $y = y$ combining (3.117), (3.118) and (3.119) we obtain the answer:

$(x,y) : (2,1), (5,3), (13,8), (34,21), (89,55), (233,144), \ldots$

If you remember the previous problem, you probably want to look at Fibonacci sequence:

$$\boxed{1}, 1, \boxed{2}, 3, \boxed{5}, 8, \boxed{13}, 21, \boxed{34}, 55, \boxed{89}, 144, \boxed{233} \ldots 1$$

Answer. $(x,y) : (2,1), (5,3), (13,8), (34,21), (89,55), (233,144), \ldots$

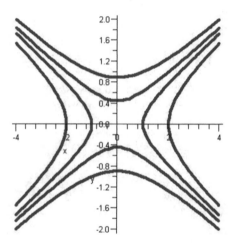

Fig. 3.1 $x^2 - 5y^2 = d$

Consider the following equations representing different hyperbolas shown in Figure 3.1:

$$x^2 - 5y^2 = 1$$
$$x^2 - 5y^2 = -1$$
$$x^2 - 5y^2 = 4$$
$$x^2 - 5y^2 = -4$$

Solutions to all of these equations can be generated by matrix (3.89) or by iteration:

$$(x,y) \rightarrow (9x+20y, 4x+9y).$$

However, if for the first two equations we have only one chain of the solutions: For $x^2 - 5y^2 = 1$ we have the following:

$$(1,0) \rightarrow (9,4) \rightarrow (161,71) \rightarrow (2889,1292) \rightarrow \ldots$$

And for $x^2 - 5y^2 = -1$

$$(2,1) \rightarrow (38,17) \rightarrow (682,305) \rightarrow (12238,5473) \rightarrow ...$$

Both the third and fourth equations have three starting solutions and hence, three different chains. Thus for $x^2 - 5y^2 = 4$ we have the following chains:

$$(2,0) \rightarrow (18,8) \rightarrow (322,144) \rightarrow (5778,2584) \rightarrow ...$$
$$(3,1) \rightarrow (47,21) \rightarrow (843,377) \rightarrow (15127,6765) \rightarrow ...$$
$$(7,3) \rightarrow (123,55) \rightarrow (2207,987) \rightarrow (39603,17711) \rightarrow ...$$

For the last equation $x^2 - 5y^2 = -4$ we also have three original nontrivial solutions that would generate all other solutions in three chains:

$$(1,1) \rightarrow (29,13) \rightarrow (521,233) \rightarrow ...$$
$$(4,1) \rightarrow (76,34) \rightarrow (1364,610) \rightarrow ...$$
$$(11,5) \rightarrow (199,89) \rightarrow (3571,1597) \rightarrow ...$$

Each time we use matrix $B = \begin{pmatrix} 9 & 20 \\ 4 & 9 \end{pmatrix}$ multiplying it by different starting solutions. Why could not we use, for example, a matrix of type (3.89) with different entries? For example, we know that the ordered pair (3,1) is a solution for $x^2 - 5y^2 = 4$. What would happen if we use $a = 3$, $c = 1$, $N = 5$ and hence, the following matrix:

$B_4 = \begin{pmatrix} 3 & 5 \\ 1 & 3 \end{pmatrix}$? Unfortunately, this new matrix would not generate correct ordered pairs because $(3x + 5y)^2 - 5(x + 3y)^2 = 4 \cdot (x^2 - 5y^2) \neq x^2 - 5y^2$. Instead, it would generate solutions to $x^2 - 5y^2 = 16$, $x = 14$, $y = 6$, then to $x^2 - 5y^2 = 64$, $x = 72$, $y = 32$, and so on by getting a point on a different hyperbola but not on the original one. Moreover, the determinant of matrix B_4 equals 4 and not one so the idea of using such a matrix would never give us correct answers for the four problems above.

Let us summarize what we have learned about Pell's equations. Consider the equation,

$$x^2 - 21y^2 = 1 \tag{3.120}$$

Let us find the representation of $\sqrt{21}$ as periodic continued fraction:

$$\sqrt{21} = \underline{4} + \frac{1}{\alpha_1}$$

$$\alpha_1 = \frac{1}{\sqrt{21}-4} = \frac{\sqrt{21}+4}{5} = 1 + \frac{1}{\alpha_2}$$

$$\alpha_3 = \frac{1}{\alpha_2 - 1} = \frac{1}{\frac{\sqrt{21}+1}{4} - 1} = \frac{4}{\sqrt{21}-3} = \frac{\sqrt{21}+3}{3} = 2 + \frac{1}{\alpha_4}$$

$$\alpha_4 = \frac{1}{\alpha_3 - 2} = \frac{1}{\frac{\sqrt{21}+3}{3} - 2} = \frac{\sqrt{21}+3}{4} = 1 + \frac{1}{\alpha_5}$$

$$\alpha_5 = \frac{1}{\alpha_4 - 1} = \frac{1}{\frac{\sqrt{21}+3}{4} - 1} = \frac{\sqrt{21}+1}{5} = 1 + \frac{1}{\alpha_6}$$

$$\alpha_6 = \frac{1}{\alpha_5 - 1} = \frac{1}{\frac{\sqrt{21}+1}{5} - 1} = \sqrt{21} + 4 = 8 + \frac{1}{\alpha_7}$$

$$\alpha_7 = \frac{1}{\alpha_6 - 8} = \frac{1}{\sqrt{21}+4-8} = \boxed{\frac{\sqrt{21}+4}{5}} = \alpha_1$$

Thus it will be $\sqrt{21} = 4; \overline{1,1,2,1,1,8}$.

By Lagrange, the period equals 6, then $m = 1 \cdot 6 - 1 = 5$ (odd number) will give us the first convergent fraction, $\frac{P_5}{Q_5}$ among all other fractions:

$$\frac{4}{1}; \frac{5}{1}, \frac{9}{2}, \frac{23}{5}, \frac{32}{7}, \frac{55}{12}$$

the numerator and denominator of which would give us a nontrivial solution to (3.120).

$$x_0 = 55, \quad y_0 = 12, \quad 55^2 - 21 \cdot 12^2 = 1.$$

All other solutions can be obtained from either raising $\alpha = 55 + 12\sqrt{21}$ to consecutive natural powers or by the iteration of the trivial solution $(1,0)$:

$$(x,y) \rightarrow (55x + 252y, 12x + 55y)$$

That will produce all its solutions by the chain:

$$(1,0) \rightarrow (55,12) \rightarrow (6049,1320) \rightarrow \ldots$$

Next, consider another equation,

$$x^2 - 29y^2 = 1 \tag{3.121}$$

and again let us find all its solutions.

$$\sqrt{29} = \underline{5} + \frac{1}{\alpha_1}$$

$$\alpha_1 = \frac{1}{\sqrt{29}-5} = \boxed{\frac{\sqrt{29}+5}{4}} = 2 + \frac{1}{\alpha_2}$$

$$\alpha_2 = \frac{1}{\alpha_1 - 2} = \frac{1}{\frac{\sqrt{29}+5}{4} - 2} = \frac{4}{\sqrt{29}-3} = \frac{\sqrt{29}+3}{5} = \underline{1} + \frac{1}{\alpha_3}$$

$$\alpha_3 = \frac{1}{\alpha_2 - 1} = \frac{1}{\frac{\sqrt{29}+3}{5} - 1} = \frac{5}{\sqrt{29} - 2} = \frac{\sqrt{29}+2}{5} = 1 + \frac{1}{\alpha_4}$$

$$\alpha_4 = \frac{1}{\alpha_3 - 1} = \frac{1}{\frac{\sqrt{29}+2}{5} - 1} = \frac{\sqrt{29}+3}{4} = 2 + \frac{1}{\alpha_5}$$

$$\alpha_5 = \frac{1}{\alpha_4 - 2} = \frac{1}{\frac{\sqrt{29}+3}{4} - 2} = \sqrt{29}+5 = 10 + \frac{1}{\alpha_6}$$

$$\alpha_6 = \frac{1}{\alpha_5 - 10} = \frac{1}{\sqrt{29}+5 - 10} = \boxed{\frac{\sqrt{29}+5}{4}} = \alpha_1$$

Thus,

$$\sqrt{29} = 5; \overline{2, 1, 1, 2, 10}.$$

The period of this continued fraction equals 5, then the first solution to (3.121) must appear at $m = 2 \cdot 5 - 1 = 9$, the smallest odd number. Writing all convergent approaching fractions, we obtain the following:

$$\frac{5}{1}, \frac{11}{2}, \frac{16}{3}, \frac{27}{5}, \frac{70}{13}, \frac{727}{135}, \frac{1524}{283}, \frac{2251}{418}, \frac{3775}{701}, \boxed{\frac{9801}{1820}} \qquad (3.122)$$

It is true that

$$9801^2 - 29 \cdot 1820^2 = 1$$

and that

$$x_0 = 9801, \ y_0 = 1820 \qquad (3.123)$$

is the smallest nontrivial solution. If, for example, we multiply the vector of this solution from the left by the inverse of the matrix B, we obtain the trivial solution $(1,0)$,

$$\begin{pmatrix} 9801 & 52780 \\ 1820 & 9801 \end{pmatrix}^{-1} \cdot \begin{pmatrix} 9801 \\ 1820 \end{pmatrix} = \begin{pmatrix} 1 \\ 0 \end{pmatrix}$$

It is interesting to note that if instead of (3.123) we substitute $x_0 = 70$, $y_0 = 13$, then we would obtain solution to the equation

$$x^2 - 29y^2 = -1 \qquad (3.124)$$

Note that all even powers of $\alpha = 70 + \sqrt{29} \cdot 13$ would produce solutions to (3.121)

$$\left(70 + \sqrt{29} \cdot 13\right)^2 = 9801 + \sqrt{29} \cdot 1820$$

and all odd powers produce consecutive solutions to (3.124), such as

$$\left(70 + \sqrt{29} \cdot 13\right)^3 = 1372210 + \sqrt{29} \cdot 254813,$$
$$1372210^2 - 29 \cdot 254813^2 = -1$$

Moreover, if we start substituting into the left side of (3.121) the numerators and denominators of all possible convergent fractions (3.122), we will obtain different integer numbers on the right-hand side that also change with some periodicity,

$$5^2 - 29 \cdot 1^2 = -4$$
$$11^2 - 29 \cdot 2^2 = 5$$
$$16^2 - 29 \cdot 3^2 = -5$$
$$27^2 - 29 \cdot 5^2 = 4$$
$$70^2 - 29 \cdot 13^2 = -1$$
$$727^2 - 29 \cdot 135^2 = 4$$
$$1524^2 - 29 \cdot 283^2 = -5$$
$$2251^2 - 29 \cdot 418^2 = 5$$
$$3775^2 - 29 \cdot 701^2 = -4$$
$$9881^2 - 29 \cdot 1820^2 = 1$$
$$\cdots$$

The following lemma is true.

> **Lemma 18**
>
> The general solution of the equation,
>
> $$x^2 - N \cdot y^2 = \pm M$$
>
> can be generated by some convergent fraction approaching $\sqrt{N}$.

Note that while a solution to the Pell's equation for any natural non-perfect square N always exists, a solution to the equation

$$x^2 - N \cdot y^2 = -1 \tag{3.125}$$

for $N = 2k$ does not always exist. We can state the following lemma.

> **Lemma 19**
>
> If $N = 2k$, then in order the equation $x^2 - Ny^2 = -1$ to be solvable in integers, it is necessary that N is not a multiple of 4 and that N is not divisible by any prime number in the form $= 4l + 3$.

Proof. If we rewrite the given equation in the equivalent form as

$$x^2 + 1^2 = N \cdot m$$

We can see that the left side of the equation is the sum of two squares, then it is obvious that $N \neq 4t$, and also $N \neq m \cdot p$, $p = 4l + 3$. The second constraint comes from the works of Lagrange and Gauss on representation of a number as the sum of two squares. You can learn more about it in the following chapter.

Remark. This lemma is only a necessary but not a sufficient condition for the equation to be solvable in integer numbers, x and y. For example, consider the even number $N = 34$. Though 34 is not a multiple of 4 and also $34 = 2 \cdot 17$, where $17 = 4 \cdot 4 + 1 \neq 4s + 3$, the corresponding to (3.125) equation, $x^2 - 34y^2 = -1$, does not have any solution in integers. I am not sure if the sufficient condition is found.

3.6 Wilson's Theorem and Equations with Factorials

Problem 201

Solve in natural numbers $x! + 4x - 9 = y^2$.

Solution. Rewriting this equation as follows:

$$x! + 4x = y^2 + 3^2$$

we can see that the right side of it is the sum of two squares. This means that the left side divided by 4 must leave a remainder of 1. See more information in Section 4.2. This can be written as

$$x! + 4x = y^2 + 3^2 = 4k + 1$$

On the left we have a multiple of 4 and factorial of a natural number x, which is also multiple of 4 for all $x \geq 4$ and hence the given equation will not have solutions. Next, we will consider only $x < 4$, $x = 1, 2, 3$.

1) Let $x = 3$

$$3! + 4 \cdot 3 = y^2 + 9$$
$$6 + 12 = 18 = y^2 + 9$$
$$y = 3$$

2) If $x = 2$, we obtain a second solution to the given equation and corresponding $y = 1$.

Answer. $(x, y) = \{(3, 3), (2, 1)\}$

Problem 202

Find integer numbers n and k satisfying the equation $12 \cdot n! + 11^n + 2 = k^2$.

Solution. Let us use the last digit of the left and right sides. We know that if $n \geq 5$, then $n!$ ends in zero. 11^n always ends in 1. Then if $n \geq 5$, $12 \cdot n! + 11^n + 2$ will always end in $0 + 1 + 2 = 3$. On the other hand k^2 cannot end in 3. Next, we will consider only cases when $n < 5$:

$$\begin{aligned} n &= 1, \quad 12 + 11 + 2 = 25 = k^2, \ k = \pm 5 \\ n &= 2, \quad 12 \cdot 2 + 11^2 + 2 = 147 \neq k^2 \\ n &= 3, \quad 12 \cdot 6 + 11^3 + 2 = 1405 \neq k^2 \\ n &= 4, \quad 12 \cdot 4! + 11^4 + 2 = 14931 \neq k^2. \end{aligned}$$

We have a solution only for $n = 1$ and it is $k = \pm 5$.
Answer. $(n = 1, k = \pm 5)$.

Next, I want to offer you a problem from Moscow City Math Olympiad, 2001.

Problem 203

Find the last two digits of the sum:

$$S = 1! + 2! + 3! + \ldots + 2015! + 2016! + 2017!$$

Solution. We know that $5! = 120$ and ends in 0. We also know that $10! = 3,628,800$ ends in two zeros. (We actually did not have to evaluate this number, we need to find only its zeros.) Hence all factorial $n!$ for $n \geq 10$ will end in two zeros! How can we use this fact? We can look at the given sum as the sum of two quantities:

$$(1! + 2! + \ldots + 9!) + [10! + 11! + \ldots + 2016! + 2017!]$$

In order to find the last two digits in the given sum, we simply have to add all factorials inside parentheses, from 1! to 9! that is

$$1! + 2! + \ldots + 9! = 1 + 2 + 6 + 24 + 720 + 5040 + 40320 + 362880 = 4091\underline{13}$$

Therefore S ends in 13.
Answer. 13 are the last two digits.

Problem 204

Find all integer x and y satisfying the equation: $1! + 2! + 3! + \cdots + x! = y^2$.

Solution. We remember that

$$n! = 1 \cdot 2 \cdot 3 \cdot 4 \cdots (n-1) \cdot n.$$

Hence,

- if $x = 1$ then $y^2 = 1$ and $y = \pm 1$. We have $(1, 1)$ and $(1, -1)$.
- if $x = 2$ then $1 + 2 = y^2$ and no solutions in integers.
- if $x = 3$ then $1 + 2 + 6 = 9 = y^2$ and $y = \pm 3$. We obtain $(3, -3)$ and $(3, 3)$.
- if $x = 4$ then $1 + 2 + 6 + 24 = 33 = y^2$ and it has no solutions.
- if $x \geq 5$ then $5! + 6! + 7! + \cdots = 10N$. Since $1! + 2! + 3! + 4! = 33$ ends in 3, $1! + 2! + 3! + \cdots + x!$ will end in a 3. However, there is no perfect square that ends in a 3. Thus, the given equation has no solutions.

Answer. $(1, 1)$, $(1, -1)$, $(3, 3)$, and $(3, -3)$.

Now it is time to introduce Wilson's Theorem that constitutes necessary and sufficient condition for a number to be prime. This theorem was first stated by Wilson but proved by Lagrange (1736–1813). We have the following statement.

Theorem 27 (Wilson's Theorem)

If p is a prime number, then p divides $(p-1)! + 1$, i.e.,

$$(p-1)! \equiv -1 \pmod{p}$$

or

$$((p-1)! + 1) \equiv 0 \pmod{p}$$

In other words, if $(p-1)! + 1$ is not divisible by p, then p is not prime.

First, we will see that this works using the following example. Consider $10!$.

$$10! = 1 \cdot 2 \cdot 3 \cdot 4 \cdot 5 \cdot 6 \cdot 7 \cdot 8 \cdot 9 \cdot 10.$$

Let us investigate how this product is related to division by 11. 11 is bigger than any number of the product. It is clear that

$$1 \equiv 1 \pmod{11}$$
$$10 \equiv 10 \pmod{11}.$$

However, since $10 - 11 = -1$, the last congruence can be written as

$$10 \equiv -1 \pmod{11}.$$

Next, we will consider interesting pairs within this product: $(2, 6)$, $(3, 4)$, $(5, 9)$, $(7, 8)$. The product of each pair divided by 11 leaves a remainder of 1. Thus,

$$12 = (2 \cdot 6) \equiv 1 \quad (\text{mod } 11)$$
$$12 = (3 \cdot 4) \equiv 1 \quad (\text{mod } 11)$$
$$45 = (5 \cdot 9) \equiv 1 \quad (\text{mod } 11) \ (45 = 11 \cdot 4 + 1)$$
$$56 = (7 \cdot 8) \equiv 1 \quad (\text{mod } 11) \ (56 = 11 \cdot 5 + 1)$$
$$1 \equiv 1 \quad (\text{mod } 11)$$
$$10 \equiv -1 \quad (\text{mod } 11).$$

Multiplying the left and the right sides of the congruences, we will obtain the required

$$10! \equiv -1 \quad (\text{mod } 11).$$

Proof. Suppose that a is one of the $(p-1)$ positive integers, i.e.,

$$1, 2, 3, 4, \ldots, (p-2), (p-1) \tag{3.126}$$

and consider the linear congruence

$$ax \equiv 1 \quad (\text{mod } p). \tag{3.127}$$

We want to find a pair for a among all $1, 2, 3, \ldots, (p-1)$ numbers such that their product divided by p leaves a remainder of 1.

Let a' be solution of (3.127), then because p is a prime, then

$$a \cdot a' \equiv 1 \quad (\text{mod } p).$$

If $a = a'$, then $a = 1$ or $a = p - 1$, because

$$a^2 \equiv 1 \quad (\text{mod } p) \text{ or } (a^2 - 1) \equiv 0 \quad (\text{mod } p)$$
$$(a - 1)(a + 1) \equiv 0 \quad (\text{mod } p).$$

From this last congruence it follows that

$$(a - 1) \equiv 0 \quad (\text{mod } p) \ a = 1$$
$$(a + 1) \equiv 0 \quad (\text{mod } p) \ a = p - 1.$$

If we omit the numbers 1 and $(p-1)$ in the sequence (3.126), then we can pair the remaining integers $2, 3, 4, 5, \ldots, (p-3), (p-2)$ into pairs $a \cdot a' \equiv 1$ (mod p). When these $(p-3)/2$ congruences are multiplied together and the factors rearranged, we obtain:

$$(2 \cdot 3 \cdot 4 \cdot \cdots \cdot (p-2)) \equiv 1 \quad (\text{mod } p),$$

or

$$(p-2)! \equiv 1 \quad (\text{mod } p).$$

Multiplying both sides of the last congruence by $(p-1)$

$$(p-2)!(p-1) \equiv (p-1) \cdot 1 \quad (\text{mod } p),$$

or

$$(p-1)! \equiv -1 \pmod{p}.$$

The proof is complete.

Here I offer you some problems on application of Wilson's Theorem.

Problem 205

What is the remainder when 97! is divided by 101?

Solution. Since 101 is prime, then the following is true:

$$100! \equiv -1 \pmod{101}$$
$$97! \cdot 98 \cdot 99 \cdot 100 \equiv -1 \pmod{101}$$

Next we can consider the following true congruences modulo 101:

$$98 \equiv -3 \pmod{101}$$
$$99 \equiv -2 \pmod{101}$$
$$100 \equiv -1 \pmod{101}$$

Multiplying these three congruences and substituting it into our congruence obtained above, we have

$$6 \cdot 97! \equiv 1 \pmod{101}$$

We need to find such multiple of 101 that by adding 1 to it, would become divisible by 6. Since

$$102 = 101 + 1 = 6 \cdot 17,$$

finally we get

$$6 \cdot 97! \equiv (6 \cdot 17) \pmod{101}$$

Dividing both sides of which we obtain the answer:

$$97! \equiv 17 \pmod{101}.$$

Answer. 17.

Problem 206 (Dudley)

Prove that $(2(p-3)! + 1) \equiv 0 \pmod{p}$.

Proof. By Theorem 27

$$(p-1)! \equiv -1 \pmod{p}, \qquad (3.128)$$

which can be written as

$$(p-1)(p-2)(p-3)! \equiv -1 \pmod{p}.$$

On the other hand we can state that

$$(p-1) \equiv -1 \pmod{p}, \tag{3.129}$$

$$(p-2) \equiv -2 \pmod{p}, \tag{3.130}$$

or

$$(p-1)(p-2) \equiv 2 \pmod{p},$$

which can be written as

$$(p-1)(p-2) = kp+2. \tag{3.131}$$

Substituting (3.131) into (3.128) we have

$$(kp+2)(p-3)! \equiv -1 \pmod{p}$$
$$(kp(p-3)! + 2(p-3)!) \equiv -1 \pmod{p}.$$

Since the first term of the last equation is a multiple of p, then it can be written as

$$2(p-3)! \equiv -1 \pmod{p},$$

or

$$(2(p-3)! + 1) \equiv 0 \pmod{p}.$$

Remark. Our solution can be much shorter if we substitute (3.129) and (3.130) into (3.128):

$$(-1)(-2)(p-3)! \equiv -1 \pmod{p}$$
$$2(p-3)! \equiv -1 \pmod{p}.$$

Problem 207

Find the least nonnegative residue of 70! $\pmod{5183}$.

Solution. Since $5183 = 71 \cdot 73$, we will use Theorem 4 along with Theorem 27. On one hand,

$$70! \equiv -1 \pmod{71}. \tag{3.132}$$

On the other hand, with the use of the previous problem the following is true:

$$2 \cdot 70! \equiv -1 \pmod{73},$$

or

$$2 \cdot 70! \equiv 72 \quad (\text{mod } 73)$$
$$70! \equiv 36 \quad (\text{mod } 73). \tag{3.133}$$

From (3.132) we have that

$$70! = 71n - 1. \tag{3.134}$$

Let us combine (3.133) and (3.134) and apply Theorem 4:

$$(71n - 1) \equiv 36 \quad (\text{mod } 73)$$
$$71n \equiv 37 \quad (\text{mod } 73)$$
$$(73n - 2n) \equiv 37 \quad (\text{mod } 73)$$
$$-2n \equiv 37 \quad (\text{mod } 73)$$
$$2n \equiv -37 \quad (\text{mod } 73) \tag{3.135}$$
$$2n \equiv 36 \quad (\text{mod } 73)$$
$$n \equiv 18 \quad (\text{mod } 73).$$

Substituting (3.135) into (3.134) we obtain that

$$70! = 71(73k + 18) - 1 = 5183k + 1277,$$

or

$$70! \equiv 1277 \quad (\text{mod } 5183).$$

Answer. 1277.

The following problem shows how Wilson's theorem can be combined with the Euclidean algorithm.

Problem 208

Find the remainder of 53! divided by 61.

Solution. By Wilson's Theorem

$$60! \equiv -1 \, (mod \, 61).$$

Method 1. Extracting 53!, this can be rewritten as

$$60 \cdot 59 \cdot 58 \cdot 57 \cdot 56 \cdot 55 \cdot 54 \cdot 53! \equiv -1 \, (mod \, 61)$$
$$(-1)(-2)(-3)(-4)(-5)(-6)(-7) \cdot 53! \equiv -1 \, (mod \, 61)$$
$$5040 \cdot 53! \equiv 1 \, (mod \, 61)$$
$$38 \cdot 53! \equiv 1 \, (mod \, 61)$$

In order to reduce this congruence and to divide both sides by 38, we will apply the Euclidean algorithm:

$$
\left.\begin{array}{l}
61 = 38 \cdot 1 + 23 \\
38 = 23 \cdot 1 + 15 \\
23 = 15 \cdot 1 + 8 \\
15 = 8 \cdot 1 + 7 \\
8 = 7 \cdot 1 + 1
\end{array}\right\} \Rightarrow
$$

Working backward, we obtain

$$
\begin{aligned}
1 &= 8 - 7 = 8 - (15 - 8) \\
&= 2 \cdot 8 - 15 = 2\,(23 - 15) - 15 \\
&= 2 \cdot 23 - 3 \cdot 15 = 2 \cdot 23 - 3\,(38 - 23) \\
&= 5 \cdot 23 - 3 \cdot 38 = 5\,(61 - 38) - 3 \cdot 38 \\
&= 5 \cdot 61 - 8 \cdot 38.
\end{aligned}
$$

Now we know that $-8 \cdot 38 = 1 - 5 \cdot 61$ and that the congruence with the factorial can be reduced by common factor of 38 as follows:

$$
53! \equiv -8 \pmod{61} \equiv 53 \pmod{61}.
$$

Method 2. We can simplify the congruence step by step as follows:

$$
\begin{aligned}
53! \cdot 2 \cdot 3 \cdot 4 \cdot 5 \cdot 6 \cdot 7 &\equiv 1 \pmod{61} \\
53! \cdot 3 \cdot 4 \cdot 7 \cdot 5 \cdot 12 &\equiv -60 \pmod{61} \\
53! \cdot 7 \cdot 4 \cdot 3 &\equiv -1 \pmod{61} \\
53! \cdot 7 \cdot 12 &\equiv 60 \pmod{61} \\
53! \cdot 7 &\equiv 5 \pmod{61} \\
53! \cdot 7 &\equiv -56 \pmod{61} \\
53! &\equiv -8 \pmod{61} \\
53! &\equiv 53 \pmod{61}.
\end{aligned}
$$

Therefore 53! divided by 61 leaves a remainder of 53.
Answer. 53.

Chapter 4
Pythagorean Triples, Additive Problems, and More

Additive problems, representing an integer number as a sum of other numbers, their squares, or cubes, have interested people for centuries. Ancient Greeks knew that every triangular number can be written by the sum of natural numbers from 1 to n,

$$
\begin{aligned}
1+2 &= 3 & &= T(2) \\
1+2+3 &= 6 & &= T(3) \\
1+2+3+4 &= 10 & &= T(4)
\end{aligned}
$$

$$\cdots$$

$$1+2+3+\cdots+n = \frac{n\cdot(n+1)}{2} = T(n)$$

Pythagoras played with the sides of a right triangle and discovered the algorithm to find all (a,b,c) triples of its sides that satisfy,

$$a^2 + b^2 = c^2$$

Some methods of finding such triples rely on pure arithmetic and were known to the Babylonians in 2000 BC. More modern methods rely on algebraic geometry and are based on finding all rational points on the unit circle or a sphere (in the case of quadruples, $a^2 + b^2 + c^2 = d^2$).

It is worth mentioning that algebraic geometry has its roots in the work of Diophantus and later Descartes and Fermat. For example, in *Arithmetica*, the greatest book of the ancient algebraists, Diophantus beautifully solves the problem of finding rational solutions to the following equation:

$$x^2 - y^2 = 1$$

and obtains parameterized formulae for all its solutions. This problem will be offered to you in this Chapter.

In general, the problem of the existence of at least one rational point on a curve of the second degree turned out to be very difficult. The first nontrivial advances

The original version of this chapter was revised: Belated correction has been incorporated. The erratum to this chapter is available at https://doi.org/10.1007/978-3-319-90915-8_6

© Springer International Publishing AG, part of Springer Nature 2018   245
E. Grigorieva, *Methods of Solving Number Theory Problems*,
https://doi.org/10.1007/978-3-319-90915-8_4

toward its solution were in the papers of the Indian mathematicians Brahmagupta
(VII century) and Bhaskara (XII century). Finally, a full solution was found much
later, in 1768, by the French mathematician J.-L. Lagrange (1736–1813).

Diophantus did not restrict himself to equations of the second degree. He successfully took up the third degree, demonstrating his general method by solving, for
example, this equation

$$y^2 - x^3 = \pm 1$$

To my knowledge, the general case of this problem, $y^2 - x^3 = c$ is still unsolved.
In this chapter, you will have a chance to prove that for a particular case (when
$c = 7$), this equation does not have a solution in natural numbers.

And of course, the most famous problem is Fermat's Last Theorem that

$$a^n + b^n = c^n$$

does not have any solution in integers for $n > 2$. In 1637 Pierre Fermat stated his
theorem as a marginal note of his copy of Diophantus' *Arithmetica*, and we all know
that unfortunately the margin was too small to write down his beautiful proof.

Fermat's Last Theorem can be also rewritten as: There is no rational solution to
the equation $x^n + y^n = 1$, $n > 2$, $n \in \mathbb{N}$. Euler proved Fermat's Last Theorem for
$n = 3$. A. Legendre and I. Dirichlet and Sofie Germen found a proof for $n = 5$.
Though this theorem was announced proven by Wiles in 1995 on 130 pages, many
mathematicians are still looking for a "beautiful" proof.

While the representation of a square by the sum of two or three other squares is
completely solved, other questions arise. Can a number that is not a perfect square
be written as a sum of squares? Pierre Fermat stated that any number is either a
triangular number; or the sum of two, three, or four triangular numbers; or the sum
of two, three, or four squares; or a pentagonal number; or the sum of five pentagonal
numbers. There is a proof for the fact that every number is a square or can be written
as the sum of two, three, or four squares in the work of Euler and Lagrange. The
general proof was given later by French mathematician A. Cauchy (1789–1857).

In 1770, the English mathematician Waring formulated without proof that any
positive integer $N > 1$ can be represented as a sum of the nth powers of k positive
integers, i.e., as $N = x_1^n + x_2^n + \cdots + x_k^n$; the number of terms k depends only on n.
There is more information about the Waring problem in Section 4.2. The special
significance of Waring's problem is that during its investigation, powerful methods
of analytic number theory have been created.

Many contest problems continue to ask for the decomposition of numbers into
squares, cubes, or other natural powers. I remember from my own Olympiad experience that I was asked to prove that any natural number can be written as the sum
of cubes of five integers. I will give you the outline of my ideas. Consider the sum
of two cubes,

$$(n+1)^3 + (n-1)^3 = 2n^3 + 6n$$

This can be easily rewritten as

$$6n = (n+1)^3 + (n-1)^3 + (-n)^3 + (-n)^3,$$

which means that any multiple of 6 is the sum of four cubes. We can always add 0 to the sum, hence, a multiple of 6 is the sum of five cubes. Since all other numbers can be written as

$$6n \pm 1, \; 6n \pm 2, \; 6n + 3$$

and working with the obtained representation, we can show that any number can be written by the sum of five cubes. For example, if a number like 31 divided by 6 leaves a remainder of 1, then the following is true:

$$6n + 1 = 1^3 + (n+1)^3 + (n-1)^3 + (-n)^3 + (-n)^3.$$

This way you can represent many prime numbers by the sum of five cubes. The complete solution to this and many other interesting additive problems, such as the following equation in fourteen variables:

$$x_1^4 + x_2^4 + \cdots + x_{14}^4 = 1000000000002015.$$

will be found here. This Chapter is also devoted to some basic concepts, such as quadratic residues, and Legendre and Jacobi symbols. It is recommended to all college students taking an introductory course in number theory.

4.1 Finding Pythagorean Triples. Problem Solved by Ancient Babylonians

In this section, we will learn how to solve the equation

$$x^2 + y^2 = z^2 \tag{4.1}$$

in order to find Pythagorean Triples. Rewriting a square of a number as the sum of squares of two other numbers was addressed and solved in ancient Babylon more than 1000 years BC. When clay tablets were found, it was also discovered that the Babylonians gave only x and z values in their tables. However, the largest Babylonian triple was $x = 12709$, $y = 13500$, $z = 18541$ where $y = 13500 = 2^2 \cdot 3^3 \cdot 5^3$ but neither x nor z are divisible by 2, 3, or 5.

The ancient Greeks solution was based on geometric interpretation of the problem and consideration of a right triangle with sides x, y, and hypotenuse z. The method of generating new triples was given by Euclid, Book X, Prop. 28, 29.

It is clear from the geometric viewpoint that if there is a triple (x, y, z) that works, then if we multiply each number (side length) by the same integer factor then next triple will also be generated.

$$(kx)^2 + (ky)^2 = (kz)^2.$$

Ancient Babylonians also revealed that sides x and y cannot be both odd and that the two numbers must either be both even or one number even and the other odd. Let us recall, for example, $3^2 + 4^2 = 5^2$. If you take and multiply each number

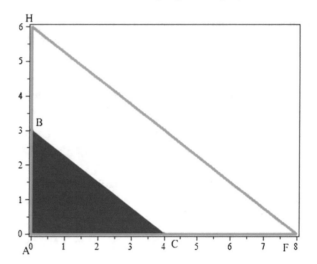

Fig. 4.1 Pythagorean Triples

by 2, we will get the sum of the squares of the two even numbers, $x = 6$ and $y = 8$.
Multiplying the original triple by 3, we will get 12 and 9, respectively, etc.

All these can be seen in the Figure 4.1 and explained by Thales Theorem and
relationships between the sides in a right triangle.

How do you prove these allegations from the modern mathematics point of view?
Let us do it using three different methods, an arithmetic, algebraic geometry, and
trigonometry.

4.1.1 Method 1. An Arithmetic Approach.

Ancient Greeks were looking for the smallest triangle with the sides satisfying the
equation. Algebraically this can be written as the following statement.

Statement 1.   The smallest solution to $x^2 + y^2 = z^2$ is a such that
$(x,y) = 1$, $(x,z) = 1$, $(y,z) = 1$.

Proof. Assume that the greatest common divisor of x and y exists and it is $(x,y) = p$,
then $x = x_1 \cdot p$, $y = y_1 \cdot p$. After substitution into the equation, we obtain $z^2 = p^2 \cdot (x_1^2 + y_1^2)$, i.e., $z = p \cdot z_1$. Then if two numbers are divisible by p, then they all are
divisible by p and this triple will not be the smallest. Therefore all variables must
be relatively prime.

Can both sides be odd numbers?

Statement 2. Assume that $x = 2k + 1$ and $y = 2n + 1$. Then $x^2 + y^2 = z^2$ has no
solution in integers.

Proof. $x^2 = 4k^2 + 4k + 1$ and $y^2 = 4n^2 + 4n + 1$, then $z^2 = 4 \cdot m + 2$, which means
that if such square exists, then divided by 4 it must leave a remainder of 2. However,
all squares divided by 4 must give only 0 (if it is a square of an even number) or 1,
and never 2. Therefore two odd numbers cannot be Pythagorean squares.

Obviously, Pythagorean triples can be all even but they will not represent the smallest triangle (solution).

Suppose that $y = 2 \cdot y_0$, an even number, and x is odd. Substituting this into (4.1), we obtain

$$4 \cdot (y_0)^2 = z^2 - x^2$$

or factor the right-hand side as follows:

$$4 \cdot (y_0)^2 = (z - x) \cdot (z + x)$$

Obviously, both factors on the right are even numbers, hence each of them can be divided by 2,

$$(y_0)^2 = \frac{z - x}{2} \cdot \frac{z + x}{2} \qquad (4.2)$$

Let us give this formula consideration. First, we can demonstrate that two factors on the right-hand side are relatively prime, so their greatest common divisor equals 1, i.e., $\left(\frac{z-x}{2}, \frac{z+x}{2}\right) = 1$.

We will prove it by contradiction. Assume that $\frac{z-x}{2} = n \cdot p$, $\frac{z+x}{2} = m \cdot p$, i.e., both are divisible by p. Then their sum $\left(\frac{z+x}{2} + \frac{z-x}{2}\right) = z = p \cdot t$ and their difference $\left(\frac{z+x}{2} - \frac{z-x}{2}\right) = x = p \cdot t$ are both divisible by p, that is contradiction.

Statement 3. Given $a^2 = b \cdot c$, $(b, c) = 1$, then b and c are also perfect squares, such as $b = \beta^2$, $c = \gamma^2$, and $a^2 = \beta^2 \cdot \gamma^2$.

Proof. If

$$a = p_1^{\alpha_1} \cdot p_2^{\alpha_2} \cdot \ldots \cdot p_n^{\alpha_n}$$

then its square can be written as

$$a^2 = p_1^{2\alpha_1} \cdot p_2^{2\alpha_2} \cdot p_3^{2\alpha_3} \cdot \ldots \cdot p_n^{2\alpha_n} = b \cdot c.$$

Assume that, for example, p_1 is a divisor of b. Obviously since $(b, c) = 1$, then p_1 cannot be divisor of c, then the entire $p_1^{2\alpha_1}$ must be divisor of b. Next, assume that p_2 is divisor of c, then using analogous approach we conclude that c must be divisible by the entire factor $p_2^{2\alpha_2}$, etc. Therefore we can state that $b = \beta^2$, $c = \gamma^2$. The proof is complete.

Hence using equation (4.2), we can denote

$$\frac{z - x}{2} = n^2, \quad \frac{z + x}{2} = m^2$$

From the formula above, by adding subtracting and multiplying the introduced quantities, we can write down the algorithm of finding all three variables-solutions of the equation (4.1):

$$\begin{cases} y = 2 \cdot y_0 = 2 \cdot m \cdot n \\ x = m^2 - n^2 \\ z = m^2 + n^2. \end{cases}$$

You can easily check the validity of these formulas. Substituting the formulas for x and y into the left side of the equation, we obtain the right-hand side:

$$(m^2 - n^2)^2 + (2mn)^2 = m^4 - 2m^2n^2 + n^4 + 4m^2n^2 = (n^2 + m^2)^2.$$

These formulas were known to ancient Babylonians and you can practice finding Pythagorean triples by solving the following problem.

Problem 209

Prove that all positive solutions of the equation $a^2 + b^2 = c^2$ can be obtained by formulas $a = 2uv$, $b = u^2 - v^2$, $c = u^2 + v^2$, where u, v are relatively prime.

Proof. It is obvious that

$$(2uv)^2 + (u^2 - v^2)^2 = (u^2 + v^2)^2.$$

There are infinitely many Pythagorean triples. For example, $65^2 = 56^2 + 33^2$ or $41^2 = 40^2 + 9^2$.

Some of them can be found in the Table 2.2. Here $a = 2uv$, $b = u^2 - v^2$, $c = u^2 + v^2$, and $(u, v) = 1$.

u	v	a	b	c
2	1	4	3	5
3	2	12	15	13
4	1	8	15	17
4	3	24	7	25
5	2	20	21	29
5	4	40	9	41
6	1	12	35	37
6	5	60	11	61
7	4	56	33	65

Table 4.1 Pythagorean Triples

The Pythagoras algorithm is simple: 1) Take any pair of relatively prime numbers, for example, $u = 5, v = 2$; 2) Using the formulas above calculate $a = 2 \cdot 2 \cdot 5 = 20$, $b = u^2 - v^2 = 5^2 - 2^2 = 21$, and $c = u^2 + v^2 = 5^2 + 2^2 = 29$. 3) Finally, $20^2 + 21^2 = 29^2$.

Problem 210

Find all right triangles with integer sides and one leg of 45.

Solution. Let one side be 45. The following is true

$$b = u^2 - v^2 = (u - v) \cdot (u + v) = 45 = 1 \cdot 45 = 3 \cdot 15 = 5 \cdot 9.$$

We can find the values of u and v by solving three systems:

1. $\begin{cases} u - v = 1 \\ u + v = 45 \end{cases} \Rightarrow u = 23,\ v = 22 \Rightarrow a = 2 \cdot 23 \cdot 22 = 1012,\ c = 23^2 + 22^2 = 1013$

$$\boxed{1012^2 + 45^2 = 1013^2}$$

2. $\begin{cases} u - v = 3 \\ u + v = 15 \end{cases} \Rightarrow u = 9,\ v = 6 \Rightarrow a = 2 \cdot 9 \cdot 6 = 108,\ c = 9^2 + 6^2 = 117$

$$\boxed{108^2 + 45^2 = 117^2}$$

3. $\begin{cases} u - v = 5 \\ u + v = 9 \end{cases} \Rightarrow u = 7,\ v = 2 \Rightarrow a = 2 \cdot 7 \cdot 2 = 28,\ c = 7^2 + 2^2 = 53$

$$\boxed{28^2 + 45^2 = 53^2}$$

Answer. $(45, 1012, 1013), (45, 108, 117), (45, 28, 53)$.

4.1.2 Method 2. Using Algebraic Geometry

Consider again equation (4.1) and divide its both sides by z^2.

$$\frac{x^2}{z^2} + \frac{y^2}{z^2} = 1,$$

which can be written as

$$p^2 + q^2 = 1, \quad p = \frac{x}{z}, \quad q = \frac{y}{z}, \quad p, q \in Q. \tag{4.3}$$

Let us think about this equation. It describes all points on the unit circle. But if (p, q) are rational numbers, then the equation describes all rational points on the unit circle.

How can we find all these points? Clearly, if we find all such points, we will have the solution to the problem. There are four points on the unit circle that are the solutions, $(0, 1)$, $(-1, 0)$, $(0, -1)$, and $(1, 0)$. Let us connect point $(0, -1)$ with any rational point (p, q) satisfying the equation.

Can we find its equation? Yes. Consider the line as $y = kx + b$. The Y-intercept, $b = -1$, and the slope of the line are also rational:

$$k = \frac{q + 1}{p}$$

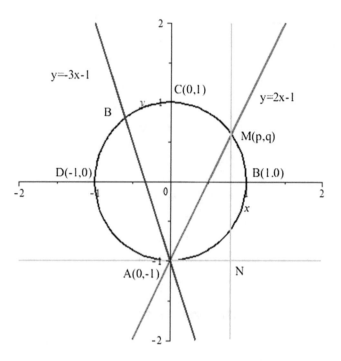

Fig. 4.2 Pythagorean Triples

Property 26

The line $y = kx - 1$ that passes through point $(0, -1)$ and another rational point of the unit circle (p, q) will always have rational slope $k \in Q$.

Property 27

Any rational point (p, q) on the unit circle can be obtained by intersection of the circle with the line through point $(0, -1)$ and rational slope.

Property 28

Any line $y = kx - 1$, where $k \in Q$ intersects the unit circle $x^2 + y^2 = 1$ in a rational point.

Because a point (p, q) is on the line and the circle simultaneously, we have the following:

$$y = kx - 1, \quad (x,y) = (x, kx - 1)$$

$$p^2 + (k \cdot p - 1)^2 = 1$$
$$p^2 + k^2 p^2 - 2kp + 1 = 1$$
$$p^2 + k^2 p^2 - 2kp = 0, \quad p \neq 0$$
$$p(k^2 + 1) = 2k$$
$$p = \frac{2k}{k^2+1} \in Q.$$

For the second coordinate, q, we will obtain

$$q = \frac{2k \cdot k}{k^2 + 1} - 1 = \frac{k^2 - 1}{k^2 + 1}.$$

Because as it was shown earlier, k is rational number, assume that $k = \frac{m}{n}$.

$$p = \frac{x}{z} = \frac{2mn}{n^2+m^2}$$
$$q = \frac{y}{z} = \frac{m^2-n^2}{m^2+n^2}$$

Finally, the formulas for (x,y,z) are given by formulas

$$\begin{aligned} k &= m/n \\ x &= 2mn \\ y &= m^2 - n^2 \\ z &= m^2 + n^2. \end{aligned} \tag{4.4}$$

Now, any Pythagorean triple can be generated using these formulas. Moreover, we can just choose any line with a rational slope and the Y intercept of -1, and for each such a line the formula above will generate a new Pythagorean triple. Assume that $y = 2x - 1$, then $k = 2$, $m = 2$, and $n = 1$, then the corresponding Pythagorean triple is $x = 4$, $y = 3$, and $z = 5$. Assume that the line equation is $y = \frac{17}{23}x - 1$, then $k = 17/23$, $m = 17$, and $n = 23$, and the corresponding Pythagorean triple is

$$x = 2 \cdot 17 \cdot 23 = 782, \ y = 240, \ z = 17^2 + 23^2 = 818.$$

You can check that this triple works and that $782^2 + 240^2 = 818^2 = 669124$.

This method of algebraic geometry proved the Pythagorean triple formula without usage of any divisibility properties. We proved that all rational points of the unit circle can be described by a line $y = kx - 1$, $k \in Q$ that passes through point $(0, -1)$ and the rational point (p,q) on the unit circle. Instead of solving the equation $x^2 + y^2 = z^2$ in integers, algebraic geometry approach solves $p^2 + q^2 = 1$ and demonstrates that all solutions to the Pythagorean problem can be obtained by drawing a line through point $(0, -1)$ and any other rational point of the unit circle! Start from any line $y = kx - 1$, $k \in Q$.

Example. For example, take line $y = (\frac{5}{8})x - 1$. Since $k = 5/8$, then evaluate $m = 5$ and $n = 8$. Evaluate $x = 2mn = 2 \cdot 5 \cdot 8 = 80$, $y = m^2 - n^2 = 5^2 = 8^2 = -39$, and $z = m^2 + n^2 = 25 + 64 = 89$.

Finally, we obtain the correct formula:

$$80^2 + 39^2 = 89^2.$$

Algebraic geometry is a new field of mathematics that has roots in the work of Diophantus who, for example, offered a beautiful and correct solution to the following problem.

Problem 211

Find all rational solutions to the following equation:

$$x^2 - y^2 = 1 \tag{4.5}$$

Solution. Equation (4.5), like any equation of the variables x, y, can be considered as a curve in the XOY plane. In this case, this is a hyperbola (Figure 4.3). An obvious solution $(1,0)$—is the intersection point A of the curve with the X axis. Let's draw a secant line through this point.

$$y = k(x - 1) \tag{4.6}$$

and find its second point of intersection with the curve (4.5), a point M.

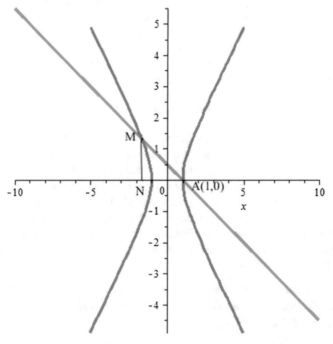

Fig. 4.3 Rational Points on Hyperbola

To do this, substitute expression (4.6) for y into equation (4.5) and solve the resulting quadratic equation with respect to x. We get

$$x^2 - k^2(x^2 - 2x + 1) = 1$$
$$(1 - k^2)x^2 + 2k^2 x - (k^2 + 1) = 0$$
$$x_{1,2} = \frac{-k^2 \pm \sqrt{k^4 + (1-k^2)(k^2+1)}}{1-k^2}$$
$$x_{1,2} = \frac{-k^2 \pm 1}{1-k^2}$$

For any rational k this formula defines a point on our curve, and hence a rational solution of the given equation. (For $k = \pm 1$, the secant crosses the curve only at the point A (see Figure 4.3). Conversely, for any rational solution, that is, a rational point M on the curve, the secant MA is given by the equation (4.6) with rational k (for which the legs of the right triangle ANM are rational).

$$(x_k, y_k) = \left(\frac{k^2 + 1}{k^2 - 1}, \frac{2k}{k^2 - 1} \right) \qquad (4.7)$$

Thus, formula (4.7) for all rational

$$k \neq \pm 1$$

gives all solutions in rational numbers of equation (4.5).

For example, for $k = -2$, the equation of the secant will be $y = -2(x - 1)$ and the coordinates of point M - $x = \frac{5}{3}, y = -\frac{4}{3}$.

Answer. $(x_k, y_k) = \left(\frac{k^2+1}{k^2-1}, \frac{2k}{k^2-1} \right), k \neq \pm 1, k \in \mathbb{Z}$

Remark. Diophantus himself, of course, did not introduce the coordinate system into consideration and did not consider the curve of the given equation. The coordinate method appeared only in the works of Descartes in the 17th century. Diophantus made the substitution (4.6) in a purely algebraic way, and, of course, he found the formula (4.7) independently.

4.1.3 Method 3. Using Trigonometry

Let us again look for the rational points on the unit circle:

$$x^2 + y^2 = 1. \qquad (4.8)$$

We know that point $C(-1,0)$ satisfies this equation. Assume that the center of the circle is point $O(0,0)$ and that there is another point on the circle, A the rational coordinates of which are also solutions to the equation. (See Figure 4.4) Let us drop a perpendicular AB to the X axis. Connect points O and A and let side CA intersect the Y axis at point D. Assume that $\angle AOB = \alpha$, then $AB = \sin \alpha$ and $OB = \cos \alpha$. Additionally because the triangle COA is isosceles, and its vertex angle is supplement to α, then $\angle ACO = \angle CAO = \alpha/2$.

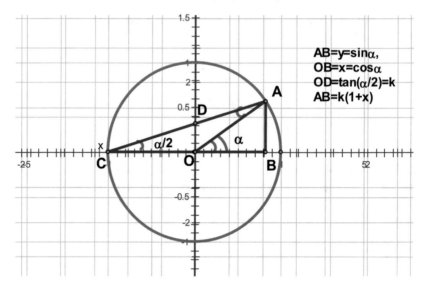

Fig. 4.4 Pythagorean Triples. Trigonometric Approach

The sine and cosine of angle α can be expressed in terms of the tangent of half angle $\frac{\alpha}{2}$:

$$\sin \alpha = \frac{2\tan(\alpha/2)}{1+\tan^2(\alpha/2)}, \quad \cos \alpha = \frac{1-\tan^2(\alpha/2)}{1+\tan^2(\alpha/2)}$$

If for any reason you do not remember these formulas, we can easily derive them. Denote $\tan(\alpha/2) = k$. From the picture, $AB = y = \sin(\alpha)$, $OB = x = \cos(\alpha)$, and $y = CB \cdot \tan(\alpha/2) = (1+x)k$. Substituting these expressions into the circle equation we obtain

$$k^2(1+x)^2 + x^2 = 1$$
$$k^2 + 2k^2 x + k^2 x^2 + x^2 - 1 = 0$$
$$(1+k^2) \cdot x^2 + 2k^2 \cdot x + (k^2 - 1) = 0.$$

The last row above is a quadratic equation in x. We will solve it using $D/4$ formula:

$$x = \frac{-k^2 \pm \sqrt{k^4 - (k^2-1)(k^2+1)}}{k^2+1}$$
$$x = \frac{-k^2 \pm 1}{k^2+1}$$

We can take only one solution with plus and also evaluate the corresponding y:

$$x = \frac{1-k^2}{1+k^2}, \quad y = \frac{2k}{k^2+1}.$$

Since $y = k(1+x)$, then $k = \frac{y}{1+x}$, hence for any rational (x,y), the value of k (slope and the tangent value) is also rational and vice versa. Let $k = \frac{u}{v}$, the following formulas are valid:

$$x = \frac{v^2 - u^2}{v^2 + u^2}, \quad y = \frac{2uv}{v^2 + u^2} \tag{4.9}$$

Substituting formulas for x and y into (4.8) and multiplying the both sides by the denominator, we obtain the required formula for Pythagorean triples,

$$(v^2 - u^2)^2 + (2uv)^2 = (v^2 + u^2)^2.$$

Remark. We just learned how to generate Pythagorean triples. Using formula (4.4) we can find simultaneously three variables x, y, z satisfying (4.1). If you make a table of the sum of many squares, you could notice that no sum of two squares can, for example, generate a perfect square of 11 or 121, so you would never decompose $23^2 = 529$ as sum of two other perfect squares. On the other hand, $17^2 = 289 = 15^2 + 8^2$ easily. Why can some numbers be represented by the sum of two other squares but many, such as 121 or 529 cannot? Please, do not think that only perfect square of a prime can be written by sum of two other squares. In fact, you can check that the square of a composite number 34 can be represented as sum of two other squares as $34^2 = 30^2 + 16^2$. Can we explain why? In order to find the answer, you must be looking for the remainders that each of the numbers z leaves when divided by 4. Thinking a little more, you could find a pattern that Fermat found 400 years ago: if a number or any of its factors divided by 4 leaves a remainder of 3, then such a number cannot be written by the sum of two squares. A more detailed explanation of this fact can be found in the Waring's problem section of the book.

4.1.4 Integer Solutions of $a^2 + b^2 + c^2 = d^2$.

Using ideas of algebraic geometry and stereographic projection, we can find all integer solutions to the equation:

$$a^2 + b^2 + c^2 = d^2.$$

Dividing both sides of the equation by d^2, we will obtain the following equivalent equation:

$$\left(\frac{a}{d}\right)^2 + \left(\frac{b}{d}\right)^2 + \left(\frac{c}{d}\right)^2 = 1.$$

Similarly to the previous section, we can introduce new variables

$$x = \frac{a}{d}, \ y = \frac{b}{d}, \ z = \frac{c}{d} \tag{4.10}$$

And rewrite the given equation as follows:

$$x^2 + y^2 + z^2 = 1. \tag{4.11}$$

The beauty of the problem now is that we will look for the rational solutions to this equation that can be seen as finding rational points on the unit sphere. Consider such a point P' on the unit sphere. If we connect the north pole $N = (0,0,1)$ and point P', this line will intersect the tangent plane at unique point P. (Figure 4.5)

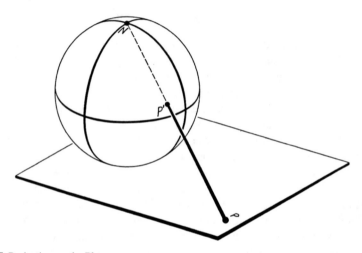

Fig. 4.5 Projection on the Plane

Instead, we can intersect this line with the plane that is parallel to the previous plane but passes through the center of the sphere (XOY) (at point A on Fig. 4.6)

Method 1.

1) We know that point $N = (0,0,1)$ satisfies the equation. Assume that we find another rational point on the sphere $P(x,y,z)$, then the line through points N and P will intersect the plane XOY in some point $A(u,v,0)$ with rational coordinates, $ON = 1$. Let its coordinates satisfy the equation:

$$u^2 + v^2 = r^2,$$

where $r = OA$, and r is some rational number.

2) Let us make an additional construction and drop perpendicular PB from P to line OA, where B is the foot of the perpendicular. Additionally, we will drop perpendicular BM in the plane XOY to the X axis and perpendicular BQ to the Y axis. Obviously $OM = x$ and $OQ = y$. Consider **the cross-section of the sphere** by the plane NOA shown in the Figure 4.6.

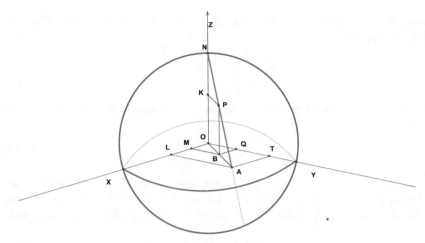

Fig. 4.6 Stereographic Projection from North Pole

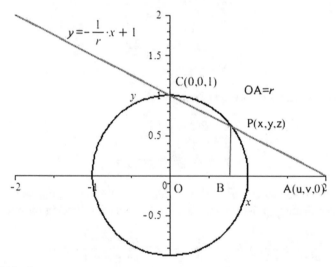

Fig. 4.7 Projection

We can find the equation of the line that intersects with the vertical axis (the Y intercept) of 1 and the X intercept of r. (Figure 4.7) These are not actual intercepts. Because PB segment in the graph is actually z coordinate of the unknown point P with rational coordinates. (Figure 4.6) Finding the equation of the line connecting points N and A is not difficult.

$$y = kx + 1$$
$$0 = kr + 1, \ k = -1/r$$
$$y = -\tfrac{1}{r} \cdot x + 1$$

Since point P is on the unit circle, then its coordinates must satisfy the circle equation. We obtain:

$$x^2 + (-\tfrac{1}{r} \cdot x + 1)^2 = 1$$
$$x + -\tfrac{x}{r^2} - \tfrac{2}{r} = 0$$
$$x(1 + \tfrac{1}{r^2}) - \tfrac{2}{r} = 0.$$

From which we obtain formulas for the $\tilde{x}$ and $\tilde{y}$ coordinates of point P in the plane NOA:

$$\tilde{x} = OB = \tfrac{2r}{r^2+1}$$
$$\tilde{y} = PB = z = \tfrac{r^2-1}{r^2+1}. \tag{4.12}$$

3) The formulas above are not the answer. We need to find the actual coordinates of point P, $x = OM$, and $y = ON$. Using Figure 4.6, consider triangles OBM and OAL. The triangles are similar, so the ratio of the corresponding sides is the same:

$$\triangle OBM \sim \triangle OLA$$
$$\tfrac{OM}{OL} = \tfrac{OB}{OA}$$
$$\tfrac{x}{u} = \tfrac{OB}{r}$$
$$x = \tfrac{OB \cdot u}{r}$$
$$x = OL \cdot \tfrac{OB}{OA} = \tfrac{2r \cdot u}{(r^2+1) \cdot r}$$
$$x = \tfrac{2u}{u^2+v^2+1}.$$

Similarly, we can write for other pair of sides:

$$\triangle OBM \sim \triangle OLA$$
$$\tfrac{MB}{LA} = \tfrac{OB}{OA}$$
$$y = LA \cdot \tfrac{OB}{OA} = \tfrac{2v}{u^2+v^2+1}.$$

Above we used the fact that $u^2 + v^2 = r^2$. Since $z = BP = \tfrac{r^2-1}{r^2+1}$, the rational parameterized coordinate of point P on the unit sphere has the following formulas:

$$x = \frac{2u}{u^2+v^2+1}, \quad y = \frac{2v}{u^2+v^2+1}, \quad z = \frac{u^2+v^2-1}{u^2+v^2+1}, \quad u,v \in Z. \tag{4.13}$$

Formulas allow us to find all possible rational points on the unit sphere and simultaneously to find solutions (a,b,c,d) to the equation in integers. The algorithm is very simple now. Choose u and v pair, evaluate corresponding to it (x,y,z) by the

(u,v)	(x,y,z)	(a,b,c,d)
$(1,1)$	$(\frac{2}{3},\frac{2}{3},\frac{1}{3})$	$(2,2,1,3)$
$(2,5)$	$(\frac{2}{15},\frac{5}{15},\frac{14}{15})$	$(2,5,14,15)$
$(17,3)$	$(\frac{34}{299},\frac{6}{299},\frac{297}{299})$	$(34,6,297,299)$
$(2,11)$	$(\frac{2}{63},\frac{11}{63},\frac{62}{63})$	$(2,11,62,63)$

Table 4.2 $a^2+b^2+c^2=d^2$

formulas above and finally reduce fraction so they would have the same denominator and record (a,b,c,d). The Table 4.2 contains several solutions to the given equation. Let us explain how we obtained, for example, the last row of the table. Let $u=2, v=11$, then

$$x=\frac{4}{2^2+11^2+1}=\frac{4}{126}=\frac{2}{63}$$

$$y=\frac{22}{2^2+11^2+1}=\frac{22}{126}=\frac{11}{63}$$

$$z=\frac{2^2+11^2-1}{1^2+11^2+1}=\frac{124}{126}=\frac{62}{63}$$

Finally, $a=63\cdot x=2$, $b=63\cdot y=11$, $c=63\cdot z=62$, $d=63$ that makes the given equation true: $2^2+11^2+62^2=63^2$.

Method 2. Let us get the same result using stereographic projection of point M on the unit sphere to the plane of XOY that passes through the diameter of the unit sphere and is perpendicular to the axis Z.

For this we will connect point on the sphere with the south pole point $S(0,0,-1)$ (Figure 4.8) that obviously satisfies the equation (4.11). Now we will have a detailed Figure 4.9. Here M_1 is the image of M on the diameter plane XOY. Stereographic projection is one- to - one mapping from the surface of the sphere to the plane, such as

$$M(x,y,z) \rightarrow M_1(u,v,0)$$

Next, we will do some geometric constructions. (See Figure 4.9)

1) First, we will draw line KM parallel to OM_1. Consider similar triangles,

$$\triangle KSM \sim \triangle OSM_1$$

$$\frac{KM}{OM_1}=\frac{KS}{OS}, \quad KS=z, \quad OS=1, \quad OM_1=r.$$

$$KM=(1+z)\cdot OM_1=(1+z)\cdot r.$$

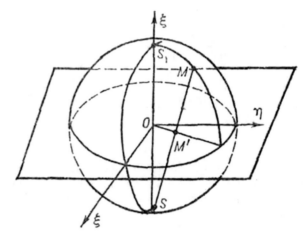

Fig. 4.8 Projection on the Plane from South Pole

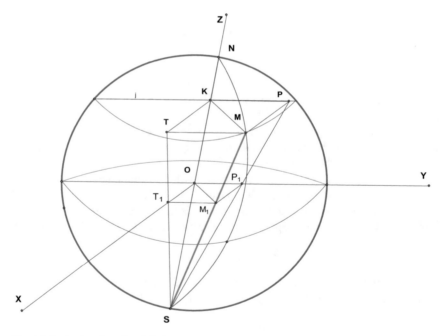

Fig. 4.9 Projection from South Pole

2) Next, let us draw the line $M_1 T_1$ parallel to the Y axis (perpendicular to the X axis) and line MP parallel to the X axis. Denote, $OT_1 = u$, $OP_1 = v$, where $(u, v, 0)$ are coordinates of point M_1. Draw a segment OM_1 by connecting the center of the sphere and point M_1. Connect point S and T_1, S and P_1. Simultaneously, let us draw two lines through point M, parallel to the X and Y axes. Also draw two lines through

point K, one parallel to the X axis and the other parallel to the Y axis. At the intersection we obtain points T and P. Plane formed by points M, T, K, P is parallel to XOY plane.

3) Consider similar triangles TKS and $T_1 OS$:

$$\triangle TKS \sim \triangle T_1 OS$$

$$\frac{TK}{T_1 O} = \frac{KS}{OS}, \quad TK = x, \quad OS = 1.$$

$$\frac{x}{u} = \frac{1+z}{1},$$

$$x = (1+z) \cdot u. \tag{4.14}$$

4) Consider similar triangles $\triangle KPS \sim \triangle OP_1 S$:

$$\frac{KP}{OP_1} = \frac{KS}{OS}, \quad KP = y, \quad OS = 1.$$

$$\frac{y}{v} = \frac{1+z}{1},$$

$$y = (1+z) \cdot v. \tag{4.15}$$

These two formulas show how coordinates x and y of point M depend on the coordinates of its image M_1 on the plane.

5) Squaring both formulas and adding their left and the right sides, we obtain the following chain of equations:

$$x^2 + y^2 = (1+z)^2 \cdot (u^2 + v^2), \quad x^2 + y^2 + z^2 = 1$$
$$1 - z^2 = (1 + 2z + z^2)(u^2 + v^2), \quad u^2 + v^2 = r^2$$
$$(1 + r^2)z^2 + 2r^2 \cdot z + (r^2 - 1) = 0.$$

Solving this quadratic equation in variable z, we obtain

$$z = \frac{-r^2 \pm \sqrt{r^4 - (r^2 - 1)(r^2 + 1)}}{r^2 + 1}$$

Taking only solution with a plus sign, we obtain $z = \frac{1-r^2}{1+r^2}$, which can be written as

$$z = \frac{1 - (u^2 + v^2)}{1 + u^2 + v^2} \tag{4.16}$$

Substituting this z into formulas for x and y (4.14), (4.15), and replacing $1+z = 1 + \frac{1-r^2}{1+r^2} = \frac{2}{1+r^2}$, finally, we obtain

$$x = \frac{2u}{1+u^2+v^2}, \quad y = \frac{2v}{1+u^2+v^2}. \tag{4.17}$$

The problem is solved. By selecting rational values for u and v, we will get corresponding rational coordinates (x,y,z) (using (4.16), (4.17)) of point M and then the answer (a,b,c,d) (4.10). Similarly, we can make a table. (Table 4.3)

(u,v)	(x,y,z)	(a,b,c,d)
$(1,1)$	$\left(\frac{2}{3},\frac{2}{3},\frac{1}{3}\right)$	$(2,2,-1,3)$
$\left(\frac{1}{3},\frac{1}{5}\right)$	$\left(\frac{150}{259},\frac{90}{259},\frac{191}{259}\right)$	$(150,90,191,259)$

Table 4.3 $a^2+b^2+c^2 = d^2$

Using formula (4.13) we can find three integers such that the sum of their squares is a perfect square itself. What if a number is not a perfect square? Can it be represented by the sum of three squares? When? The following problem will help you to think in the right direction.

Problem 212

Prove that if $n = x^2+y^2+z^2$, where $x,y,z \in \mathbb{Z}$, then its square is also sum of three squares,

$$n^2 = m^2+k^2+l^2.$$

Proof. Let $n = x^2+y^2+z^2$, such that $x \geq y \geq z$, then $x^2+y^2-z^2 > 0$ and we have the following chain of correct transformations:

$$n^2 = \left(x^2+y^2+z^2\right)^2$$
$$= x^4+y^4+z^4+2x^2y^2+2y^2z^2+2x^2z^2$$
$$= \left(x^2+y^2-z^2\right)^2 + (2yz)^2 + (2xz)^2,$$

where

$$m = x^2+y^2-z^2, \ k = 2yz, \ l = 2xz.$$

Proof is complete.

Example. Let $x = 4$, $y = 3$, and $z = 2$, then

$$n = 4^2 + 3^2 + 2^2 = 29.$$

Set $m = 4^2 + 3^2 - 22 = 21$, $k = 2 \cdot 3 \cdot 2 = 12$, and $l = 2 \cdot 4 \cdot 2 = 16$. We can see that the following is true:

$$n^2 = 29^2 = 21^2 + 12^2 + 16^2.$$

Remark. Think about this: Is the converse of the statement above true? (If a square of a number is sum of three other squares, does it mean that the number itself is also sum of three squares?) In the following section we will learn that all natural numbers in the form:

$$n = (8k + 7) \cdot 4^m$$

cannot be represented as sum of three squares. Obviously the converse statement is false. Thus, $15^2 = 2^2 + 5^2 + 14^2$, and $63^2 = 62^2 + 11^2 + 2^2$ but neither $15 = 8 \cdot 1 + 7$ nor $638 \cdot 7 + 7 =$ can be written by sum of three squares.

4.2 Waring's and other Related Problems

The Waring's problem (1770, English mathematician) formulated without proof: Any positive integer $N > 1$ can be represented as a sum of n-th powers of k positive integers, i.e., as $N = x_1^n + x_2^n + \cdots + x_k^n$; the number of terms k depends only on n. A particular case of the Waring's problem is the Lagrange theorem, which concerns the fact that each N is the sum of at most four squares. The first general solution (for any n) of the Waring's problem was given by D. Hilbert (1909) with a very rough estimate of the number of terms k in relation to n. More exact estimates of k were obtained in the 1920s by G. Hardy and J. Littlewood; and in 1934, L. M. Vinogradov, by the method of trigonometric sums that he created, obtained results that were close to definitive. An elementary solution of the Waring's problem was given in 1942 by Russian mathematician L. V. Linnik. At present the Waring's problem for $n = 2$, which concerns with representation of a natural number by two, three, four, or more squares, is entirely solved. The special significance of the Waring's problem consists in the fact that during its investigation powerful methods of the analytic number theory were created.

4.2.1 Representing a Number by Sum of Squares

Representation of a number by sum of two or more squares generated many interesting problems. Some problems have the form of the known theorems (properties). An especially interesting problem is what prime numbers can be written by the sum of two squares. Olympiad and math contest problems are often inspired by this field of number theory.

Problem 213 (USSR Olympiad 1986)

Prove that for any quadratic equation of the type $x^2 + ax + 1 = b$ with integer coefficients and natural roots, the expression $a^2 + b^2$ cannot be prime number.

Solution. Method 1. (Using Vieta's Theorem). We can rewrite the equation as

$$x^2 + ax + (1 - b) = 0$$

Assume that this equation has the following natural roots, x_1, x_2, then these roots must satisfy the Vieta's theorem:

$$x_1 \cdot x_2 = 1 - b$$
$$x_1 + x_2 = -a.$$

Solving this system for a and b, we will square each of them and then obtain the following true relationships:

$$b = 1 - x_1 \cdot x_2$$
$$a = -(x_1 + x_2)$$
$$b^2 = (1 - x_1 \cdot x_2)^2$$
$$a^2 = (x_1 + x_2)^2$$
$$
\begin{aligned}
a^2 + b^2 &= x_1^2 + x_2^2 + (x_1^2 x_2^2) \\
&= x_2^2 \cdot (1 + x_1^2) + (1 + x_1^2) \\
&= (1 + x_1^2)(1 + x_2^2).
\end{aligned}
$$

Because the sum of the squares is a product of two quantities, then $a^2 + b^2$ is a composite, not prime number.

Method 2. Using complex numbers.

Consider a quadratic polynomial $p(x) = x^2 + ax + (1 - b)$. Let us substitute for x an imaginary unit i and evaluate $p(i)$ remembering that the following is true:

$$i^2 = -1,$$
$$i^3 = -i,$$
$$i^4 = 1,$$
$$i^5 = i...$$

Thus, $p(i) = i^2 + a \cdot i + 1 - b = -1 + i \cdot a + 1 - b = b + i \cdot a$. Similarly, $p(-i) = b - i \cdot a$. Hence, $p(i), p(-i)$ are **complex conjugate numbers** and because of the condition of the problem their product is a natural number:

$$p(i) \cdot p(-i) = b^2 + a^2. \tag{4.18}$$

On the other hand, the given quadratic equation has two natural roots, i.e., x_1 and x_2. Then the quadratic polynomial can be factored as

$$p(x) = (x - x_1)(x - x_2).$$

This allows us to find $p(i)$ and $p(-i)$ in a different way substituting i and $-i$ in this formula. The following in true:

$$
\begin{aligned}
p(i) &= (i - x_1)(i - x_2), \\
p(-i) &= (-i - x_1)(-i - x_2), \\
p(i)p(-i) &= (i - x_1)(i - x_2)(-i - x_1)(-i - x_2) \\
&= [(i - x_1)(-i - x_1)] \cdot [(i - x_2)(-i - x_2)] \\
&= (1 + x_1^2)(1 + x_2^2).
\end{aligned}
\tag{4.19}
$$

Equating the right sides of two formulas, we obtain that

$$a^2 + b^2 = (1 + x_1^2)(1 + x_2^2).$$

This also proves that the sum of the squares representing parameters of the given quadratic equations is not prime number.

Remark. By solving the problem above using complex numbers method, we can expand the ideas to the case of polynomial of the nth degree and state that for any polynomial $p_n(x) = a_n x^n + a_{n-1} x^{n-1} + \cdots + a_2 x^2 + a_1 x + a_0$, the expression $p(i)$ is a complex number $p(i) = A + i \cdot B$, where the values of A and B depend only on the coefficients of the given polynomial. Obviously, $p(-i) = A - i \cdot B$ as its complex conjugate. Hence, their product is $p(i) \cdot p(-i) = A^2 + B^2$.

Example. Let $p(x) = x^{14} + 2x^{13} - 5x^4 + 3x + 9$. Replacing x by i, we get $p(i) = i^{14} + 2i^{13} - 5i^4 + 3i + 9$. Using properties of the imaginary one and its powers, we can simplify it as

$$p(i) = (i^2)^7 + 2(i^2)^6 \cdot i - 5 \cdot 1 + 3i + 9 = -1 + 2i + 4 + 3i = 3 + 5 \cdot i$$

and

$$p(-i) = 3 - 5 \cdot i.$$

| Problem 214 |

Given n natural numbers $x_1, x_2, x_3, ..., x_n$. Prove that the number

$$N = (1+x_1^2) \cdot (1+x_2^2) \cdot ... \cdot (1+x_n^2)$$

can be represented by the sum of two squares.

Proof. Consider a polynomial of nth degree with n integer roots, $x_1, x_2, x_3, ..., x_n$. Then it can be factored as

$$(x-x_1)(x-x_2)(x-x_3) \cdot ... \cdot (x-x_n).$$

Evaluating $p(i)$ and its complex conjugate $p(-i)$ we obtain:

$$p(i) = (i-x_1)(i-x_2)(i-x_3) \cdot ... \cdot (i-x_n),$$

$$p(-i) = (-i-x_1)(-i-x_2)(-i-x_3) \cdot ... \cdot (-i-x_n),$$

Then their product is

$$p(i) \cdot p(-i) = (i-x_1)(i-x_2) \cdot ... \cdot (i-x_n) \cdot (-i-x_1)(-i-x_2) \cdot ... \cdot (-i-x_n),$$

which, by pairing corresponding conjugate factors together, such as

$$(i-x_k)(-i-x_k) = 1+x_k^2$$

can be written as

$$p(i)p(-i) = (1+x_1^2)(1+x_2^2) \cdot ... \cdot (1+x_n^2). \tag{4.20}$$

On the other hand, a polynomial of nth degree with integer coefficients will have the leading coefficient 1, but as we showed earlier, for any polynomial, $p(i) = A + iB$, $p(-i) = A - iB$ are simply complex conjugate numbers, and hence their product is a sum of two squares,

$$p(i)p(-i) = A^2 + B^2. \tag{4.21}$$

By equating the right-hand sides of two equations, we obtain

$$(1+x_1^2)(1+x_2^2) \cdot ... \cdot (1+x_n^2) = A^2 + B^2.$$

that completes the proof.

Question. Can a sum of two squares be a prime number? What prime numbers can be always written by sum of two squares?

We have some important statements.

Property 29 (Fermat Theorem on Prime Representation as Sum of Two Squares)

Any prime of the form $(4k+1)$ can be written as the sum of two squares.

The prime numbers for which this is true are called Pythagorean Primes. Thus, $5, 13, 17, 29, 37,$ and 41 are such primes and we can state that $5 = 1^2 + 2^2$, $37 = 1^2 + 6^2$ or $41 = 4^2 + 5^2$. On the other hand, the primes, $3, 7, 1, 19, 23, 31, \ldots$ or any other prime of type $p = 4k+3$ cannot be written as the sum of two squares. The first proof of this property was given by Euler.

Let us solve the following problems.

Problem 215

Prove that prime in the form $p = 4k+3$ cannot be written by sum of two squares.

Proof. Assume the contradiction that

$$p = 4k+3 = x^2 + y^2.$$

it means that if, for example, x is an even and y is odd natural number, then the following would be true

$$x^2 = 4k, \quad y^2 = (2n+1)^2 = 4n^2 + 4n + 1 = 4m + 1, \quad x^2 + y^2 = 4l + 1 \neq 4k + 3 = p.$$

The line above means that square of an even number divided by 4 leaves no reminder, while the square of an odd number leaves a remainder of 1, then the sum of such numbers divided by 4 must also leave a remainder of 1. Obviously, any prime in the form $4k+3$ divided by 4 leaves a remainder of 3. We obtained the contradiction and hence, have proven the statement.

Lemma 20

Representation of a prime $p = 4k+1$ by the sum of two squares is unique.

We will prove this statement by solving the following problem.

Problem 216

Prove that if a prime can be written as the sum of two squares, then this representation is unique.

Proof. Assume that prime can be represented by sum of two squares in two different ways:

$$p = a^2 + b^2 = c^2 + d^2, \quad a \neq c, b \neq d, \tag{4.22}$$

i.e., this presentation is not unique.

Consider the following true equality,

$$a^2 \cdot c^2 - b^2 \cdot d^2 = a^2[c^2 + d^2] - d^2[b^2 + a^2] \tag{4.23}$$

Since the expressions inside brackets of (4.23) equal to p (see (4.22)) and denoting $a^2 - d^2 = n$, it follows that it can be rewritten as

$$a^2 \cdot c^2 - b^2 \cdot d^2 = p \cdot n \tag{4.24}$$

On the other hand, applying the difference of squares formula to the left side of (4.24), we obtain

$$(ac + bd)(ac - bd) = p \cdot n.$$

From this equality it follows that one of the factors on the left-hand side is a multiple of p. Without loss of generality, we can assume that

$$ac + bd = p \cdot k \tag{4.25}$$

Using double representation of prime p (4.22) and replacing each of them by the product of corresponding complex numbers, we can write down its square as follows:

$$\begin{aligned}
p^2 &= (a + ib)(a - ib)(c + id)(c - id) \\
&= [(a - ib)(c + id)] \cdot [(a + ib)(c - id)] \\
&= [(ac + bd) + i(ad - bc)] \cdot [(ac + bd) - i(ad - bc)] \\
&= (ac + bd)^2 + (ad - bc)^2.
\end{aligned}$$

or

$$p^2 = (ac + bd)^2 + (ad - bc)^2 \tag{4.26}$$

Considering (4.25) and (4.26), we have two cases:

Case 1. Let $ad - bc \neq 0$. Then $(ad - bc)^2 = \alpha^2 > 0$, $(ac + bd)^2 = p^2 \cdot k^2$, and then

$$(ac + bd)^2 + (ad - bc)^2 = p^2 \cdot k^2 + \alpha^2 > p^2,$$

which is false.

Case 2. Assume that $ad - bc = 0$, then $ad = bc$, $\Rightarrow a = c$, $b = d$.

Therefore representation of a prime $p = 4k + 1$ by the sum of two squares is unique.

Property 30

The following equation $x^2 + y^2 = np$, where $n \in \mathbb{N}$, $p = 4k + 1$, and p is prime always has a solution.

Example.

$$8^2 + 2^2 = 68 = 4 \cdot 17, \quad 17 = 4 \cdot 4 + 1, \ n = 4,$$
$$9^2 + 2^2 = 85 = 5 \cdot 17, \quad 17 = 4 \cdot 4 + 1, \ n = 5,$$
$$121 + 49 = 170 = 10 \cdot 17, \quad 170 = 11^2 + 7^2, \ n = 10.$$

Property 31

If $x^2 + y^2 = p$, where $p = 4n + 1$, and prime, then $(kx)^2 + (ky)^2 = k^2 p$.

Example. For example,

$$13 = 9 + 4 = 3^2 + 2^2$$

let $k = 7$, the following is true:

$$(7 \cdot 3)^2 + (7 \cdot 2)^2 = 7^2 \cdot 13 = 637,$$
$$21^2 + 14^2 = 637.$$

Property 32

If a number n can be written as a sum of two squares, $n = a^2 + b^2$, then it defines a unique decomposition of

$$k^2 \cdot n = (ka)^2 + (kb)^2.$$

Example.

$13 = 3^2 + 2^2$ rewrite 325 as sum of two squares,

$$325 = 65 \cdot 5 = 5^2 \cdot 13 = (5 \cdot 3)^2 + (5 \cdot 2)^2 = 15^2 + 10^2.$$

Proofs of some of the properties will be given in the following sections.

Lemma 21

Any number m, such that $m = 4k + 1 = 4c^2 + 1$ can be written as the sum of two squares, one of which is 1.

The proof of this statement is obvious because

$$m = 4c^2 + 1 = (2c)^2 + 1^2.$$

Example. Let $m = 145$. $145 = 4 \cdot 36 + 1 = 4 \cdot 6^2 + 1$. Therefore $c = 6$ and $145 = 1^2 + 12^2$. (true).

Problem 217

Rewrite numbers 13, 17, 23, 25, 29, 145 as sums of two squares. Create the general rule by which an integer number can be written as the sum of squares of two other integers?

Solution. We can find mentally that

$$13 = 3^2 + 2^2$$
$$17 = 4^2 + 1^2$$
$$25 = 3^2 + 4^2$$
$$29 = 5^2 + 2^2$$
$$145 = 12^2 + 1^2 = 8^2 + 9^2.$$

Unfortunately, no combination of squares will give us 23. Certainly not every integer can be written as the sum of two squares. Do you know why?

Let us find something common about these numbers: 13, 17, 29 are primes, each divided by 4 leaves a reminder of 1, $(4k + 1)$. Number 23 is a prime but in the different form, $(4k + 3)$. Number 25 is not a prime, and it is a perfect square of 5. Number 145 can be factored as $5 \cdot 29$, where 29 and 5 are both primes in the form of $(4k + 1)$.

In 1775, Lagrange used quadratic forms to prove the generalization of the Fermat Theorem (Property 29). The following generalizes the possibility of representation of a number by sum of two squares.

> ### Property 33
>
> An integer number $n > 1$ can be written as sum of two squares, $n = x^2 + y^2$, if its prime factorization contains no prime in the form $p = 4n + 3$, raised to an odd power.

Example.

$$2450 = 2 \cdot 5^2 \cdot 7^2$$
$$5 = 4 \cdot 1 + 1, \ 7 = 4 \cdot 1 + 3.$$

Although 7 divided by 4 gives a remainder 3, it is raised to an even power 2. Therefore 2450 can be written as sum of two squares, $2450 = 7^2 + 49^2$.

On the other hand, next example will show why 3430 cannot be written as sum of two squares.

Example.

$$3430 = 2 \cdot 5 \cdot 7^3$$
$$5 = 4 \cdot 1 + 1, \ 7 = 4 \cdot 1 + 3.$$

Since 7 divided by 4 gives a remainder 3, and it is raised to an odd power 3, then 3430 cannot be represented by sum of two squares.

Representing a number by the sum of two squares generated interesting problems. Many problems have the form of the known theorems (properties). We just learned that no prime in the form $4n + 3$ can be represented as sum of two squares. An obvious question is "Can the sum of two squares be divisible by a prime of form $p = 4n + 3$?" Let us solve the following problem.

> ### Problem 218
>
> Prove that if $a^2 + b^2$ is divisible by prime number $p = 4n + 3$, then both numbers a and b are divisible by p.

Proof. Assume that neither a nor b is multiple of $p = 4n + 3$ and so

$$a \neq a_1 \cdot p, \ b \neq b_1 \cdot p$$

Based on the condition of the problem we can write it as

$$a^2 + b^2 = k \cdot p$$

or using congruence notation in an equivalent form:

$$a^2 \equiv -b^2 \pmod{p = 4n + 3} \tag{4.27}$$

Let us raise (4.27) into an odd power of $(2n+1)$:

$$\left(a^2\right)^{2n+1} \equiv \left(-b^2\right)^{2n+1} \quad (\text{mod } p = 4n+3)$$

After simplification we will obtain

$$a^{4n+2} \equiv -b^{4n+2} \quad (\text{mod } 4n+3) \tag{4.28}$$

On the other hand, applying Fermat's Little Theorem, and because

$$4n+2 = p-1, \; p = 4n+3,$$

we can write,

$$\begin{aligned} a^{4n+2} &\equiv 1 \quad (\text{mod } 4n+3) \\ b^{4n+2} &\equiv 1 \quad (\text{mod } 4n+3) \end{aligned} \tag{4.29}$$

Combining (4.29) and (4.28), we would obtain the following strange result:

$$a^{4n+2} \equiv 1 \equiv -b^{4n+2} \equiv -1 \quad (\text{mod } p = 4n+3)$$

However, $1 \equiv -1 \pmod{p = 4n+3}$ is false for all prime greater than 2. Therefore we obtained contradiction and both numbers a and b must be divisible by $p = 4n+3$.

> **Property 34**
>
> A product of two numbers, each of which is the sum of two squares is also the sum of two squares.

Let $n_1 = a^2 + b^2$ and $n_2 = c^2 + d^2$. The Theorem states that the product of these two numbers $n_1 \cdot n_2 = m^2 + n^2$.

The proof of this Property is asked in the following problem.

> **Problem 219**
>
> Prove that numbers in the form of $(a^2 + b^2)(c^2 + d^2)$ can be always written as the sum of a) four squares b) two squares.

Proof. Notice that

$$(a^2 + b^2)(c^2 + d^2) = (ac)^2 + (ad)^2 + (bc)^2 + (bd)^2. \tag{4.30}$$

This proves the first part of the problem that each such a number can be written by the sum of four squares.

The second part of the problem can be proven in several ways.

Method 1. By completing a square in (4.30) in two different ways, we can obtain two formulas

$$\begin{aligned}
& [(ac)^2 + (bd)^2] + [(ad)^2 + (bc)^2] \\
&= (ac)^2 + (bd)^2 + 2abcd + (ad)^2 + (bc)^2 - 2abcd \\
&= (ac+bd)^2 + (ad-bc)^2 \\
&= (ac-bd)^2 + (ad+bc)^2.
\end{aligned}$$

Therefore,

$$\begin{aligned}
(a^2+b^2)(c^2+d^2) &= (ac+bd)^2 + (ad-bc)^2 \\
(a^2+b^2)(c^2+d^2) &= (ac-bd)^2 + (ad+bc)^2.
\end{aligned} \tag{4.31}$$

Method 2. (Using complex numbers) Discovery of complex numbers simplified many proofs. In high school you learn that $i^2 = -1$ and that the product of two complex conjugate numbers $((A+iB)$ and $(A-iB))$ is a real number. Let us demonstrate it by using the difference of squares formula:

$$(A+iB) \cdot (A-iB) = A^2 - (i \cdot B)^2 = A^2 + B^2.$$

On one hand, using the formula above we can rewrite the left side as follows

$$(a^2+b^2) \cdot (c^2+d^2) = ((a+ib)(a-ib)) \cdot ((c+id)(c-id))$$

On the other hand, the same product of four complex numbers can be paired as follows:

$$((a+ib)(c-id)) \cdot ((a-ib)(c+id))$$

or as

$$((a+ib)(c+id)) \cdot ((a-ib)(c-id))$$

Multiplying complex numbers inside each parentheses we obtain

$$\begin{aligned}
(ac - iad + ibc - i^2bd)(ac + iad - ibc - i^2bd) \\
((ac+bd) - i \cdot (ad-bc)) \cdot ((ac+bd) + i \cdot (ad-bc)),
\end{aligned}$$

and

$$\begin{aligned}
(ac + iad + ibc + i^2bd)(ac - iad - ibc + i^2bd) \\
((ac-bd) - i \cdot (ad+bc)) \cdot ((ac-bd) + i \cdot (ad+bc)).
\end{aligned}$$

The last row of each equation above is a product of two complex conjugate numbers, which gives us the immediate result and proves the statement,

$$(a^2+b^2) \cdot (c^2+d^2) = (ac+bd)^2 + (ad-bc)^2 = (ac-bd)^2 + (ad+bc)^2.$$

Formula (4.31) generates numbers that can be written as sum of two squares for selected a, b, c, and d. This formula was found in the work of Diophantus.

Example. Numbers $a = 1$, $b = 2$, $c = 3$, and $d = 4$ will generate the following

$$(1^2+2^2)(3^2+4^2) = (1\cdot 3+2\cdot 4)^2+(1\cdot 4-2\cdot 3)^2 = (1\cdot 3-2\cdot 4)^2+(1\cdot 4+2\cdot 3)^2$$
$$125 = 5\cdot 25 = 11^2+2^2 = 5^2+10^2.$$

Example. Number $145 = 5\cdot 29 = (1+2^2)\cdot (5^2+2^2)$, so $a = 1$, $b = 2$, $c = 5$, and $d = 2$ will create two different representations of 145 as sum of two squares:

$$145 = (1\cdot 5+2\cdot 2)^2+(1\cdot 2-2\cdot 5)^2 = 9^2+8^2$$
$$145 = (1\cdot 5-2\cdot 2)^2+(1\cdot 2+2\cdot 5)^2 = 1^2+12^2.$$

In order to find representations of other numbers, we can use the following ideas.

Problem 220

Rewrite 52 and 400 as sums of two squares.

Solution. Notice that $52 = 4\cdot 13$, and $400 = 16\cdot 25$. Regarding Property 31, since $13 = 4\cdot 3+1$, $25 = 4\cdot 6+1$ we know that

$$13 = 3^2+2^2 \Rightarrow 52 = 2^2\cdot 13 = (2\cdot 3)^2+(2\cdot 2)^2 = 6^2+4^2$$
$$25 = 3^2+4^2 \Rightarrow 400 = 4^2\cdot 25 = (4\cdot 3)^2+(4\cdot 4)^2 = 12^2+16^2.$$

Problem 221

Given two numbers, not multiples of 7. Prove that the sum of their squares cannot be a multiple of 7.

Proof. Any non-multiple of 7 and its square can be written as

$$k = 7n\pm 1 \Rightarrow k^2 = 49n^2\pm 14n+1$$
$$k = 7n\pm 2 \Rightarrow k^2 = 49n^2\pm 28n+4$$
$$k = 7n\pm 3 \Rightarrow k^2 = 49n^2\pm 42n+9.$$

We can see that no combination of the numbers will give us a multiple of 7. Therefore, the sum of two squares cannot be a multiple of 7.

It is interesting to know that if a number can be written by a sum of two squares, then some its multiples also can be written as a sum of squares. Let us solve some problems now.

> **Problem 222**
>
> Given $n = x^2 + y^2$. Prove that $2n$, $5n$, $13n$ are also sums of two squares.

Proof. 1) Let us consider the following sum and expand it using the given equality.

$$(x+y)^2 + (x-y)^2 = x^2 + 2xy + y^2 + x^2 - 2xy + y^2 = 2 \cdot (x^2 + y^2) = 2 \cdot n.$$

2) In order to extract 5 we can add the following binomial squares:

$$(2x+y)^2 + (x-2y)^2 = 4x^2 + 4xy + y^2 + x^2 - 4xy + 4y^2 = 5x^2 + 5y^2 = 5n.$$

3) Similarly, we can proof that $13n$ is a sum of two squares of the form:

$$(3x+y)^2 + (2x-3y)^2 = 13(x^2 + y^2) = 13n.$$

Example. Because $11^2 + 3^2 = 130$, then 260 also is sum of two squares, so as $650 = 5 \cdot 130$ and $1690 = 13 \cdot 130$. Using the formula from Case 1 we obtain

$$260 = (11+3)^2 + (11-3)^2 = 14^2 + 3^2$$

Using the formula from Case 2 we have

$$650 = (2 \cdot 11 + 3)^2 + (11 - 2 \cdot 3)^2 = 25^2 + 5^2$$

Finally, using the formula from Case 3, we have

$$1690 = (3 \cdot 11 + 2 \cdot 3)^2 + (2 \cdot 11 - 3 \cdot 3)^2 = 39^2 + 13^2.$$

> **Property 35** (Legendre's Three Squares Theorem)
>
> If a natural number can be represented as a sum of three squares,
>
> $$N = x^2 + y^2 + z^2,$$
>
> then N cannot be written as
>
> $$N = 4^a \cdot (8b+7), \quad a, b \in \mathbb{Z}^+.$$

Proof. Consider first, the case when $a = 0$ and let us show that no number in the form $N = 8b + 7$ can be represented by three squares. Assume the contradiction and that it is possible. As we proved in Problem 8 of Chapter 1, a square of any even number is either a multiple of 8 or leaves a remainder of 4, and a square of any odd number divided by 8 leaves a remainder of 1. For convenience, we can additionally represent all possible odd numbers by the following group:

$$(8k \pm 1), \ (8k \pm 3), \ (8k \pm 5); (8k \pm 7)$$

Squares of any of these numbers will leave a remainder of 1 when divided by 8.
  All possible even numbers can be described by the following group:

$$8k, \ (8k \pm 2), \ (8k \pm 4), \ (8k \pm 6),$$

and their squares divided by 8 can leave either 0 or 4 as remainder.
  Therefore, a square of any natural number divided by 8 leaves a remainder of 0, 1 or 4. Then any combination of three squares divided by 8 would leave a remainder of 0 or 1 or 2 or 3...or 6, but never it would be 7. Therefore, a number $N = 8b + 7$ cannot be written by sum of three squares.
  Second, assume that for some integer value $a = n > 0$, the number

$$N_n = 4^n \cdot (8b + 7)$$

is represented by sum of three squares, i.e.,

$$N_n = 4^n \cdot (8b + 7) = m^2 + k^2 + l^2,$$

then the sum of three squares must be divisible by 4, which means that all numbers on the right-hand side must be even. Denote $m = 2m_1$, $k = 2k_1$, and $l = 2l_1$. Because $8b + 7$ is not a multiple of 4, dividing left and right sides by 4 we obtain a new equation:

$$N_{n-1} = 4^{n-1} \cdot (8b + 7) = \left(m_1^2 + k_1^2 + l_1^2\right) < m^2 + k^2 + l^2 = N_n.$$

Again, this is possible if $m_1 = 2m_2$, $k_1 = 2k_2$, and $l_1 = 2l_2$. Dividing both sides by 4 we would obtain a new equation:

$$N_{n-2} = 4^{n-2} \cdot (8b + 7) = \left(m_2^2 + k_2^2 + l_2^2\right) < m_1^2 + k_1^2 + l_1^2 = N_{n-1}.$$

Continuing the same procedure, we would finally get a situation that after s consecutive operations, the power of 4 after division would become 0 and we would obtain an odd sum on the right,

$$N_{n-s} = (8b + 7) = \left(m_s^2 + k_s^2 + l_s^2\right),$$

that is false as we proved above.
  Contradiction and so no number of the type $N = 4^a \cdot (8b + 7)$ can be represented by a sum of three squares.

Thus, a sequence of the numbers

$$7, 15, 23, 28, 31, 39, 47, 55, 60, 63, 71, \ldots$$

cannot be written as sum of three squares.

Remark. Pierre Fermat gave a criterion for a representation of a number $n = 3a + 1$ by three squares, but did not provide the proof. In 1796, Gauss proved the **Eureka Theorem** and stated that every positive integer n is the sum of three triangular numbers. That is equivalent to the statement that any number $N = 8n + 3$ is a sum of three squares.

For example,

$$11 = 1^2 + 1^2 + 3^2, \quad n = 1$$
$$19 = 1^2 + 3^2 + 3^2, \quad n = 2$$
$$26 = 3^2 + 4^2 + 1^2, \quad n = 3.$$

Theorem 28 (Gauss Eureka Theorem)

If N is a sum of three triangular numbers, then $8N + 3$ is the sum of three squares.

Proof. If N is the sum of three triangular numbers, then the following is valid:

$$N = \frac{n(n+1)}{2} + \frac{m(m+1)}{2} + \frac{k(k+1)}{2}.$$

Multiplying both sides by 8 we obtain

$$8N = 4n^2 + 4n + 4m^2 + 4m + 4k^2 + 4k.$$

Finally, if we add 3 to both sides of the equation and complete the squares on the right-hand side, we will prove the statement:

$$8N + 3 = (2n + 1)^2 + (2m + 1)^2 + (2k + 1)^2$$

Note. The Reversed Theorem is also true and if a number represented by the sum of three squares has the form $8N + 3$, then N is the sum of three triangular numbers.

Example. Consider $11^2 + 5^2 + 9^2 = 227 = 8 \cdot 28 + 3$, then $N = 28$. Because

$$(2 \cdot 5 + 1)^2 + (2 \cdot 2 + 1)^2 + (2 \cdot 4 + 1)^2 = 8 \cdot 28 + 3,$$

we can find that

$$28 = T(5) + T(2) + T(4) = \frac{5 \cdot (5+1)}{2} + \frac{2 \cdot (2+1)}{2} + \frac{4 \cdot (4+1)}{2} = 15 + 3 + 10.$$

Property 36 (Euler's Four-Square Identity)

The product of two numbers, each of which is a sum of four squares, is itself a sum of four squares.

Proof.

$$(a_1^2 + a_2^2 + a_3^2 + a_4^2)(b_1^2 + b_2^2 + b_3^2 + b_4^2)$$
$$= (a_1 b_1 + a_2 b_2 + a_3 b_3 + a_4 b_4)^2 + (a_1 b_2 - a_2 b_1 + a_3 b_4 - a_4 b_3)^2 \qquad (4.32)$$
$$+ (a_1 b_3 - a_2 b_4 - a_3 b_1 + a_4 b_2)^2 + (a_1 b_4 + a_2 b_3 - a_3 b_2 - a_4 b_1)^2.$$

Let us prove a different statement but related to the previous property by solving the following problem.

Problem 223

Prove that if each of two given numbers is a sum of fours squares then the product of the numbers can be written as sum of eight squares and as sum of sixteen squares.

Proof. Assume that the given number is

$$N = (a_1^2 + a_2^2 + a_3^2 + a_4^2)(b_1^2 + b_2^2 + b_3^2 + b_4^2)$$

Multiplying two quantities, we can easily prove the second part of the problem and show that such a number is the sum of sixteen squares:

$$\begin{aligned}
N = &(a_1 \cdot b_1)^2 + (a_1 \cdot b_2)^2 + (a_1 \cdot b_3)^2 + (a_1 \cdot b_4)^2 \\
&+ (a_2 \cdot b_1)^2 + (a_2 \cdot b_2)^2 + (a_2 \cdot b_3)^2 + (a_2 \cdot b_4)^2 \\
&+ (a_3 \cdot b_1)^2 + (a_3 \cdot b_2)^2 + (a_3 \cdot b_3)^2 + (a_3 \cdot b_4)^2 \\
&+ (a_4 \cdot b_1)^2 + (a_4 \cdot b_2)^2 + (a_4 \cdot b_3)^2 + (a_4 \cdot b_4)^2 \\
= &\sum\sum (a_i \cdot b_j)^2.
\end{aligned}$$

Next, we will prove that such a number can be also written by the sum of eight squares. Using complex numbers we can rewrite this product as

$$((a_1 + ia_2)(a_1 - ia_2) + (a_3 + ia_4)(a_3 - ia_4) \cdot ((b_1 + ib_2)(b_1 - ib_2) + (b_3 + ib_4)(b_3 - ib_4)$$

Next, we will multiply to factors in a different the following manner:

$$(A+B) \cdot (C+D) = A \cdot C + A \cdot D + B \cdot C + B \cdot D.$$

Each product on the right we can find separately.

$$
\begin{aligned}
AC &= (a_1 + ia_2)(a_1 - ia_2) \cdot (b_1 + ib_2)(b_1 - ib_2) \\
&= ((a_1 + ia_2)(b_1 + ib_2)) \cdot ((a_1 - ia_2)(b_1 - ib_2)) \\
&= ((a_1 b_1 - a_2 b_2) + i(a_1 b_2 + a_2 b_1)) \cdot ((a_1 b_1 - a_2 b_2) - i(a_1 b_2 + a_2 b_1)) \\
&= (a_1 b_1 - a_2 b_2)^2 + (a_1 b_2 + a_2 b_1)^2.
\end{aligned}
$$

Similarly, we obtain other terms of representation.

$$
\begin{aligned}
AD &= ((a_1 + ia_2)(b_3 + ib_4)) \cdot ((a_1 - ia_2)(b_3 - ib_4)) \\
&= (a_1 b_3 - a_2 b_4)^2 + (a_1 b_4 + a_2 b_3)^2.
\end{aligned}
$$

$$
\begin{aligned}
BC &= ((a_3 + ia_4)(b_1 + ib_2)) \cdot ((a_3 - ia_4)(b_1 - ib_2)) \\
&= (a_3 b_1 - a_4 b_2)^2 + (a_3 b_2 + a_4 b_1)^2.
\end{aligned}
$$

$$
\begin{aligned}
BD &= ((a_3 + ia_4)(b_3 + ib_4)) \cdot ((a_3 - ia_4)(b_3 - ib_4)) \\
&= (a_3 b_3 - a_4 b_4)^2 + (a_3 b_4 + a_4 b_3)^2.
\end{aligned}
$$

Because $N = AC + AD + BC + BD$, finally, we obtain

$$
\begin{aligned}
N &= (a_1^2 + a_2^2 + a_3^2 + a_4^2)(b_1^2 + b_2^2 + b_3^2 + b_4^2) \\
&= (a_1 b_1 - a_2 b_2)^2 + (a_1 b_2 + a_2 b_1)^2 + (a_1 b_3 - a_2 b_4)^2 + (a_1 b_4 + a_2 b_3)^2 \quad (4.33) \\
&\quad + (a_3 b_1 - a_4 b_2)^2 + (a_3 b_2 + a_4 b_1)^2 + (a_3 b_3 - a_4 b_4)^2 + (a_3 b_4 + a_4 b_3)^2.
\end{aligned}
$$

This completes the proof.

Let us show a numerical example for the proven statement.

Example. Assume that $n_1 = 1^2 + 2^2 + 3^2 + 4^2 = 30$ and $n_2 = 5^2 + 6^2 + 7^2 + 8^2 = 174$, then $n_1 \cdot n_2 = 30 \cdot 174 = 5220$. Additionally, $a_1 = 1, a_2 = 2, a_3 = 3, a_4 = 4, b_1 = 5, b_2 = 6, b_3 = 8$, and $b_4 = 8$ that using formulas from Problem 223 will generate the following numbers.

$$
\begin{aligned}
m_1 &= a_1 b_1 - a_2 b_2 = 1 \cdot 5 - 2 \cdot 6 = -7, \quad m_1^2 = 49 \\
m_2 &= a_1 b_2 + a_2 b_1 = 1 \cdot 6 + 2 \cdot 5 = 16, \quad m_2^2 = 256 \\
m_3 &= a_1 b_3 - a_2 b_4 = 1 \cdot 7 - 2 \cdot 8 = -9, \quad m_3^2 = 81 \\
m_4 &= a_1 b_4 + a_2 b_3 = 1 \cdot 8 + 2 \cdot 7 = 22, \quad m_4^2 = 484 \\
m_5 &= a_3 b_1 - a_4 b_2 = 3 \cdot 5 - 4 \cdot 6 = -9, \quad m_5^2 = 81 \\
m_6 &= a_3 b_2 + a_4 b_1 = 3 \cdot 6 + 4 \cdot 5 = 38, \quad m_6^2 = 1444 \\
m_7 &= a_3 b_3 - a_4 b_4 = 3 \cdot 7 - 4 \cdot 8 = -11, \quad m_7^2 = 121 \\
m_8 &= a_3 b_4 + a_4 b_3 = 3 \cdot 8 + 4 \cdot 7 = 52, \quad m_8^2 = 2708.
\end{aligned}
$$

Therefore using positive numbers and (4.33) we obtain

$$N = n_1 \cdot n_2 = (1^2 + 2^2 + 3^2 + 4^2)(5^2 + 6^2 + 7^2 + 8^2)$$
$$= 7^2 + 16^2 + 9^2 + 22^2 + 9^2 + 38^2 + 11^2 + 52^2 = 5220.$$

Theorem 29 (Lagrange's Four-Square Theorem)

Any natural number can be represented as the sum of four integer squares.

We will omit its rigorous algorithmic proof given by Lagrange. Part of the proof considers the case given by Euler's four-square identity formula. For all other numbers, the idea is that if a number can be represented as a sum of three squares, such as 3, then by adding a square of zero, it will be the sum of four squares,

$$3 = 1^2 + 1^2 + 1^2 + 0^2$$

If a number is in the form of $4^a \cdot (8n+7)$, such as 31 or 310, and cannot be represented by a sum of three squares by Legendre's Three Squares Theorem, then it can be fixed as follows:

$$31 = 5^1 + 2^2 + 1^2 + 1^2$$
$$310 = 17^2 + 4^2 + 2^2 + 1^2.$$

Consider the following product

$$\left(a_1^2 + a_2^2 + \cdots + a_n^2\right) \cdot \left(b_1^2 + b_2^2 + \cdots + b_n^2\right) = c_1^2 + c_2^2 + \cdots + c_n^2, \qquad (4.34)$$

It was proven that if a number is a product of two other numbers, each of them is the sum of n squares, then the number can be represented by the sum of n squares only for $n = 1$, 2, 4, and 8. The first three cases are

- $n = 1$, $a^2 \cdot b^2 = (ab)^2$
- $n = 2$, (4.31) (Diophantus formula)
- $n = 4$. (4.32) (Euler's formula)

Problems involving representation of a natural number as the sum of two, three, four, or more squares can be challenging and often appear on different math contests. Below we summarize some rules that we have learned in this Section.

- Prime p can be written as the sum of two squares, $p = x^2 + y^2$, iff $p = 4n + 1$.
- A natural number $N = x^2 + y^2$ iff any its prime factors of type $4n + 3$ occur to even power.
- If $N = 8n + 7$, then N cannot be represented as the sum of three squares.
- If $N = x^2 + y^2 + z^2$, then $N \neq 4^k \cdot (8n + 7)$ (by Legendre's Three Squares Theorem).
- Any natural number can be represented as the sum of four squares (Lagrange).

- If $N = n \cdot m$, where m and n are the sums of either of two or four squares, then so is their product N.

Let us see how you understand the material by solving the following problems.

Problem 224

If possible, rewrite the number 32,045 by the sum of two, three, four, eight, and sixteen squares. Explain.

Solution. Using prime factorization this number can be written as follows:

$$32045 = 5 \cdot 13 \cdot 17 \cdot 29 = (4+1)(9+4)(16+1)(25+4)$$
$$= (2^2+1)(3^2+2^2)(4^2+1^2)(5^2+2^2).$$

We can attack this problem in different ways. Because each prime factor of 32045 is in the form of $(4k+1)$, then there are many decompositions available for 32045.

1. Can we rewrite $N = 32045$ as the sum of two squares? Yes, we can because its prime factorization does not contain any factor of $(4k+3)$. Estimating the square root of the given number, we obtain $[\sqrt{32045}] = 179 = x$, then $N - x^2 = N - 179^2 = 32045 - 179^2 = 4 = 2^2$, hence, $N = 32045 = 179^2 + 2^2$. Next, we can subtract from N the square of 178 (this number is one less than 179) and obtain $N = 32045 - 178^2 = 361 = 19^2$. Hence, $N = 32045 = 178^2 + 19^2$. By trying for x all integers less than the previous one, we get that if $x = 173$, then

$$N - 173^2 = 2116 - 46^2, \quad N = 173^2 + 46^2.$$

We can continue to subtract, but using formula (4.31) we will consider a better way to represent 32045 as sum of two squares, which will be demonstrated below, in part 4.

2. Can this number be written by a sum of three squares? Yes, it can because $N = 32045 = 8 \cdot 4005 + 5 \neq 8n + 7$.

Assume that $N = x^2 + y^2 + z^2$. We can similarly to the previous example, start subtracting a square of a number, x, then check if the difference $n = N - x^2$ can be written by the sum of two squares. If our n does not have any prime factor of type $4k + 3$ raised to an odd power, then we will find a combination of squares of two other numbers, such that $n = y^2 + z^2$.

Thus, let $x = 174$, then $n = N - x^2 = 32045 - 174^2 = 1769 = 4 \cdot 442 + 1 = 4k + 1$, then 1769 can be represented by the sum of two squares. In order to find it, we can do one of the following:

Estimate a square root of $n = 1769$, denote $y = [\sqrt{1769}] = 42$, subtract the answer from n, decrease y by 1, repeat subtraction until your get a perfect square (z). Note that if $y = 40$, then it works as follows

$$n - 40^2 = 1769 - 40^2 = 169 = 13^2, \quad N = 32045 = x^2 + y^2 + z^2 = 174^2 + 40^2 + 13^2.$$

Alternatively, we can factor 1769 as follows

$$1769 = 29 \cdot 61 = (5^2 + 22) \cdot (5^2 + 6^2)$$

and apply formula (4.31) for $a = 5$, $b = 2$, $c = 5$, and $d = 6$:

$$1769 = y^2 + z^2 = (5 \cdot 4 - 2 \cdot 6)^2 + (5 \cdot 6 + 2 \cdot 5)^2 = 13^2 + 40^2$$
$$= (5 \cdot 4 + 2 \cdot 6)^2 + (5 \cdot 6 - 2 \cdot 5)^2 = 37^2 + 20^2.$$

Thus, two of possible representations of 32045 as sum of three squares are

$$32045 = 174^2 + 13^2 + 40^2 = 174^2 + 37^2 + 20^2.$$

3. How can we rewrite 32045 as a sum of four squares?
    Note that $N - 176^2 = 32045 - 176^2 = 1069 = 8k + 5$. Next, $[\sqrt{1069}] = 32$, subtracting it from 1069, we obtain

$$1069 - 32^2 = 45 = 6^2 + 3^2.$$

Hence,

$$N = 32045 = 176^2 + 32^2 + 6^2 + 3^2.$$

In order to rewrite 32,045 by the sum of other four squares, we can do the following.

$$32045 = (5 \cdot 13) \cdot (17 \cdot 29) = 65 \cdot 493$$
$$= (5 \cdot 17) \cdot (13 \cdot 29) = 85 \cdot 377$$
$$= (5 \cdot 29) \cdot (13 \cdot 17) = 145 \cdot 221.$$

Using complex numbers it can be shown that each of the factors can be represented by a sum of two squares in two different ways:

$$65 = 8^2 + 1^2 = 4^2 + 7^2$$
$$493 = 22^2 + 3^2 = 18^2 + 13^2.$$

Thus, for the second factor we have

$$17 \cdot 29 = (4 + i)(4 - i)(5 + 2i)(5 - 2i)$$
$$493 = [(4 + i)(5 - 2i)] \cdot [(4 - i)(5 + 2i)] = (22 - 3i)(22 + 3i) = 22^2 + 3^2$$
$$493 = [(4 + i)(5 + 2i)] \cdot [(4 - i)(5 - 2i)] = (18 + 13i)(18 - 13i) = 18^2 + 13^2.$$

Multiplying both representations of 65 and 493 by each other, we obtain the following four possible representations of 32045 by a sum of four squares.

$$(8^2 + 1^2)(22^2 + 3^2) = (8 \cdot 22)^2 + (8 \cdot 3)^2 + (1 \cdot 22)^2 + (1 \cdot 3)^2 = 176^2 + 24^2 + 22^2 + 3^2$$
$$(8^2 + 1^2)(18^2 + 13^2) = (8 \cdot 18)^2 + (8 \cdot 13)^2 + (1 \cdot 18)^2 + (1 \cdot 13)^2 = 144^2 + 104^2 + 18^2 + 13^2$$
$$(4^2 + 7^2)(22^2 + 3^2) = (4 \cdot 22)^2 + (4 \cdot 3)^2 + (7 \cdot 22)^2 + (7 \cdot 3)^2 = 88^2 + 12^2 + 154^2 + 21^2$$
$$(4^2 + 7^2)(18^2 + 13^2) = (4 \cdot 18)^2 + (4 \cdot 13)^2 + (7 \cdot 18)^2 + (7 \cdot 13)^2 = 72^2 + 52^2 + 126^2 + 91^2.$$

Above are not the only representations by four squares, and each time we pair new factors, we will get a new representation. For example, 32045 can be also rewritten as

$$32045 = (5 \cdot 17) \cdot (13 \cdot 29) = 85 \cdot 377.$$

Using complex numbers it can be shown that each of the factors can be represented by a sum of two squares in two different ways. We have for 85:

$$\begin{aligned}
85 &= 5 \cdot 17 = (2-i)(2+i)(4-i)(4+i) \\
&= [(2-i)(4+i)] \cdot [(2+i)(4-i)] = (7-6i)(7+6i) = 7^2 + 6^2 \\
&= [(2-i)(4-i)] \cdot [(2+i)(4+i)] = (9-2i)(9+2i) = 9^2 + 2^2.
\end{aligned}$$

Thus, for the second factor of 377 we have

$$\begin{aligned}
377 &= 13 \cdot 29 = (3+2i)(3-2i)(5+2i)(5-2i) \\
&= [(3+2i)(5+2i)] \cdot [(3-2i)(5-2i)] = (11-16i)(11+16i) = 11^2 + 16^2 \\
&= [(3+2i)(5-2i)] \cdot [(3-2i)(5+2i)] = (19+4i)(19-4i) = 19^2 + 4^2.
\end{aligned}$$

Multiplying both representations of 85 and 377 by each other, we obtain the following four possible representations of 32045 by a sum of new four squares.

$$\begin{aligned}
32045 &= (7^2 + 6^2)(11^2 + 16^2) = 77^2 + 112^2 + 66^2 + 96^2 \\
32045 &= (7^2 + 6^2)(19^2 + 4^2) = 133^2 + 28^2 + 114^2 + 24^2 \\
32045 &= (9^2 + 2^2)(11^2 + 16^2) = 99^2 + 144^2 + 22^2 + 32^2 \\
32045 &= (9^2 + 2^2)(19^2 + 4^2) = 171^2 + 36^2 + 38^2 + 8^2.
\end{aligned}$$

Finally, 32045 can be also rewritten as

$$32045 = (5 \cdot 29) \cdot (13 \cdot 17) = 145 \cdot 221.$$

Using complex numbers it can be shown that each of the factors can be represented by a sum of two squares in two different ways. We have for 145,

$$\begin{aligned}
145 &= 5 \cdot 29 = (2+i)(2-i)(5+2i)(5-2i) \\
&= [(2+i)(5-2i)] \cdot [(2-i)(5+2i)] = (12+i)(12-i) = 12^2 + 1^2 \\
&= [(2+i)(5+2i)] \cdot [(2-i)(5-2i)] = (8+9i)(8-9i) = 8^2 + 9^2.
\end{aligned}$$

Thus, for the second factor of 221 we have

$$\begin{aligned}
221 &= 13 \cdot 17 = (3+2i)(3-2i)(4-i)(4+i) \\
&= [(3+2i)(4-i)] \cdot [(3-2i)(4+i)] = (14-5i)(14+5i) = 14^2 + 5^2 \\
&= [(3+2i)(4+i)] \cdot [(3-2i)(4-i)] = (10+11i)(10-11i) = 10^2 + 11^2.
\end{aligned}$$

Multiplying both representations of 145 and 221 by each other, we obtain the following four possible representations of 32045 by a sum of four squares.

$$32045 = (14^2 + 5^2)(12^2 + 1^2) = 168^2 + 14^2 + 60^2 + 5^2$$
$$32045 = (14^2 + 5^2)(8^2 + 9^2) = 112^2 + 126^2 + 40^2 + 45^2$$
$$32045 = (10^2 + 11^2)(12^2 + 1^2) = 120^2 + 10^2 + 132^2 + 11^2$$
$$32045 = (10^2 + 11^2)(8^2 + 9^2) = 80^2 + 90^2 + 88^2 + 99^2.$$

Above we found twelve different representations of $32045 = 5 \cdot 13 \cdot 17 \cdot 29$ as a sum of four squares. Are there any more?

The answer was found by Jacobi in 1870 and can be formulated as **Jacobi's Four-Square Theorem**: The number of representations of an odd natural number n as the sum of four squares (denoted as $r_4(n)$) is 8 times the sum of the divisors of n. Since $\tau(32045) = 16$, then $r_4(n) = 8 \cdot 16 = 128$.

4. Let us find other representations of 32045 as a sum of two squares. 32045 can be represented as a product of two numbers, each of which is sum of two squares, in several ways. For example, as follows

$$32045 = 145 \cdot 221 = 65 \cdot 493 = 85 \cdot 377$$

using formula (4.31) can be rewritten by sum of two squares in two different ways.

In the example above, consider $32045 = 145 \cdot 221$, where

$$145 = 1^2 + 12^2 = 9^2 + 8^2$$
$$221 = 10^2 + 11^2 = 14^2 + 5^2.$$

Combining representations of two factors of 32045 and applying formula (4.31) we have

$(1^2 + 12^2) \cdot (10^2 + 11^2)$	$= (1 \cdot 10 - 12 \cdot 11)^2 + (1 \cdot 11 + 12 \cdot 10)^2 = 122^2 + 131^2$
$a = 1,\ b = 12,\ c = 10,\ d = 11$	$= (1 \cdot 10 + 12 \cdot 11)^2 + (1 \cdot 11 - 12 \cdot 10)^2 = 142^2 + 109^2$
$(1^2 + 12^2) \cdot (14^2 + 5^2)$	$= (1 \cdot 14 - 12 \cdot 5)^2 + (1 \cdot 5 + 12 \cdot 14)^2 = 46^2 + 173^2$
$a = 1,\ b = 12,\ c = 14,\ d = 5$	$= (1 \cdot 14 + 12 \cdot 5)^2 + (1 \cdot 5 - 12 \cdot 14)^2 = 74^2 + 163^2$
$(9^2 + 8^2) \cdot (10^2 + 11^2)$	$= (9 \cdot 10 - 8 \cdot 11)^2 + (9 \cdot 11 + 8 \cdot 10)^2 = 2^2 + 179^2$
$a = 9,\ b = 8,\ c = 10,\ d = 11$	$= (9 \cdot 10 + 8 \cdot 11)^2 + (9 \cdot 11 - 8 \cdot 10)^2 = 178^2 + 19^2$
$(9^2 + 8^2) \cdot (14^2 + 5^2)$	$= (9 \cdot 14 - 8 \cdot 5)^2 + (9 \cdot 5 + 8 \cdot 14)^2 = 86^2 + 157^2$
$a = 9,\ b = 8,\ c = 14,\ d = 5$	$= (9 \cdot 14 + 8 \cdot 5)^2 + (9 \cdot 5 - 8 \cdot 14)^2 = 166^2 + 72^2.$

Above we obtained 24 different ways to represent 32045 by two squares. Further, if we factor 32045 as follows:

$$32045 = 5 \cdot 6409 = (1^2 + 2^2) \cdot (80^2 + 3^2), \quad a = 1,\ b = 2,\ c = 80,\ d = 3,$$

and then again apply formula (4.31):

$$32045 = 86^2 + 157^2 = 74^2 + 163^2,$$

we obtain two representations that have been already listed before. Are there new representations of 32045 as a sum of two squares? Try it yourself.

5. Can we rewrite 32045 as a sum of 8 squares? Consider factorization of 32045 and multiplication of the factors inside brackets:

$$32045 = \left[(2^2+1^2)\cdot(3^2+2^2)\right]\cdot\left[(4^2+1^2)\cdot(5^2+2^2)\right]$$
$$= \left((2\cdot3)^2+(2\cdot2)^2+(1\cdot3)^2+(1\cdot2)^2\right)\cdot\left((4\cdot5)^2+(4\cdot2)^2+(1\cdot5)^2+(1\cdot2)^2\right)$$
$$= \left(6^2+4^2+3^2+2^2\right)\cdot\left(20^2+8^2+5^2+2^2\right).$$

In order to use formula (4.33), we define $a_1 = 6$, $a_2 = 4$, $a_3 = 3$, $a_4 = 2$, $b_1 = 20$, $b_2 = 8$, $b_3 = 5$, and $b_4 = 2$
and obtain

$$32045 = 88^2 + 128^2 + 22^2 + 32^2 + 44^2 + 64^2 + 11^2 + 16^2.$$

If we multiply other pairs of factors, such as

$$32045 = \left[(2^2+1^2)\cdot(4^2+1^2)\right]\cdot\left[(3^2+2^2)\cdot(5^2+2^2)\right]$$

Or as

$$32045 = \left[(2^2+1^2)\cdot(5^2+2^2)\right]\cdot\left[(3^2+2^2)\cdot(4^2+1^2)\right].$$

We will obtain other representations of 32045 by sum of eight squares. Please try it yourself.

6. Now let us rewrite 32,045 by a sum of 16 squares:

$$32045 = [(2^2+1)(3^2+2^2)]\cdot[(4^2+1^2)(5^2+2^2)]$$
$$= [(2\cdot3)^2+(2\cdot2)^2+(1\cdot3)^2+(1\cdot2)^2]\cdot[(4\cdot5)^2+(4\cdot2)^2+(1\cdot5)^2+(1\cdot2)^2]$$
$$= (2\cdot3\cdot4\cdot5)^2+(2\cdot3\cdot4\cdot2)^2+(2\cdot3\cdot1\cdot5)^2+(2\cdot3\cdot1\cdot2)^2$$
$$+(2\cdot2\cdot4\cdot5)^2+(2\cdot2\cdot4\cdot2)^2+(2\cdot2\cdot1\cdot5)^2+(2\cdot2\cdot1\cdot2)^2$$
$$+(1\cdot3\cdot4\cdot5)^2+(1\cdot3\cdot4\cdot2)^2+(1\cdot3\cdot1\cdot5)^2+(1\cdot3\cdot1\cdot2)^2$$
$$+(1\cdot2\cdot4\cdot5)^2+(1\cdot2\cdot4\cdot2)^2+(1\cdot2\cdot1\cdot5)^2+(1\cdot2\cdot1\cdot2)^2$$
$$= 120^2+48^2+30^2+12^2+80^2+32^2+20^2+8^2+60^2$$
$$+24^2+15^2+6^2+40^2+16^2+10^2+4^2.$$

Representation of a number by a sum of several squares is solved. It is very interesting to find out if such a sum of perfect squares can be also a perfect square. There are many problems that can be created regarding this topic. Let us try to solve some of them now.

Problem 225

Find eleven consecutive natural numbers of which the sum of squares is also a perfect square.

Solution. Let

$$N = (n-5)^2 + (n-4)^2 + (n-3)^2 + (n-2)^2 + (n-1)^2 + n^2 + (n+1)^2$$
$$+ (n+2)^2 + (n+3)^2 + (n+4)^2 + (n+5)^2.$$

It is easy to simplify this number by pairing corresponding squares as follows:

$$\begin{aligned}
(n-5)^2 + (n+5)^2 &= 2n^2 + 2 \cdot 25 \\
(n-4)^2 + (n+4)^2 &= 2n^2 + 2 \cdot 16 \\
(n-3)^2 + (n+3)^2 &= 2n^2 + 2 \cdot 9 \\
(n-2)^2 + (n+2)^2 &= 2n^2 + 2 \cdot 4 \\
(n-1)^2 + (n+1)^2 &= 2n^2 + 2 \cdot 1 \\
n^2 &= n^2.
\end{aligned}$$

Adding the left and the right sides of the equations, and applying the formula for the sum of k squares, we obtain

$$\begin{aligned}
N &= 11n^2 + 2 \cdot (25 + 16 + 9 + 4 + 1) = 11n^2 + 2 \cdot \tfrac{5 \cdot (5+1) \cdot (2 \cdot 5 + 1)}{6} \\
&= 11n^2 + 10 \cdot 11 = 11 \cdot (n^2 + 10) \\
N &= 11 \cdot (n^2 + 10).
\end{aligned}$$

Next, we need to make sure that N is a perfect square. Since 11 is prime, then the following must be true:

$$n^2 + 10 = 11 \cdot k^2. \tag{4.35}$$

Under the condition above, we will have

$$N = 11 \cdot 11 \cdot k^2 = (11k)^2 = m^2, \quad m = 11k.$$

Let us solve equation (4.35) in natural numbers. First, we want to rewrite this equation in factorized form. For this, we we will represent it as follows:

$$\begin{aligned}
n^2 + 10 &= 10k^2 + k^2 \\
n^2 - k^2 &= 10 \cdot (k^2 - 1) \\
(n-k)(n+k) &= 10(k-1)(k+1) \\
(n-k)(n+k) &= 2 \cdot 5 \cdot (k-1) \cdot (k+1).
\end{aligned}$$

The last equation can be solved by rewriting it as possible systems of equations. Only the following system will give us natural values for variable n and k:

$$\begin{cases} n - k = 2(k+1) \\ n + k = 5(k-1) \end{cases} \Leftrightarrow k = 7, \ n = 23, \ m = 77.$$

Finally, we obtain that the sum of eleven consecutive squares from 18 to 28 is a square of 77:

$$18^2 + 19^2 + 20^2 + 21^2 + 22^2 + 23^2 + 24^2 + 25^2 + 26^2 + 27^2 + 28^2 = 5929 = 77^2.$$

Answer. The sum of eleven consecutive squares from 18 to 28 is a square of 77.

4.2.2 Sum of Cubes and More

Can we represent a natural number by the sum of two, three, four, five, or more cubes of some integer numbers?

You can find many examples that indicate such possibility for certain numbers. For example, 52 can be written as sum of two cubes $52 = 64 - 8 = 4^3 + (-2)^3$ or as sum of five cubes as $52 = 4^3 + (-3)^3 + 2^3 + 2^3 + (-1)^3$ or as $52 = 11^3 + (-10)^3 + (-10)^3 + 9^3 + (-2)^3$. On the other hand, 54 can be written as sum of four cubes as $54 = 10^3 + 8^3 + (-9)^3 + (-9)^3$. If we are allowed to use zero then 54 can be written as sum of five cubes easily as

$$54 = 10^3 + 8^3 + (-9)^3 + (-9)^3 + 0^3.$$

In Chapter 1 you learned about perfect numbers. It is interesting to know that all perfect numbers, except 6, can be written by a sum of consecutive cubes!

Example.
$$28 = 1^3 + 3^3$$
$$496 = 1^3 + 3^3 + 5^3 + 7^3$$
$$8128 = 1^3 + 3^3 + 5^3 + 7^3 + 9^3 + 11^3 + 13^3 + 15^3$$

In this section we are going to find some general rules of such decomposition and learn if and when it is possible to represent a natural number by the sum of several cubes. First, let us prove some statements by solving the following problems.

Problem 226

Prove that any integer that can be represented as a sum of the cubes of two other integers must be divisible by the sum of these integers.

Proof. Assume that $N = n^3 + m^3$, applying the formula of sum of two cubes, we obtain
$$N = n^3 + m^3 = (n+m) \cdot (n^2 - nm + m^2),$$

For example, $3^3 + 5^3 = 27 + 125 = 152 = 8 \cdot 19 = (3+5) \cdot 19$.
The formula above is often used in the following form:

$$n^3 + m^3 = (n+m)^3 - 3nm(n+m) = (n+m) \cdot L.$$

Problem 227

Prove that any multiple of 6 can be written as a sum of four cubes.

Proof. Notice that
$$(n+1)^3 + (n-1)^3 - n^3 - n^3 = 6n,$$
which means that any multiple of 6 can be represented as
$$6n = (n+1)^3 + (n-1)^3 + (-n)^3 + (-n)^3.$$

The proof is complete.

In the example above 54 was rewritten as sum of four cubes. Using this formula, you can make your own examples.

Problem 228 (MGU Russian Students' Olympiad)

Prove that any integer can be written as a sum of five cubes.

Proof. Consider the following obvious equality
$$(n+1)^3 + (n-1)^3 + (-n)^3 + (-n)^3 + 0^3 = 6n.$$

From this we know that every multiple of 6 can be written as the sum of five cubes. As we learned earlier in this chapter, all integer numbers are either multiples of 6 or leave a remainder of 1, 2, 3, 4, or 5 when divided by 6. Therefore,

$$
\begin{aligned}
6n+1 &= 6n+1^3 & &= (n+1)^3 + (n-1)^3 + (-n)^3 + (-n)^3 + 1^3 \\
6n+2 &= 6(n-1)+2^3 & &= n^3 + (n-2)^3 + (-n+1)^3 + (-n+1)^3 + 2^3 \\
6n+3 &= 6(n-4)+3^3 & &= (n-3)^3 + (n-5)^3 + (-n+4)^3 + (-n+4)^3 + 3^3 \\
6n+4 &= 6(n+2)+(-2)^3 & &= (n+3)^3 + (n+1)^3 + (-n-2)^3 + (-n-2)^3 + (-2)^3 \\
6n+5 &= 6(n+1)+(-1)^3 & &= (n+2)^3 + n^3 + (-n-1)^3 + (-n-1)^3 + (-1)^3.
\end{aligned}
$$

The proof is complete.

Let us explain how each formula was obtained using, for example, the representation of $6n+3$. First, we will rewrite as $6n+3 = 6 \cdot (n-4) + 3^3$, second, we simply replace all n in the formula by $(n-4)$, etc. Obviously, now we can rewrite any number as sum of five cubes. Take any number, for example, $N = 111$, divide it by 6 and obtain that $111 = 6 \cdot 18 + 3 = 6n+3$, $n = 18$, replace n by 18 in the third formula above:
$$111 = 15^3 + 13^3 + (-14)^3 + (-14)^3 + 3^3.$$

Modern number theory has solved almost every possible problem except some famous unsolved mentioned in the preface. However, people cannot remember all of them with the solutions and often such problems appear on different contests. Let us consider and prove the following statement.

Problem 229

Prove that a square of the arithmetic mean of two consecutive numbers, n and $n+1$ can be represented by the sum of n consecutive cubes from 1 to n.

Proof. Even ancient Greeks knew the formula of the sum of the first n consecutive cubes:

$$1^3 + 2^3 + 3^3 + \cdots + n^3 = \left[\frac{n(n+1)}{2}\right]^2$$

that proves the statement.

From this we can always know what numbers can be certainly rewritten by the sum of $7, 20$, or 2018 cubes. Thus, 44100 can be written by 20 consecutive cubes as follows:

$$1^3 + 2^3 + 3^3 + \cdots + 20^3 = \left(\frac{20 \cdot 21}{2}\right)^2 = 44100.$$

Problem 230

Find all positive integers n such that the equation $x^3 + y^3 + z^3 + t^3 = 100^{100^n}$ has an integer positive solution x, y, z, and t.

Solution. Note that

$$1^3 + 2^3 + 3^3 + 4^3 = 100. \tag{4.36}$$

Therefore, if $n = 0$, then solution exists. However, we need a solution for all natural n.

Let us modify (4.36) as follows:

$$\left(1 \cdot 100^k\right)^3 + \left(2 \cdot 100^k\right)^3 + \left(3 \cdot 100^k\right)^3 + \left(4 \cdot 100^k\right)^3$$
$$= 100 \cdot 100^{3k} = 100^{3k+1} = 100^{100^n}. \tag{4.37}$$

From (4.37) we can conclude that this equation has solutions if $3k + 1 = 100^n$, that can be written as $3k = 100^n - 1$.

Then the solution exists for any natural n since $100^n - 1$ is always divisible by 3. The corresponding value of k can be evaluated by the formula:

$$k = \frac{(100^n - 1)}{3},$$

Answer. $x = 1 \cdot 100^k$, $y = 2 \cdot 100^k$, $z = 3 \cdot 100^k$, and $t = 4 \cdot 100^k$, where $k = \frac{100^n - 1}{3}$.

Problem 231 (USSR Olympiad 1981)

Solve in natural numbers $x^3 - y^3 = xy + 61$.

Solution. Since $x > y$, let $x = y + a$, $a > 0$.

$$(y+a)^3 - y^3 = (y+a)y + 61$$
$$y^3 + 3y^2a + 3ya^2 + a^3 - y^3 - y^2 - ay = 61$$
$$(3a-1)y^2 + (3a^2 - a)y + a^3 = 61$$

From the last quadratic equation with positive coefficients and natural vales of y, it is obvious that

$$a^3 < 61, \quad 1 \le a \le 3.$$

Considering $a = 1$, $a = 2$, and $a = 3$, we found that at $a = 1$ the equation has integer solution.

$$2y^2 + 21y + 1 - 61 = 0$$
$$(y-5)(y+6) = 0$$
$$y = 5, x = 6$$

Answer. $x = 6$, $y = 5$.

As we are about to complete this section, I want to offer you two interesting problems, the first one was mentioned in the work of Diophantus.

Problem 232

Prove that the equation $x^2 - y^3 = 7$ has no solutions in integer numbers.

Proof. 1. Assume that y is even, $y = 2n$, then

$$x^2 - 8n^3 = 7$$
$$x^2 = 8n^3 + 4 + 3$$
$$x^2 = 4k + 3.$$

Because the square of a natural number cannot leave a remainder of 3 when divided by 4, then y cannot be an even number.

2. Assume that y is odd and that $y = 2n + 1$, then $y - 1 = 2n$ and we can rewrite our equation as

$$x^2 + 1 = y^3 + 8$$
$$x^2 + 1 = y^3 + 2^3 = (y+2)(y^2 - 2y + 4)$$
$$x^2 + 1 = (y+2)((y-1)^2 + 3)$$
$$x^2 + 1 = (y+2)(4n^2 + 3) = m \cdot (4k + 3)$$

as we know $x^2 + 1^2$ cannot contain any factor of $4k + 3$. Therefore, the given equation has no solutions.

Problem 233

Solve the equation in integers:

$$x_1^4 + x_2^4 + x_3^4 + \cdots + x_{13}^4 + x_{14}^4 = 100000000002015.$$

Solution. The left side is positive and represents the sum of the fourth powers of even and odd numbers. Obviously the fourth power of any even number is divisible by 16 and we need to find out what happens to any odd power of number x. Consider the following difference:

$$x^4 - 1 = (x - 1) \cdot (x + 1) \cdot (x^2 + 1)$$

If $x = 2n - 1$, then $x + 1 = 2n$, $x - 1 = 2n - 2$, and $x^2 + 1 = 4n^2 - 4n + 2$, and hence,

$$x^4 - 1 = (2n - 2) \cdot 2n \cdot (4n^2 - 4n + 2) = 2 \cdot 2 \cdot (n - 1) \cdot n \cdot 2 \cdot (2n^2 - 2n + 1)$$

Because the product of two consecutive integers is a multiple of 2, then the following is true.
$$x^4 - 1 = 16k,$$

which means that the fourth power of any odd number divided by 16 leaves a remainder of 1. Thus, the left side of the equation divided by 16 can have maximum remainder as 14 that can be written as

$$r_{left} = r_1 + r_2 + \cdots + r_{14} \leq 14.$$

If we divide the right side by 16, we will obtain 15. Therefore, the given equation does not have solutions.
Answer. No solutions.

4.3 Quadratic Congruence and Applications

In this section we will use congruence in order to prove important statements, some from the previous sections.

Theorem 30

The quadratic congruence $(x^2 + 1) \equiv 0 \pmod{p}$, where p is odd prime, has solutions if and only if $p \equiv 1 \pmod 4$.

Proof. Let a be any solution of $(x^2 + 1) \equiv 0 \pmod p$, then

$$a^2 \equiv -1 \pmod p. \tag{4.38}$$

Since p does not divide a, then the outcome of Theorem 13 is

$$a^{p-1} \equiv 1 \pmod p. \tag{4.39}$$

Raising both sides of (4.38) into power of $2n$ we obtain:

$$\begin{aligned} 1 &\equiv (a^2)^{2n} \equiv (-1)^{2n} \pmod p \\ a^{4n} &\equiv 1 \pmod p. \end{aligned} \tag{4.40}$$

For the same values of a, (4.39) and (4.40) are simultaneously true if and only if $4n = p - 1$. Therefore, $p = 4n + 1$ or $p \equiv 1 \pmod 4$.

Theorem 31

Solutions of $(x^2 + 1) \equiv 0 \pmod p$ satisfy $\left(\frac{p-1}{2}\right)! \equiv x \pmod p$.

Proof. In order to prove this theorem, let us apply again Wilson Theorem 27.

First, we will consider $(p-1)!$. If p is an odd prime, then $(p-1)!$ consists of $\frac{p-1}{2}$ factors that can be paired:

$$(p-1)! = 1 \cdot 2 \cdot 3 \cdots \cdots \frac{p-1}{2} \cdot \frac{p+1}{2} \cdots \cdots (p-2)(p-1). \tag{4.41}$$

For example, for $p = 13$, $p - 1 = 12$ we obtain

$$\begin{aligned} 12! &= 1 \cdot 2 \cdot 3 \cdot 4 \cdot 5 \cdot 6 \cdot \quad \big| 7 \cdot 8 \cdot 9 \cdot 10 \cdot 11 \cdot 12 \\ (p-1)! &= 1 \cdot 2 \cdot 3 \cdots \cdots \frac{p-1}{2} \cdot \Big| \frac{p+1}{2} \cdots \cdots (p-2) \cdot (p-1) \end{aligned}$$

Starting from the right to the left of (4.41), consider obvious congruences:

$$(p-1) \equiv -1 \pmod{p}$$
$$(p-2) \equiv -2 \pmod{p}$$
$$(p-3) \equiv -3 \pmod{p}$$

· · · · · · · · · · · · · · · · · · · ·

$$\frac{p+1}{2} \equiv -\frac{(p-1)}{2} \pmod{p}$$
$$\frac{p-1}{2} \equiv \frac{p-1}{2} \pmod{p}$$

· · · · · · · · · · · · · · · · · · · ·

$$2 \equiv 2 \pmod{p}$$
$$1 \equiv 1 \pmod{p}.$$

Here we use obvious formulas:

$$\frac{p+1}{2} - p = \frac{p+1-2p}{2} = -\frac{p-1}{2}.$$

Multiplying the left and the right sides of the congruences we get

$$(p-1)! = 1 \cdot (-1) \cdot 2 \cdot (-2) \cdot 3 \cdot (-3) \cdot \ldots \cdot \frac{p-1}{2} \cdot \left(-\frac{p-1}{2}\right) \pmod{p}.$$

The relationship above can be simplified as

$$(p-1)! \equiv (-1)^{\frac{p-1}{2}} \left(1 \cdot 2 \cdot 3 \cdot \ldots \cdot \frac{p-1}{2}\right)^2 \pmod{p},$$

or

$$(p-1)! \equiv (-1)^{\frac{p-1}{2}} \left(\left(\frac{p-1}{2}\right)!\right)^2 \pmod{p}. \tag{4.42}$$

Congruence (4.42) must satisfy Theorem 27:

$$(p-1)! \equiv -1 \, (\bmod \, p)). \tag{4.43}$$

Combining (4.42) and (4.43) we obtain

$$-1 \equiv (-1)^{\frac{p-1}{2}} \left(\left(\frac{p-1}{2}\right)!\right)^2 \pmod{p}. \tag{4.44}$$

Formula (4.44) can stay along but we want to consider only such prime p that satisfies the equation below

$$(x^2 + 1) \equiv 0 \pmod{p}. \tag{4.45}$$

Regarding Theorem 30, if such p exists then they must satisfy

$$p = 4k + 1. \tag{4.46}$$

Substituting (4.46) into (4.44) and using the fact that $(p-1)/2 = 2k$, we have

$$-1 \equiv \left(\left(\frac{p-1}{2} \right)! \right)^2 \quad (\text{mod } p),$$

which can be written as

$$\left[\left(\frac{p-1}{2} \right)! \right]^2 \equiv -1 \quad (\text{mod } p). \tag{4.47}$$

From (4.45) and (4.47) it follows that

$$\left(\frac{p-1}{2} \right)! \equiv x \quad (\text{mod } p). \tag{4.48}$$

is the solution of (4.45). Congruence (4.48) means that prime p divides the difference between x and $\left(\frac{p-1}{2} \right)!$.

Let us see how these two theorems can be applied to solving the following problems.

Problem 234

Decide whether or not the congruence $(x^2 + 1) \equiv 0 \ (\text{mod } 13)$ has a solution. If a solution exists, then solve the congruence.

Solution. This congruence has a solution in integers because $p = 13 = 4 \cdot 3 + 1 = 4k + 1$. Let us recall our congruence

$$(x^2 + 1) \equiv 0 \quad (\text{mod } 13). \tag{4.49}$$

The answer can be obtained directly as

$$(x^2 + 1) \equiv 0 \quad (\text{mod } 13)$$
$$x^2 \equiv -1 \quad (\text{mod } 13)$$
$$x^2 \equiv (13 \cdot 2 - 1) \quad (\text{mod } 13)$$
$$x^2 \equiv 25 \quad (\text{mod } 13)$$
$$x \equiv 5 \quad (\text{mod } 13), \ x \equiv -5 \quad (\text{mod } 13).$$

This gives us two possible answers for x,

$$x = 13k + 5, \text{ or } x = 13k + 8. \tag{4.50}$$

Both x will satisfy the congruence

$$(x^2 + 1) \equiv 0 \quad (\text{mod } 13), \ k = 0, 1, 2, 3, \ldots$$

For example,

$$5^2 + 1 = 26 = 2 \cdot 13$$
$$8^2 + 1 = 65 = 5 \cdot 13$$

In general (4.50) generates any solution for (4.49)

$$x = 13k + 5, \qquad\qquad x = 13k + 8$$
$$(13k+5)^2 + 1 = 13n, \qquad (13k+8)^2 + 1 = 13m$$
$$169k^2 + 2 \cdot 13k \cdot 5 + 25 + 1 = 13n, \quad 169k^2 + 2 \cdot 13k \cdot 8 + 64 + 1 = 13m$$
$$13(13k^2 + 10k + 2) = 13n, \qquad 13(13k^2 + 16k + 5) = 13m$$
$$n = 13k^2 + 10k + 2, \qquad\qquad m = 13k^2 + 16k + 5.$$

We can make the following table for $k = 0, 1$, and 2.

k	$n = 13k^2 + 10k + 2$	$x = 13k+5$	$x^2 + 1 = 13n$	$m = 13k^2 + 16k + 5$	$x = 13k+8$	$x^2 + 1 = 13m$
0	2	5	$5^2 + 1 = 13 \cdot 2$	5	8	$8^2 + 1 = 13 \cdot 5$
1	25	18	$18^2 + 1 = 13 \cdot 25$	34	21	$21^2 + 1 = 13 \cdot 34$
2	74	31	$31^2 + 1 = 13 \cdot 74$	89	34	$34^2 + 1 = 13 \cdot 89$

Table 4.4 Solution to quadratic congruence

Because of the symmetry of the equation (4.49) for x, we know that if a pair (x, l) satisfies it, then $(-x, l)$ is a solution as well. Thus, $(5, 2)$ solution makes $(-5, 2)$, $(8, 5)$ will make $(-8, 5)$, etc.

Regarding Theorem 31, a positive solution to (4.49) can be written in the form,

$$\left(\tfrac{p-1}{2}\right)! \equiv x \pmod{p}$$
$$720 = 6! \equiv 5 \pmod{13} \ (720 = 13 \cdot 55 + 5)$$
$$x \equiv 5 \pmod{13}$$
$$26 = (5^2 + 1) \equiv 0 \pmod{13}$$

Having one of the solutions

$$x \equiv 5 \pmod{13}, \tag{4.51}$$

how can we obtain all other solutions generated by $x = 13m + 8$? Formula (4.51) can be rewritten as $x \equiv -8 \pmod{13}$. However, if $x = -8$ is a solution then $x = 8$ is related to it solution, etc.

Answer. $x \equiv 5 \pmod{13}$ or $x \equiv 8 \pmod{13}$.

In each case decide whether or not the congruence has a solution, solve it.

(a)$x^2 \equiv -1 \pmod{29}$,
(b)$x^2 \equiv -1 \pmod{37}$.

Solution. Regarding Theorem 30 solutions to (a) and (b) exist because 29 and 37 divided by 4 leave a remainder of 1. Now we can find solution to equation (a) and (b).

Let us solve (a) first.

$$x^2 \equiv -1 \pmod{29}$$
$$x^2 \equiv (29 \cdot 5 - 1) \pmod{29}$$
$$x^2 \equiv 144 \pmod{29}$$
$$x \equiv \pm 12 \pmod{29}$$
$$x \equiv 12 \pmod{29}, \text{ or } x \equiv 17 \pmod{29},$$

and

$$12^2 + 1 = 145 = 5 \cdot 29 \equiv 0 \pmod{29}$$
$$17^2 + 1 = 290 = 10 \cdot 29 \equiv 0 \pmod{29}.$$

On the other hand, from Theorem 27 it follows that

$$x^2 \equiv -1 \pmod{29}$$
$$\left(\tfrac{29-1}{2}\right)! = 14! \equiv x \pmod{29}$$
$$x = 12$$
$$14! \equiv 12 \pmod{29}.$$

Note. We see that $14! = 8.71782912E + 10$ is a big number, but we know that when divided by 29 it gives a remainder of 12.

Now, let us solve (b).

$$x^2 \equiv -1 \pmod{37}$$
$$x^2 \equiv (-1 + 37) \pmod{37}$$
$$x^2 \equiv 36 \pmod{37}$$
$$x \equiv 6 \pmod{37}, \text{ or } x \equiv -6 \pmod{37},$$

which gives us solutions generated by the following rule: $x = 37k + 6$ or $x = 37m + 31$.

For example,

$$x = 80 = 37 \cdot 2 + 6 \ x^2 = 6400 \ x^2 + 1 = 6401 = 37 \cdot 173,$$
$$x = 31 = 37 \cdot 1 - 6 \ x^2 = 961 \ x^2 + 1 = 962 = 37 \cdot 26.$$

On the other hand, from Theorem 27

$$\left(\tfrac{37-1}{2}\right)! = 18! \equiv x \quad (\text{mod } 37)$$
$$18! \equiv 6 \quad (\text{mod } 37).$$

Remainder here can be only even number because 18! is even. $18! = 6.40237306E + 15$ divided by 37 leaves a remainder of 6.

Answer.

(a) $x \equiv 12 \pmod{29}$ or $x \equiv 17 \pmod{29}$,

(b) $x \equiv 6 \pmod{37}$ or $x \equiv 31 \pmod{37}$.

We know that every prime number of the form $(4m+1)$ is the sum of two squares in a unique way. For example, $13 = 2^2 + 3^2$, $29 = 2^2 + 5^2$, $41 = 4^2 + 5^2$, $37 = 1^2 + 6^2$, etc. Further, for primes of the form $p = 4m+1$, if we know a solution for

$$(x^2 + 1) \equiv 0 \quad (\text{mod } p),$$

then we will find all other solutions of the following equation:

$$(x^2 + y^2) \equiv 0 \quad (\text{mod } p).$$

> **Problem 236**
>
> Find all integer solutions of $(x^2 + y^2) \equiv 0 \pmod{13}$.

Solution. Above we found that

$$(x^2 + 1) \equiv 0 \quad (\text{mod } 13)$$
$$x^2 \equiv -1 \quad (\text{mod } 13)$$
$$x^2 \equiv (-1 + 26) \quad (\text{mod } 13)$$
$$x^2 \equiv 25 \quad (\text{mod } 13).$$

Therefore,

$$(x^2 + y^2) \equiv (x - 5y)(x + 5y) \quad (\text{mod } 13)$$
$$x = \pm 5y.$$

This generates infinitely many solutions for the problem, such as $y = 1$, $x = 5$, $y = 2$, $x = 10$, $y = 3$, $x = 15$, etc. Then,

$$\begin{aligned}
5^2 + 1^2 &= 26 &&= 2 \cdot 13 \equiv 0 \quad (\text{mod } 13) \\
10^2 + 2^2 &= 104 &&= 8 \cdot 13 \equiv 0 \quad (\text{mod } 13) \\
15^2 + 3^2 &= 234 &&= 18 \cdot 13 \equiv 0 \quad (\text{mod } 13) \\
20^2 + 4^2 &= 416 &&= 32 \cdot 13 \equiv 0 \quad (\text{mod } 13).
\end{aligned} \tag{4.52}$$

Answer. $(x, y) = \{(5t, t), \ t = \pm 1, \pm 2, \pm 3, \dots\}$.

Problem 237

Solve the following equation in integers: $x^2 + y^2 - 29z = 0$.

Solution. 29 is a prime of the form $(4k+1)$. Consider the corresponding congruence:

$$(x^2 + y^2) \equiv 0 \quad (\text{mod } 29). \tag{4.53}$$

Earlier we found solutions of $x^2 \equiv -1 \ (\text{mod } 29)$ and that $x^2 \equiv 144 \ (\text{mod } 29)$. Therefore, all other solutions of (4.53) can be found from

$$(x^2 + y^2) \equiv (x - 12y)(x + 12y) \quad (\text{mod } 29).$$

We have $x = \pm 12y$. Next, for each pair of (x, y) we will calculate the corresponding z:

$$144y^2 + y^2 = 29z$$
$$145y^2 = 29z$$
$$z = 5y^2,$$

and

$$y = \pm 1, \ x = \pm 12, \ 12^2 + 1^2 = 29z, \ z = 5, \quad (x,y,z) = (\pm 12, \pm 1, 5)$$
$$y = \pm 2, \ x = \pm 24, \ 24^2 + 2^2 = 29z, \ z = 20, \ (x,y,z) = (\pm 24, \pm 2, 20)$$
$$y = \pm 3, \ x = \pm 36, \ 36^2 + 3^2 = 29z, \ z = 45, \ (x,y,z) = (\pm 36, \pm 3, 45).$$

Answer. $(x,y,z) = \{(\pm 12t, t, 5t^2), \ t = \pm 1, \pm 2, \pm 3, \ldots\}$.

Problem 238

Solve the congruence: $2x^2 + 3x + 1 \equiv 0 \ (\text{mod } 7)$.

Solution. The left side can be factored as follows:

$$(2x + 1) \cdot (x + 1) \equiv 0 \quad (\text{mod } 7)$$

Then either

$$2x + 1 \equiv 0 \quad (\text{mod } 7) \rightarrow x \equiv 3 \quad (\text{mod } 7)$$

or

$$x + 1 \equiv 0 \quad (\text{mod } 7) \rightarrow x \equiv 6 \quad (\text{mod } 7)$$

Answer. $x = 7n + 3$ or $x = 7m + 6$.

What should we do if a quadratic expression cannot be easily factored? For example, how can we solve the following congruence?

$$3x^2 + 6x + 5 \equiv 0 \pmod{7}$$

Do you remember what we did earlier, when solving quadratic equations, such as

$$3x^2 + 6x + 5 - 7y = 0?$$

We would consider it as quadratic in variable x and y to be a parameter and by writing a quadratic formula we would find out for what values of y the discriminant of the quadratic equation is a perfect square, that can be written as

$$21y - 6 = k^2$$

Then we would subtract the unit from both sides and factor the left and right sides as

$$21y - 7 = k^2 - 1$$
$$7(3y - 1) = (k - 1)(k + 1)$$

Then we would consider that the right side is the product of two consecutive odd or consecutive even numbers and use the fact that 7 is a prime number. However, the next theorem will help us to solve quadratic congruences in an easy way.

Theorem 32

Let p be an odd prime and p does not divide a. Then

$$(ax^2 + bx + c) \equiv 0 \pmod{p}$$

has an integer solution if and only if $z^2 \equiv d \pmod{p}$, where $d = b^2 - 4ac$ is the discriminant of a quadratic equation. For each solution z of

$$z^2 \equiv d \pmod{p},$$

a solution x of

$$(2ax + b) \equiv z \pmod{p}$$

is a solution of $(ax^2 + bx + c) \equiv 0 \pmod{p}$.

The proof of this theorem will be given in Section 4.3.2. Below we will solve the problem mentioned above in two different ways. The second solution is an application of Theorem 32.

Problem 239

Solve $3x^2 + 6x - 7y + 5 = 0$ in integers and find minimal positive values for (x, y).

Solution. Method 1. We can rewrite this as

$$(3x^2 + 6x + 5) \equiv 0 \pmod 7.$$

In order to extract a multiple of 7 we will multiply both sides of the equation by 5:

$$(15x^2 + 30x + 25) \equiv 0 \pmod 7.$$

Extracting the largest multiple of 7 in each term we have:

$$(14x^2 + x^2 + 28x + 2x + 21 + 4) \equiv 0 \pmod 7$$
$$(x^2 + 2x + 4) \equiv 0 \pmod 7.$$

Completing the square on the left:

$$((x+1)^2 + 3) \equiv 0 \pmod 7$$
$$(x+1)^2 \equiv -3 \pmod 7.$$

Since $-3 + 7 = 4$, the last relationship can be written as $(x+1)^2 \equiv 4 \pmod 7$, which is equivalent to two congruence equations:

$$(x+1) \equiv 2 \pmod 7 \text{ or } (x+1) \equiv -2 \pmod 7$$
$$x \equiv 1 \pmod 7 \qquad \text{or } x \equiv 4 \pmod 7.$$

Substituting x into original equation, we obtain two formulas for y:

$$x = 7k + 1 \qquad \text{or } x = 7m + 4$$
$$y = 21k^2 + 12k + 2 \text{ or } y = 21m^2 + 30m + 11.$$

Setting $k = m = 0$ we obtain minimal values of x and y as ordered pairs, $(1,2)$ and $(4,11)$.

Method 2. Let us apply Theorem 32 of this section. We have:

$$(3x^2 + 6x + 5) \equiv 0 \pmod 7$$
$$d = 6^2 - 4 \cdot 3 \cdot 5 = -24$$
$$z^2 \equiv d \pmod p \qquad\qquad z \equiv -2 \pmod 7 \text{ or } z \equiv 2 \pmod 7$$
$$z^2 \equiv -24 \pmod 7 \qquad\qquad z = 5 \qquad\qquad \text{or } z = 2.$$
$$z^2 \equiv 4 \pmod 7$$

For each value of z we obtain corresponding x:

$$(2 \cdot 3x + 6) \equiv 2 \pmod 7 \text{ or } (2 \cdot 3x + 6) \equiv 5 \pmod 7$$
$$(6x + 6) \equiv 30 \pmod 7 \quad \text{ or } (6x + 6) \equiv 5 \pmod 7$$
$$(x + 1) \equiv 5 \pmod 7 \quad\quad \text{ or } (6x + 6) \equiv 12 \pmod 7$$
$$x \equiv 4 \pmod 7 \quad\quad\quad\quad \text{ or } x \equiv 1 \pmod 7.$$

Answer. $x = 4, y = 11$ or $x = 1, y = 2$.

Problem 240

Solve the equation $2x^2 - 71y = 5$.

Solution. Let us solve the congruence instead:

$$2x^2 \equiv 5 \pmod{71}$$

Multiplying both sides by 2 it can be rewritten as more convenient

$$4x^2 \equiv 10 \pmod{71}$$

that can be solved as

$(2x)^2 \equiv 81 \pmod{71}$
$2x \equiv 9 \pmod{71}$ or $2x \equiv -9 \pmod{71}$
$2x \equiv 80 \pmod{71}$ or $2x \equiv 62 \pmod{71}$
$x \equiv 40 \pmod{71}$ or $x \equiv 31 \pmod{71}$
$x = 71n + 40$ or $x = 31 + 71k$
$y = 142n^2 + 160n + 45$ or $y = 142k^2 + 124k + 27$, $n, k \in \mathbb{Z}$.

Answer. $(x, y) : (71n + 40, 142n^2 + 160n + 45), (31 + 71k, 142k^2 + 124k + 27)$

4.3.1 Euler's Criterion on Solution of Quadratic Congruence

By solving problems from this book, you noticed that many quadratic equations do not have solutions in integers. It was helpful to use properties of congruence to reduce an equation to a simple one of the type:

$$x^2 \equiv a (mod p), (a, p) = 1, p \text{ is prime} \tag{4.54}$$

When we have a congruence in the simplest form (4.54), it would be nice to know if it has solution or not. Leonard Euler stated the following criterion.

> **Theorem 33** (Euler's Theorem on Solution of Quadratic Congruence)
>
> In order for (4.54) to have a solution, or in order for a to be residual of (4.54), it is necessary and sufficient for a to satisfy
>
> $$a^{\frac{p-1}{2}} \equiv 1 \quad (\text{mod } p). \tag{4.55}$$
>
> On the other hand, if
>
> $$a^{\frac{p-1}{2}} \equiv -1 \quad (\text{mod } p), \tag{4.56}$$
>
> then a is not a quadratic residual of congruence of (4.54).

Proof. From Fermat's Little Theorem, it follows that $a^{p-1} \equiv 1\,(mod\,p)$, which can be rewritten as $a^{\left(\frac{p-1}{2}\right)\cdot 2} - 1^2 \equiv 0\,(mod\,p)$ and then apply to this the difference of squares formula:

$$\left(a^{\frac{p-1}{2}} - 1\right) \cdot \left(a^{\frac{p-1}{2}} + 1\right) \equiv 0\,(mod\,p)$$

Two factors on the left-hand side differ by 2 and simultaneously cannot be divisible by p. This means that the following is true:

$$a^{\frac{p-1}{2}} \equiv 1(mod\,p) \Rightarrow a \text{ is a quadratic residual of } (4.54)$$

$$a^{\frac{p-1}{2}} \equiv -1(mod\,p) \Rightarrow a \text{ is not quadratic residual of } (4.54)$$

Another form of formula (4.54) is $x^2 \equiv a\,(mod\,p)$ can be written as

$$x^2 = a + n \cdot p \Rightarrow \begin{bmatrix} x^2 \equiv a\,(mod\,p) \\ a \equiv x^2\,(mod\,p) \end{bmatrix}$$

Therefore if we want to know if a solution to (4.54) exists then we can check Euler Criterion. If p is not big, then it is easy to solve quadratic congruence and decide on its solution.

> **Problem 241**
>
> Does the equation $x^2 \equiv 3 \pmod{11}$ have a solution?

Solution. Because the given equation can be rewritten as $x^2 \equiv 25 \pmod{11}$, $25 = 3 + 2 \cdot 11$, then a solution exists and can be easily found as $x \equiv \pm 5(mod\,11) \Rightarrow x \equiv 5 \pmod{11}$, $x \equiv 6 \pmod{11}$. Obviously $a = 3$ must satisfy the Euler Criterion:

$$3^{\frac{11-1}{2}} = 3^5 = 243 \equiv 1 \pmod{11}.$$

Therefore a solution exists.

Problem 242

Prove that $x^2 \equiv 2 \pmod{11}$ does not have any solutions.

Solution. Let us check Euler's criterion: $2^{\frac{11-1}{2}} = 2^5 = 32 \equiv -1 \pmod{11} \Rightarrow$ solution does not exist.

Theorem 34

Numbers $1^2, 2^2, 3^2, 4^2, ..., (\frac{p-1}{2})^2$ form a system of representatives of all classes of quadratic residuals.

Problem 243

Given $p = 17$. Find all classes of quadratic residuals.

Solution.
$$1^2 = 1, \ 2^2 = 4, \ 3^2 = 9, \ 4^2 = 16,$$
$$5^2 = 25 \equiv 8 \,(mod\,17),$$
$$6^2 = 36 \equiv 2 \,(mod\,17)$$
$$7^2 = 49 \equiv 15 \,(mod\,17)$$
$$8^2 = 64 \equiv 13 \,(mod\,17)$$

Here $\frac{p-1}{2} = \frac{17-1}{2} = 8$ and all other numbers will produce the same set of residuals $1, 2, 4, 8, 9, 13, 15, 16$.

All of these numbers make quadratic congruence $x^2 = a \,(mod\,17)$ solvable. On the other hand, for $a = 3, 5, 6, 7, 10, 11, 12, 14$ the given quadratic congruence does not have solutions.

Problem 244

Given $p = 7$. Find all classes of quadratic residuals.

Solution. All possible quadratic residuals will be produced by $1, 2$ and $3 = \frac{7-1}{2}$.

$$1^2 \equiv 1 \,(mod\,7)$$
$$2^2 = 4 \,(mod\,7)$$
$$3^2 = 9 \equiv 2 \,(mod\,7)$$
$$4^2 = 16 \equiv 2 \,(mod\,7)$$
$$5^2 = 25 \equiv 4 \,(mod\,7)$$
$$6^2 = 36 \equiv 1 \,(mod\,7)$$

Therefore, $a = 1$, 2, and 4 are residuals of $x^2 \equiv a(mod\,7)$ and $a = 3$, 5, and 6 are non-residuals and hence any of the equations below do not have solution:

$$x^2 \equiv 3 \,(mod\,7) \quad x^2 \equiv 5 \,(mod\,7) \quad x^2 \equiv 6 \,(mod\,7).$$

You can check it yourself by applying Euler criterion. For example, consider $x^2 \equiv 3 \,(mod\,7)$ and let us check $3^{\frac{7-1}{2}} = 3^3 = 27 \equiv -1 \,(mod\,7)$.

The Euler criterion is an excellent way to check whether quadratic congruence has solutions but sometimes it is hard to apply it because of big exponents.

Problem 245

Does $x^2 \equiv 8(mod\,53)$ have a solution?

Solution. Because $\frac{53-1}{2} = 26$, we need to find the remainder of 8^{26} divided by 53. In order to get the answer, we can go through the following consecutive squaring procedure:

$$8^2 = 64 \equiv 11 \quad (mod\ 53)$$
$$8^4 \equiv 121 \quad (mod\ 53) \equiv 15 \quad (mod\ 53)$$
$$8^8 \equiv 225 \quad (mod\ 53) \equiv 13 \quad (mod\ 53)$$
$$8^{16} \equiv 169 \quad (mod\ 53) \equiv 10 \quad (mod\ 53)$$
$$8^{26} = 8^2 \cdot 8^8 \cdot 8^{16} \equiv 11 \cdot 10 \cdot 13 \quad (mod\ 53) \equiv -1 \quad (mod\ 53).$$

Because $8^{\frac{53-1}{2}} = 8^{26} \equiv -1(mod\,53)$, then the given congruence does not have a solution.

Answer. No solutions.

4.3.2 Legendre and Jacobi Symbols

Solvability of quadratic congruence can be established easier by using the **Legendre symbol**.

Definition. Let p be an odd prime and $(a, p) = 1$, then $L(a, p) = \left(\frac{a}{p}\right)$ is the Legendre symbol.

The following is true:

$$\text{If } L(a,p) = \begin{cases} \left(\frac{a}{p}\right) = 1 & \Rightarrow x^2 \equiv a(mod\,p) \text{ has two solutions} \\ \left(\frac{a}{p}\right) = 0 & \Rightarrow x^2 \equiv a(mod\,p) \text{ has one solution } x \equiv 0(mod\,p) \\ \left(\frac{a}{p}\right) = -1 & \Rightarrow x^2 \equiv a(mod\,p) \text{ is not solvable in integers.} \end{cases}$$

Example. Because $x^2 \equiv 3(mod\,11)$ has two solutions, $x \equiv \pm 5(mod\,11)$, then $L(3,11) = 1$. On the other hand, $L(2,11) = -1$ because $x^2 \equiv 2(mod\,11)$ does not have a solution.

Property 37 (Properties of Legendre Symbol)

1. If $a \equiv b(mod\,p)$ then $L(a,p)=L(b,p)$
2. $L(a_1 \times a_2 \times \cdots \times a_k, p) = L(a_1,p) \times L(a_2,p) \times \cdots \times L(a_k,p)$
3. $L(a^2,p) = 1$
4. $L(1,p) = 1$
5. $L(a,p) = a^{\frac{p-1}{2}} (mod\,p)$ Euler criterion
6. $L(-1,p) = (-1)^{\frac{p-1}{2}}$
7. $L(2,p) = (-1)^{\frac{p^2-1}{8}}$
8. $L(q,p) = (-1)^{\frac{p-1}{2} \cdot \frac{q-1}{2}} \cdot L(p,q)$

The last property was first proven by Carl F. Gauss in 1796.

The above makes it possible to formulate algorithm for computing the Legendre symbol.

1. If the number is a negative, then select factor of $L(-1,p)$ and evaluate it as

$$L(-1,p) = \left(\frac{-1}{p}\right) = (-1)^{\frac{p-1}{2}} = \begin{cases} 1, & \text{if } p \equiv 1\,(mod\,4) \\ -1, & \text{if } p \equiv 3\,(mod\,4) \end{cases}$$

2. Replace a by the remainder under division of a by p.
3. Factor number a into a product of simple factors (prime factorization)

$$a = a_1^{\alpha_1} \cdot a_2^{\alpha_2} \cdot \ldots \cdot a_k^{\alpha_k}$$

If the number cannot be decomposed into primes, then go to step 7.
4. Decompose Legendre symbol
$$L(a,p) = (L(a_1,p))^{\alpha_1} \cdot (L(a_2,p))^{\alpha_2} \cdot \ldots \cdot (L(a_k,p))^{\alpha_k}$$

$$L(a_1 \cdot a_2, p) = L(a_1,p) \cdot L(a_2,p)$$

The proof is simple. Using the definition of Legendre symbol and properties of exponents, we have

$$\left(\frac{a_1 \cdot a_2}{p}\right) \equiv (a_1 \cdot a_2)^{\frac{p-1}{2}} \,(mod\,p) \equiv a_1^{\frac{p-1}{2}} \cdot a_2^{\frac{p-1}{2}} \,(mod\,p) = \left(\frac{a_1}{p}\right) \cdot \left(\frac{a_2}{p}\right)$$

5. Discard the factors having an even power α_j. Note that $\left(\frac{a^{2n}}{p}\right) = \left(\frac{(a^n)^2}{p}\right) = 1$.
6. Evaluate Legendre symbol $L(a_j, p)^{\alpha_j}$ for $a_j = 2$:

$$L(2,p) = \left(\frac{2}{p}\right) = (-1)^{\frac{p^2-1}{8}} = \begin{cases} 1, & p \equiv \pm 1 \,(mod\,8) \\ -1, & p \equiv \pm 3 \,(mod\,8) \end{cases}$$

7. If all Legendre symbols are known $L(a,p) = (L(a_1,p))^{\alpha_1} \cdot (L(a_2,p))^{\alpha_2} \cdot \ldots \cdot (L(a_k,p))^{\alpha_k}$, then calculations of Legendre symbol $L(a, p)$ is complete. Otherwise, for factors $(L(a_j,p))^{\alpha_j}$, for which $a_j \neq 2$, we apply the law of reciprocity

$$L(q,p) = L(p,q)) \cdot (-1)^{\frac{(p-1)}{2} \cdot \frac{(q-1)}{2}}$$

$$\left(\frac{q}{p}\right) = \begin{cases} \left(\frac{p}{q}\right), & p \equiv 1 \,(mod\,4) \text{ or } q \equiv 1 \,(mod\,4) \\ -\left(\frac{p}{q}\right), & p \equiv 3 \,(mod\,4) \text{ and } q \equiv 3 \,(mod\,4) \end{cases}$$

Let us solve $x^2 \equiv 8 \pmod{53}$ again applying Legendre symbol. Denote $a = 8$ and $p = 53$. Hence, we need to evaluate $L(8,53)$:

$$L(8,53) = L(2^3, 53) = (L(2,53))^3$$
$$L(2,53) = (-1)^{\frac{53^2-1}{8}} = (-1)^{351} = -1$$
$$L(8,53) = (-1)^3 = -1.$$

We did not have to evaluate $\frac{53^2-1}{8}$, it was enough to check just this,

$$53 = 8 \cdot 7 - 3 = 8n - 3$$

Therefore the given congruence does not have solutions.

Problem 246

Solve the equation
$$x^2 - 97y = 85. \tag{4.57}$$

Solution. Let us rewrite this equation in terms of quadratic congruence

$$x^2 \equiv 85 \pmod{97}$$

and first investigate whether or not this congruence has a solution. Applying properties of Legender symbol, we obtain

$$L(85,97) = L(17 \cdot 5, 97) = L(17,97) \cdot L(5,97)$$

First we will evaluate $L(17,97)$:

$$L(17,97) = (-1)^{\frac{17-1}{2} \cdot \frac{97-1}{2}} \cdot L(97,17) = (-1)^{384} \cdot L(97,17)$$
$$97 \equiv 12 \pmod{17} \rightarrow L(97,17) = L(12,17)$$
$$L(12,17) = L(3 \cdot 2^2, 17) = L(3,17) \cdot (L(2,17))^2 = L(3,17).$$

Let us evaluate $L(3,17)$,

$$L(17,97) = L(3,17) = (-1)^{\frac{3-1}{2} \cdot \frac{17-1}{2}} \cdot L(17,3) = (-1)^8 \cdot L(2,3) = (-1)^{\frac{3^2-1}{8}} = -1.$$

Hence, $L(17,97) = -1$. Now we can evaluate $L(5,97)$:

$$L(5,97) = (-1)^{\frac{5-1}{2} \cdot \frac{97-1}{2}} \cdot L(97,5)$$
$$97 = 5 \cdot 19 + 2 \rightarrow L(97,5) = L(2,5)$$
$$L(5,97) = (-1)^{2 \cdot 48} \cdot L(2,5) = (-1)^{\frac{5^2-1}{8}} = (-1)^3 = -1.$$

Finally,

$$L(85,97) = (-1) \cdot (-1) = 1,$$

which means that the given congruence has a solution. Notice that

$$97 \cdot 20 + 85 = 2025 = 45^2,$$

then the given congruence can be rewritten and solved as

$$x^2 \equiv 2025 \pmod{97}$$
$$x \equiv 45 \pmod{97} \quad \text{or } x \equiv -45 \pmod{97}$$
$$x \equiv 45 \pmod{97} \quad \text{or } x \equiv 52 \pmod{97}$$
$$x = 45 + 97n \quad \text{or } x = 52 + 97k$$
$$y = 97n^2 + 90n + 20 \quad \text{or } y = 97k^2 + 104k + 27, \ n, k \in \mathbb{Z}.$$

Both values of y were obtained after substituting the corresponding x into (4.57). Thus, solving the equation for y:

$$(45 + 97n)^2 = 85 + 97y,$$

we would obtain that $y = 97n^2 + 90n + 20$.

Obviously the time spent evaluating the Legendre symbols becomes much shorter with practice.

Answer. $(x,y) = (45 + 97n, 97n^2 + 90n + 20)$, $(52 + 97k, 97k^2 + 104k + 27)$, $n, k \in \mathbb{Z}$.

Problem 247

Solve the congruence $x^2 \equiv 506 \pmod{1103}$.

Solution. Since 1103 is prime number, we can use Legendre symbol operation to find if the given congruence is solvable. Notice that $506 = 2 \cdot 253 = 2 \cdot 11 \cdot 23$.

$$L(506, 1103) = L(2 \cdot 11 \cdot 23, 1103) = L(2, 1103) \cdot L(11, 1103) \cdot L(23, 1103)$$

Next, we will find each factor separately.

$$L(2, 1103) = (-1)^{\frac{1103^2 - 1}{8}} = 1. \tag{4.58}$$

$$L(11, 1103) = (-1)^{\left(\frac{11-1}{2}\right) \cdot \left(\frac{1103-1}{2}\right)} \cdot L(1103, 11) = -1 \cdot L(1103, 11) \tag{4.59}$$

Dividing 1103 by 11 we can state that $1103 \equiv 3 (mod\,11)$ and then

$$L(1103, 11) = L(3, 11) = (-1)^{\frac{3-1}{2}} \cdot (-1)^{\frac{11-1}{2}} = 1. \tag{4.60}$$

Hence, substituting (4.60) into (4.59) we have

$$L(11, 1103) = -1. \tag{4.61}$$

Next, we will evaluate

$$L(23, 1103) = L(1103, 23) \cdot (-1)^{\frac{23-1}{2} \cdot \frac{1103-1}{2}} = -L(1103, 23),$$

Dividing 1103 by 23, we obtain that

$$L(1103, 23) = L(23, 1103) = L(22, 23) = L(-1, 23) = (-1)^{\frac{23-1}{2}} = -1.$$

and then

$$L(23, 1103) = 1. \tag{4.62}$$

Finally, multiplying (4.58), (4.61), and (4.62), we obtain

$$L(506, 1103) = -1.$$

Which means that the given congruence has no solutions.
Answer. No solutions.

Remark. This problem can be solved easier. Because

$$1103 = 1004 - 1 = 8 \cdot 138 - 1 \equiv -1 \, (mod\,8),$$

then

$$L(2, 1103) = \left(\frac{2}{1103}\right) = (-1)^{\frac{1103^2 - 1}{8}} = 1$$

$$L(11,1103) = \left(\tfrac{11}{1103}\right) = \left(\tfrac{1103}{11}\right) \cdot (-1)^{\frac{11-1}{2} \cdot \frac{1103-1}{2}}$$
$$= \left(\tfrac{3}{11}\right) = \left(\tfrac{11}{3}\right) \cdot (-1)^{\frac{11-1}{2} \cdot \frac{3-1}{2}}$$
$$= \left(\tfrac{2}{3}\right) \cdot (-1) \cdot (-1) = (-1)^{\frac{3^2-1}{8}} = -1.$$

$$L(23,1103) = \left(\tfrac{23}{1103}\right) = \left(\tfrac{1103}{23}\right) \cdot (-1)^{\frac{23-1}{2} \cdot \frac{1103-1}{2}}$$
$$= \left(\tfrac{22}{23}\right) \cdot (-1) = \left(\tfrac{-1}{23}\right) \cdot (-1) = 1.$$

$$\left(\frac{506}{1103}\right) = \left(\frac{2}{1103}\right) \cdot \left(\frac{11}{1103}\right) \cdot \left(\frac{23}{1103}\right) = -1.$$

Next, let us prove Theorem 32 using Legendre symbol.

Proof. Consider a quadratic congruence,

$$ax^2 + bx + c \equiv 0 (mod\ p). \tag{4.63}$$

Because p is an odd prime, we can multiply both sides of the congruence by $4a$,

$$4a \cdot ax^2 + 4abx + 4ac \equiv 0 (mod\ p)$$

Next we will complete a square on the left-hand side,

$$(2ax + b)^2 - b^2 + 4ac \equiv 0 (mod\ p),$$

which can be rewritten as

$$(2ax + b)^2 \equiv (b^2 - 4ac)\ (mod\ p) \tag{4.64}$$

Recognizing the discriminant of a quadratic equation on the right-hand side, we can denote the following:

$$D = b^2 - 4ac, \quad z = 2ax + b,$$

that allows us to rewrite (4.64) in a new form stated by Theorem 32:

$$z^2 \equiv D (mod\ p) \tag{4.65}$$

Using Legendre symbol we can state the following
1. Equation (4.65) has two solutions if $L(D, p) = \left(\tfrac{D}{p}\right) = 1$.
2. Equation (4.65) has no solutions if $L(D, p) = \left(\tfrac{D}{p}\right) = -1$.
3. Equation (4.65) has one solution if $L(D, p) = \left(\tfrac{D}{p}\right) = 0,\ D = p \cdot k$.
   The proof is complete.

Let us solve the following problems.

Problem 248

Does the equation $5x^2 - 17x + 12 = 11y$ have solution in integers? If yes, then find the least positive solutions for variable x.

Solution. First, we will rewrite it using congruence notation:

$$5x^2 - 17x + 12 \equiv 0 (mod\,11)$$

And evaluate the discriminant,

$$D = 17^2 - 4 \cdot 5 \cdot 12 = 49$$

Let us evaluate Legendre symbol,

$$L(49, 11) = \left(\frac{49}{11}\right) = \left(\frac{7^2}{11}\right) = 1,$$

then the given equation has two solutions:

$$(2 \cdot 5x - 17)^2 \equiv 7^2 (mod\,11)$$

$$10x - 17 \equiv 7 (mod\,11) \qquad 10x - 17 \equiv -7 (mod\,11)$$

$$10x \equiv 13 (mod\,11) \qquad 10x \equiv 10 (mod\,11)$$

$$10x \equiv (13 + 7 \cdot 11)(mod\,11) \qquad x \equiv 1 (mod\,11)$$

$$x \equiv 9 (mod\,11)$$

$$x = 9 \qquad\qquad x = 1$$

Answer. $x = 1$, $x = 9$.

Question. Do we always need to complete the square when solving a quadratic congruence?

Problem 249

Solve $168x^2 + 169x + 84 \equiv 0 (mod\,503)$, where 503 is prime.

Solution. Let us multiply both sides by 3, so the numbers on the left-hand side will be easy to work with.

$$504x^2 + 507x + 252 \equiv 0 (mod\,503)$$

Next, we will extract a multiple of 503,

$$x^2 + 4x + 252 \equiv 0 (mod\,503), \tag{4.66}$$

and then complete the square, and simplify:

$$(x+2)^2 \equiv 255 (mod\,503) \tag{4.67}$$

Let us check whether this congruence has a solution. Thus, Legendre symbol is

$$L(255,503) = \left(\frac{255}{503}\right) = \left(\frac{3}{503}\right) \cdot \left(\frac{5}{503}\right) \cdot \left(\frac{17}{503}\right) = ? \tag{4.68}$$

Of course, we could proceed with (4.68) by using reciprocity relationship. However, it is much easy to start working with (4.66) instead. Let us multiply both sides of it by 2:

$$2x^2 + 8x + 504 \equiv 0 (mod\,503),$$

Subtracting 503 from both sides make it much easier than (4.66):

$$2x^2 + 8x + 1 \equiv 0 (mod\,503). \tag{4.69}$$

Since (4.69) is equivalent to (4.66), we can apply Theorem 32, evaluate the discriminant and then evaluate Legendre symbol for it as

$$D = 8^2 - 4 \cdot 2 \cdot 1 = 56$$
$$L(D,503) = \left(\frac{56}{503}\right) = \left(\frac{7}{503}\right) \cdot \left(\frac{2^3}{503}\right)$$
$$= \left(\frac{503}{7}\right)(-1)^{\frac{503-1}{2} \cdot \frac{7-1}{2}} \cdot \left(\frac{2}{503}\right) \cdot \left(\frac{2^2}{503}\right)$$

The last factor is one. The power of (-1) is odd, and

$$503 = 7 \cdot 71 + 6 \;\Rightarrow\; \left(\frac{503}{7}\right) = \left(\frac{6}{7}\right) = \left(\frac{-1}{7}\right) = -1.$$

Finally, we have

$$\left(\frac{2}{503}\right) = (-1)^{\frac{503^2-1}{8}} = 1, \;\; \text{because} \;\; 503 = 8 \cdot 63 - 1 \equiv -1 (mod\,8).$$

We know that the given congruence is solvable and has two solutions. Working with (4.67) please notice that

$$503 \cdot 3 + 255 = 1764 = 42^2.$$

Then (4.67) can be solved as

$$\begin{array}{ll} x + 2 \equiv 42 (mod\,503) & x + 2 \equiv -42 (mod\,503) \\ x \equiv 40 (mod\,503) & x \equiv 459 (mod\,503) \end{array}$$

Answer. $x \equiv 40; \; 459 (mod\,503)$.

Problem 250

Solve the congruence $x^2 - 6x + 7 \equiv 0 (mod 31)$.

Solution. Sometimes it is easy to solve the equation, and then apply Theorem 32. Completing the square on the left-hand side, we obtain:

$$(x-3)^2 - 2 \equiv 0 (mod 31) \text{ or } (x-3)^2 \equiv 2 (mod 31).$$

It is not hard to see that $2 + 2 \cdot 31 = 64 = 8^2$ and solve the congruence as

$$(x-3)^2 \equiv 8^2 (mod 31)$$
$$x - 3 \equiv 8 (mod 31) \text{ or } x - 3 \equiv -8 (mod 31)$$
$$x \equiv 11 (mod 31) \quad \text{or} \quad x \equiv -5 (mod 31) \equiv 26 (mod 31).$$

Answer. $x \equiv 11; 26 (mod 31)$.

Definition. Let n be an odd number greater than 1 and its prime factorization be

$$n = p_1^{k_1} \cdot p_2^{k_2} \cdot \ldots \cdot p_s^{k_s},$$

then the **Jacobi symbol** is defined as

$$J(a,n) = \left(\frac{a}{n}\right) = \left(\frac{a}{p_1}\right)^{k_1} \times \left(\frac{a}{p_2}\right)^{k_2} \times \cdots \times \left(\frac{a}{p_s}\right)^{k_s},$$

This means that the Jacobi symbol is decomposed into the product of Legendre symbols.

Property 38 (Properties of Jacobi Symbol)

1. $a_1 \equiv a (mod n) \Rightarrow \left(\frac{a_1}{n}\right) = \left(\frac{a}{n}\right)$.
2. $J(ab,n) = \left(\frac{ab}{n}\right) = \left(\frac{a}{n}\right) \cdot \left(\frac{b}{n}\right)$.
3. If $(a,n) = 1 \Rightarrow \left(\frac{a^2 b}{n}\right) = \left(\frac{b}{n}\right)$.
4. $J(1,n) = 1, \ J(-1,n) = (-1)^{\frac{n-1}{2}}$.
5. $J(2,n) = \left(\frac{2}{n}\right) = (-1)^{\frac{n^2-1}{8}}$.
6. $J(m,n) = J(n,m) \cdot (-1)^{\frac{m-1}{2} \cdot \frac{n-1}{2}}$. This property is valid for odd n and m.
7. If $J(a,n) = -1$, then $x^2 \equiv a \pmod{n}$ has no solutions.

If $J(a,n) = 1$ there is no definite answer on whether or not $x^2 \equiv a \pmod{n}$ would have a solution. Let us solve the following problem.

Problem 251

Solve the congruence $x^2 \equiv 131 \, (mod\, 255)$.

Solution. To solve this congruence is equivalent to solving the equation $x^2 = 131 + 3 \cdot 5 \cdot 17 \cdot y$. Because 255 is not prime, we can decompose Jacobi symbol into three Legendre symbols,

$$J(131, 255) = L(131, 3) \times L(131, 5) \times L(131, 17).$$

Let us evaluate

$$\left(\frac{131}{255}\right) = \left(\frac{131}{3}\right) \cdot \left(\frac{131}{5}\right) \cdot \left(\frac{131}{17}\right).$$

$$\left(\tfrac{131}{3}\right) = \left(\tfrac{2}{3}\right) = -1$$
$$\left(\tfrac{131}{5}\right) = \left(\tfrac{1}{3}\right) = 1$$
$$\left(\tfrac{131}{17}\right) = \left(\tfrac{12}{17}\right) = \left(\tfrac{2^2 \cdot 3}{17}\right)$$
$$= \left(\tfrac{3}{17}\right) = \left(\tfrac{17}{3}\right)(-1)^{\frac{17-1}{2} \cdot \frac{3-1}{2}} = -1.$$

Multiplying all factors we obtain that $J(131, 255) = \left(\frac{131}{255}\right) = 1$. However, the given congruence does not have a solution because two of the Legendre symbols in the product equal -1. (For example, $x^2 \equiv 131 \pmod 3 \equiv 2 \pmod 3$ does not have a solution)

Answer. No solutions.

Problem 252

Solve the congruence $x^2 \equiv 52 (mod\, 159)$.

Solution. Because 159 is not prime ($159 = 3 \cdot 53$), first, we will evaluate Jacobi symbol:

$$J(52, 159) = L(52, 3) \cdot L(52, 53) = \left(\frac{2^2 \cdot 13}{3}\right) \cdot \left(\frac{2^2 \cdot 13}{33}\right) = \left(\frac{13}{3}\right) \cdot \left(\frac{13}{53}\right) = 1 \cdot 1 = 1.$$

The given congruence can be written as $x^2 = 52 + 3 \cdot 53 \cdot y$ or in an equivalent form as the system of two congruences:

$$\begin{cases} x^2 \equiv 52 (mod\, 3) \\ x^2 \equiv 52 (mod\, 53) \end{cases} \Leftrightarrow \begin{cases} x^2 \equiv 1 (mod\, 3) \\ x^2 \equiv -1 (mod\, 53) \end{cases}$$

The first congruence of the system always has a solution. The second congruence $x^2 \equiv -1 (mod\, 53)$ has a solution because $53 = 4 \cdot 13 + 1 = 4k + 1$. Moreover, because

$53 \cdot 10 - 1 = 529 = 23^2$, it can be rewritten as $x^2 \equiv 529 (mod\,53) \equiv 23^2 (mod\,53)$. Hence, we need to solve the following system:

$$\begin{cases} \begin{bmatrix} x \equiv 1(mod3) \\ x \equiv -1(mod3) \end{bmatrix} \\ \begin{bmatrix} x \equiv 23(mod53) \\ x \equiv -23(mod53) \end{bmatrix} \end{cases} \Leftrightarrow \begin{cases} \begin{bmatrix} x \equiv 1(mod3) \\ x \equiv 2(mod3) \end{bmatrix} \\ \begin{bmatrix} x \equiv 23(mod53) \\ x \equiv 30(mod53) \end{bmatrix} \end{cases}$$

Therefore the given congruence has four solutions that can be obtained by solving the following four systems of linear congruences. Each such a system can be solved using Chinese Remainder Theorem.

$$1) \begin{cases} x \equiv 1(mod3) \\ x \equiv 23(mod53) \end{cases} \Rightarrow x = 76 + 159n$$

$$2) \begin{cases} x \equiv 1(mod3) \\ x \equiv 30(mod53) \end{cases} \Rightarrow x = 136 + 159m$$

$$3) \begin{cases} x \equiv 2(mod3) \\ x \equiv 23(mod53) \end{cases} \Rightarrow x = 23 + 159k$$

$$4) \begin{cases} x \equiv 2(mod3) \\ x \equiv 30(mod53) \end{cases} \Rightarrow x = 83 + 159l$$

Answer. $x \equiv 23; 76; 83; 136 \,(mod\,159)$.

4.4 Word Problems Involving Integers

It is not a secret that our students consider word problems as the most difficult. The reasons of this are different in each student's case. However, word problems usually are closely connected to real- life problems. Word problems in integers are also interesting because while solving them, we need to use common sense in order to select valid answers. For example, if one solved a problem and obtained the interval $3 < x < 7$ as a solution, where x is the number of apples in the basket, one must understand that the actual answer is 4, 5, and 6 only. Since only whole numbers of apples make sense.

I will start from simple problems that each of my readers can solve.

Problem 253

A middle school teacher decided to introduce number theory to her 6th graders. Each student got a piece of paper and scissors. The teacher asked them to cut each paper into four pieces (not necessarily equal) and then take only one of the four pieces to cut into four more pieces. Following the second cutting, one of those four pieces is cut again into four pieces, and so on. After that, she wondered if her students could cut the original paper into 97 pieces. Some students continued to cut, but a few knew the answer right away. Do you know the answer too?

Solution. If one piece is broken into 4 pieces then the total number of pieces is increased by 3 pieces. At the beginning each student had one piece, and after the first cut each student had $1 + 3 = 4$ pieces. After the second cut, each student would have 7 pieces ($7 = 1 + 3 \cdot 2$). For example, after the third cut, one would obtain 10, which also can be written as $10 = 3 \cdot 3 + 1$. If he / she continued, after the kth cut, they would have $(1 + 3k)$ pieces. Because 97 divided by 3 gives a remainder of 1 ($97 = 1 + 3 \cdot 32$), it would be possible to get 97 pieces eventually.

It is interesting that the answer does not depend on how you cut a piece. It has to be four pieces, but there is no need for scissors; the four pieces can be unequal by size and even with fuzzy, not straight edges. (For example, one could rip the paper by hand.) It is clear that an ability to get n pieces depends on whether or not n divided by 3 gives a remainder of 1.

Answer. Yes. It is possible to get 97 pieces of paper.

Problem 254

Mary wants to celebrate the 4th of July and invite her friends to a party. She spent a lot of money and now has only one hundred dollars to buy red and blue helium balloons at four and five dollars each, respectively. How many options does she have if she wants to spend all her money and to have at least one balloon of each kind.

Solution. Let x be the number of red and y the number of blue balloons. We have the equation:

$$4x + 5y = 100.$$

using extracting an integer part of the numerator method, let us solve this equations for y:

$$y = 20 - 4 \cdot \frac{x}{5}$$

In order to have integer solutions, x must be divisible by 5, then

$$x = 5k, \ k \in N.$$

Then substituting this into formula for y, and using the fact that $y > 0$ (natural number), we obtain that

$$y = 20 - 4k > 0, \implies 0 < k < 5, \ k = 1, 2, 3, 4.$$

Therefore Mary has not many combinations. All of the are:

$$k = 1, x = 5, y = 16$$
$$k = 2, x = 10, y = 12$$
$$k = 3, x = 15, y = 8$$
$$k = 4, x = 20, y = 4.$$

Answer. 5 red and 16 blue, 10 red and 12 blue, 15 red and 8 blue, or 20 red and 4 blue balloons.

Problem 255

13 pirates divide the treasure of gold coins on the ship's deck. When they tried to divide the treasure equally, it turned out that there were eight coins left. Later, two of the pirates fell overboard in a storm. When the remaining pirates again divided the coins equally, three gold coins were left. Then in the exchange of gunfire three more pirates were lost. When the surviving pirates again began to divide the treasure, this time it turned out that there were five coins left. The treasure chest can contain, at most, 500 coins. How many gold coins were in the treasure?

Solution. Method 1. Assume that the treasure consisted of n gold coins, then

$$n \leq 500$$

and the following equations are true:

$$n = 13l + 8$$
$$n = 11m + 3$$
$$n = 8k + 5.$$

Solving the first and the second equations and the second and the third equations together, we can obtain the following new equations:

$$13(l - 3) = 11(m - 4) \implies m - 4 = 13s, \ m = 13s + 4$$
$$8(k + 3) = 11(m + 2) \implies m + 2 = 8t, \ m = 8t - 2.$$

Next we need to solve a new equation:

$$13s + 4 = 8t - 2$$

Using modular arithmetics we can solve it as

$$8t \equiv 6 (mod\, 13)$$
$$8t \equiv (6 + 2 \cdot 13)(mod\, 13)$$
$$8t \equiv 32 (mod\, 13)$$
$$t \equiv 4 (mod\, 13), \ t = 13p + 4$$

Finally, since $m = 8t - 2$, we obtain for the number of gold coins the following expression with the constraint:

$$n = 11m + 3 = 11 \cdot (8t - 2) + 3 = 11\,[8(13p + 4) - 2] + 3 = 11 \cdot 104p + 333 \leq 500$$

If $p = 0$, then $n = 333$ coins.

Method 2. Using congruence and Theorem 20, we also can solve this problem. We have the system of congruences:

$$\begin{cases} n \equiv 8 \,(mod\, 13) \\ n \equiv 3 \,(mod\, 11) \\ n \equiv 5 \,(mod\, 18) \end{cases} \Rightarrow a_1 = 8, m_1 = 13, \ a_2 = 3, m_2 = 11, \ a_3 = 5, \ m_3 = 8.$$

Next, we need to evaluate corresponding M_1, M_2, M_3 and solving the following congruences to find corresponding N_1, N_2, N_3 :

$$M = 13 \cdot 11 \cdot 8 = 1144$$
$$M_1 = \frac{M}{m_1} = 88 \qquad 88 \cdot N_1 \equiv 1 \,(mod\, 13)$$
$$M_2 = \frac{M}{m_2} = 104 \qquad 104 \cdot N_2 \equiv 1 \,(mod\, 11)$$
$$M_3 = \frac{M}{m_3} = 143 \qquad 143 \cdot N_3 \equiv 1 \,(mod\, 8)$$

As soon we know N_1, N_2, N_3, the answer will be found from the formula:

$$n \equiv (a_1 N_1 M_1 + a_2 N_2 M_2 + a_3 N_3 M_3)\,(mod\, M).$$

Solving each congruence, we obtain

$88 \cdot N_1 \equiv 1 \,(mod\, 13)$	$104 \cdot N_2 \equiv 1 \,(mod\, 11)$	$143 \cdot N_3 \equiv 1 \,(mod\, 8)$
$13 \cdot 27 + 1 = 88 \cdot 4$	$85 \cdot 11 + 1 = 104 \cdot 9$	$8 \cdot 125 + 1 = 143 \cdot 7$
$88 \cdot N_1 \equiv 88 \cdot 4 \,(mod\, 13)$	$104 \cdot N_2 \equiv 104 \cdot 9 \,(mod\, 11)$	$143 \cdot N_3 \equiv 143 \cdot 7 \,(mod\, 8)$
$N_1 \equiv 4 \,(mod\, 13)$	$N_2 \equiv 9 \,(mod\, 11)$	$N_3 \equiv 7 \,(mod\, 8)$

Now, we find the answer as

$$n \equiv (8 \cdot 4 \cdot 88 + 3 \cdot 9 \cdot 104 + 5 \cdot 7 \cdot 143)\,(mod\, 1144),$$
$$n \equiv 10629 \,(mod\, 1144) \quad (10629 = 9 \cdot 1144 + 333)$$
$$n \equiv 333 \,(mod\, 1144)$$
$$\boxed{n = 333}.$$

Although the second method is a direct application of formulae given by Theorem 20, in my opinion, the first method was nicer.

Answer. 333 coins were in the treasure.

> **Problem 256**
>
> A mail worker wants to distribute 90 advertising letters into 13 mailboxes in such a way that each box has at least one letter. Show that at least two boxes would have the same number of letters.

Solution. Let each of 13 slots have a different number of letters, then the total number of letters must be

$$1 + 2 + 3 + \cdots + 12 + 13 = \frac{(1 + 13) \, 13}{2} = 91.$$

Consider the following table, in which each cell represents a slot with a different number of letters.

2	12	8	4	1	6	7	3	5	10	13	11	9

Table 4.5 91 letters in 13 slots

Since the total number of letters in the first table is 91, one more that the number of letters the mail worker has, then we need to take one letter from any of the slots, except one that has only one letter. (This is a sort of the Pigeonhole Principle as well.) For example, we can take a letter from the slot that now contains six letters and we will obtain the following distribution of letters:

2	12	8	4	1	5	7	3	5	10	13	11	9

Table 4.6 90 letters in 13 slots

Now we have two slots with the same number of letters (five letters in each). Therefore, given 90 letters, you cannot fill 13 slots with a different number of letters in each slot!

> **Problem 257**
>
> Find two different two-digit numbers, such that the sum of all other numbers is fifty times one of the two given numbers.

Solution. Let x and y be two unknown numbers. The sum of all two-digit numbers from 10 to 99 is the sum of an arithmetic sequence with the first term 10 and the last term of 99, that is $\frac{(10+99)\cdot 90}{2} = 4905$. Without loss of generality we can assume that x satisfies the condition of the problem and hence, we have the following equation to solve:

$$4905 - (x+y) = 50 \cdot x$$
$$4905 = 51x + y$$
$$3 \cdot 1635 = 3 \cdot 17 \cdot x + y.$$

Dividing both sides by 3, we have the following new equation:

$$17x = 1635 - \frac{y}{3}$$

Obviously, in order this equation to have integer solution, y must be a multiple of 3:

$$y = 3m \tag{4.70}$$

After replacing y by $3m$ we obtain the new equation in two variables to solve:

$$17x = 1635 - m$$

This equation can be solved in many ways. For example, we can divide both sides by 17 and then extract the largest multiple of 17 in the numerator of the fraction:

$$x = \frac{1635-m}{17} = \frac{17 \cdot 96 + 3 - m}{17}$$
$$= 96 + \frac{3-m}{17}. \tag{4.71}$$

It is easy to see that $3 - m$ must be divisible by 17, which can be written as

$$3 - m = 17k, \implies m = 3 - 17k \tag{4.72}$$

Substituting this into previous formula for x and setting boundaries for this number as two-digit number, we obtain the formula for x and the inequality for allowed values of k, respectively:

$$x = 96 + k \tag{4.73}$$

$$10 \le x = 96 + k \le 99 \implies -86 \le k \le 3. \tag{4.74}$$

Number $y = 3m$ is also two-digit number and hence using the formula for y, the following boundaries are valid:

$$10 \le y = 3(3 - 17k) = 9 - 51k \le 99; \implies -\frac{91}{51} \le k \le -\frac{1}{51} \Leftrightarrow k = -1. \tag{4.75}$$

The intersection of two inequalities give us a unique value of a parameter k, that is $k = -1$. Next, we will evaluate the corresponding $m = 20$, $x = 95$, and $y = 60$.
Answer. Two numbers are 95 and 60.

Problem 258

Mike and Marie are playing a game. Both are given different 2016-digit natural numbers by their math teacher and are asked to do the following: erase the last digit of the number, then multiply the remaining number by 3 and add to it the doubled erased number, continue by doing the same with the new number and so on. After applying several such operations with the given number, Marie noticed that her numbers stopped changing, which was not the case with the new numbers obtained by Mike. Marie has stopped, same as Mike, even though he continued obtaining new numbers. He said that he had no interest to continue this game. The teacher looked at Mike's calculations and noticed that Mike not only predicted Marie's final number, but even guessed correctly her possible 2016 digit number. What was Marie's "stopping" number? What was the minimal possible 2016-digit number that Marie received from the teacher?

Solution. First, Mike found Marie's last number, 17. Thus, if n is the unknown unchanged number and if its last digit is y, then $n = 10x + y$; after the next operation it will become $3x + 2y$. From $10x + y = 3x + 2y$, we obtain $7x = y$, which leaves only one choice as $y = 7$ and $x = 1$, and $n = 17$. Second, because the original number had 2016 digits (and actually if a number just longer than 17), then it will always become smaller after the described procedure, $10x + y > 3x + 2y$.

From $2(10x + y) = 17x + (3x + 2y)$, it follows that $(10x + y)$ is divisible by 17 if and only if $(3x + 2y)$ is divisible by 17. Obviously, her original number also must be divisible by 17.

Assume that the original number is divisible by 17 but has only five digits and that it was 10013. Let us conduct the algorithm proposed by the teacher:

$$3 \cdot 1001 + 2 \cdot 3 = 3009$$
$$3 \cdot 300 + 2 \cdot 9 = 918$$
$$3 \cdot 91 + 2 \cdot 8 = 289$$
$$3 \cdot 28 + 2 \cdot 9 = 102$$
$$3 \cdot 10 + 2 \cdot 2 = 34$$
$$3 \cdot 3 + 2 \cdot 4 = 17.$$

We can see that each time we obtain a number divisible by 17, until finally stop at 17. Using similar arguments, let us find the minimal possible 2016-digit number divisible by 17. Consider remainders of different powers of 10 divided by 17:

n	1	2	3	4	5	6	7	8	9	10	11	12	13	14	15	16	17
r	10	15	14	4	6	9	5	16	7	2	3	13	11	8	12	1	10

Table 4.7 Remainders of 10^n divided by 17

$$100 = 10^2 = 5 \cdot 17 + 15$$
$$1000 = 10^3 = 58 \cdot 17 + 14$$
$$10^4 = 10000 = 588 \cdot 17 + 4$$
$$10^5 = 5882 \cdot 17 + 6,$$

From Fermat's Little Theorem we know that

$$a^{p-1} \equiv 1 (mod\, p)$$

is true if p is prime. 17 is prime, then $10^{16} \equiv 1 (mod\, 17)$. We have the following true equations.

$$10 \equiv 10 (mod\, 17), \quad 10^2 \equiv 15 (mod\, 17)$$
$$10^{16} \equiv 1 (mod\, 17),$$
$$10^{17} \equiv 10 (mod\, 17), \quad 10^{18} \equiv 15 (mod\, 17).$$

Then the remainder of 17th power of 10 divided by 17 will be the same as the first power of 10. Thus 10^{17} gives that same remainder as 10, 10^{18} will give the same remainder as $10^2 = 100 (r = 15)$ and 10^{19} will have the same remainder as 10^3 divided by 17 ($r = 14$), etc.

Next, the minimal n-th digit number can be written as 10^{n-1} (for example, $1000 = 10^3$). Then the minimal 2016 digit number will be written as

$$\underbrace{100...00}_{2016 \text{ digits}}.$$

We can increase it as much as we want to satisfy the condition. For example,

$$n = 10^{2015} + a = \underbrace{100...00a}_{2016 \text{ digits}}, \ 0 < a \le 9,$$

will describe all numbers between 1000...001 and 1000...009. Or the number

$$m = 10^{2015} + ba = \underbrace{100...00ba}_{2016 \text{ digits}}, \ 0 < a \le 9, \ 0 < b \le 9$$

will describe all 2016 digit numbers between 1000...010 and 1000...99. Since $2015 = 125 \cdot 16 + 15$, using Fermat's Little Theorem again, we obtain that

$$10^{16} \qquad \equiv 1(mod\,17),$$
$$10^{2015} \qquad \equiv (10^{16})^{125} \cdot 10^{15}$$
$$10^{15} \qquad \equiv 12(mod\,17),$$
$$10^{2015} \qquad \equiv 12(mod\,17)$$
$$5 \qquad\qquad \equiv 5(mod\,17)$$
$$10^{2015} + 5 \equiv 17(mod\,17) = 17k.$$

Therefore, $n = 10^{2015} + 5 = \underbrace{100...005}_{2016\ \text{digits}}$ is the minimal 2016-digit number that could be given to Marie by the teacher and 005 are its three last digits.

Answer. $n = 10^{2015} + 5 = \underbrace{100...005}_{2016\ \text{digits}}$.

Problem 259

Three groups of fishermen caught 113 fish. The average number of fish per fisherman in the first group is 13, in the second 5, and in the third 4. How many fishermen were in each group if there were a total of 16 men?

Solution. We noticed that this problem is again a problem involving integers. We cannot have 5.5 fishermen or 7.3 fish. Let us introduce three variables corresponding to unknowns. Let x be the number of fishermen in the first group, y—the number of fishermen in the second group, z—the number in the third group. There were 16 men in total or

$$x + y + z = 16. \tag{4.76}$$

They caught 113 fish in total and using the conditions of the problem, we have the second equation:

$$13x + 5y + 4z = 113. \tag{4.77}$$

Usually two equations in three variables cannot have a unique solution. Using the fact that x, y, and z must be integers, we will find a solution.

From equation (4.76) we can express z in terms x and y and then find all integer solutions of equation (4.77)

$$z = 16 - y - x, \tag{4.78}$$

then $13x + 15y + 4(16 - y - x) = 113$ or

$$y + 9x = 49. \tag{4.79}$$

Solving (4.79) for x and extracting an integer part of the quotient we obtain:

$$x = \frac{49 - y}{9} = \frac{45 + 4 - y}{9} = 5 + \frac{4 - y}{9}. \tag{4.80}$$

In order that x be a natural number, $\frac{4-y}{9}$ must be an integer as well. Therefore $(4 - y)$ should be divisible by 9 or we can assume that

$$4 - y = 9n, \tag{4.81}$$

where n is an integer.

Replacing (4.81) into (4.78) and (4.80) gives us the following:

$$x = 5 + n, \tag{4.82}$$

$$y = 4 - 9n, \tag{4.83}$$

$$z = 7 + 8n. \tag{4.84}$$

Using common sense all x, y, and z must be positive integers that can be written as a system of three inequalities:

$$\begin{cases} y = 4 - 9n > 0 \\ x = 5 + n > 0 \\ 7 + 8n > 0 \end{cases} \tag{4.85}$$

Solving each inequality of (4.85) for n we obtain:

$$\begin{cases} n < \frac{4}{9} \\ n > -\frac{7}{8} \end{cases} \Leftrightarrow -\frac{7}{8} < n < \frac{4}{9}. \tag{4.86}$$

There is only one integer n, $n = 0$, that satisfies inequality (4.86). Plugging $n = 0$ into (4.82), (4.83), and (4.84) we finally obtain:

$$x = 5, \ y = 4, \ z = 7.$$

Answer. There were 5 fishermen in the first group, 4 in the second, and 7 in the third.

You can try to solve each problem in a different way using congruence.

Problem 260 (MGU Entrance Exam)

A fruit farmer wants to plant trees. He has fewer than 1000. If he plants them in rows, 37 trees per row, then there will be 8 trees remaining. If he plants 43 per row, there will be 11 remaining. How many trees does he have?

Solution. Let x be the number of trees. Using the condition of the problem and assuming that the farmer has either n rows of 37 planted or m rows of 43 planted, we obtain the following:

$$0 < x < 1000, \tag{4.87}$$

$$x = 37n + 8, \tag{4.88}$$

$$x = 43m + 11. \tag{4.89}$$

Combining (4.88) and (4.89) we have

$$37n + 8 = 43m + 11. \tag{4.90}$$

Presenting $43m$ as $(37m + 6m)$ we can factor the left and the right sides of (4.90):

$$37n - 37m = 6m + 3,$$

or

$$37(n - m) = 3(2m + 1). \tag{4.91}$$

We want to find only integer solutions of (4.91) because we cannot plant half of a tree. Because 3 and 37 are primes in order for (4.91) to have any integer solutions, $(n - m)$ must be divisible by 3 and $(2m + 1)$ by 37. But there are a lot of numbers that are multiples of 37: 37, 111, 148, ...

Let us use condition (4.87) that restricts possible values of x. (The fruit farmer has less than 1000 trees). Now combining (4.87) and (4.89) we obtain

$$\begin{aligned} x = {}& 43m + 11 < 1000 \\ 43m & \qquad < 989 \\ m & \qquad < 23. \end{aligned}$$

How can we use this inequality?

Recall, that $(2m + 1)$ should be a multiple of 37?. But if $m < 23$, then

$$\begin{aligned} 2m + 1 < 2 \cdot 23 + 1 = 47 \\ 2m + 1 < 47 \end{aligned}. \tag{4.92}$$

Looking at the last equation of (4.92) we can see that there is only one multiple of 37 that is less than 47, this is 37 itself. Therefore $2m + 1 = 37$, and $m = 18$. Replacing $m = 18$ into equation (4.89) we obtain $x = 43 \cdot 18 + 11 = 785$.
Answer. The fruit farmer has 785 trees.

Problem 261 (MGU Entrance Exam)

One box contains only red balls, another only blue. The number of red balls is 15/19 of the number of blue balls. When 3/7 of the red balls and 2/5 of the blue are removed from the boxes, there are less than 1000 balls in the first box and greater than 1000 balls in the second. How many balls were originally in each box?

Solution. Let x be the original number of red balls in the first box, and y be the original number of blue balls in the second box. Now we can write the following system of equations and inequalities:

$$\begin{cases} \frac{4}{7}x < 1000 \\ \frac{3}{5}y > 1000 \\ \frac{15}{19}y = x. \end{cases} \qquad (4.93)$$

Some of you could try to solve this system in a standard way and would obtain:

$$\begin{cases} \frac{4}{7} \cdot \frac{15}{19}y < 1000 \\ y > 1666\frac{2}{3} \\ \frac{15}{19}y = x, \end{cases}$$

or

$$\begin{cases} y < 2216\frac{2}{3} \\ y > 1666\frac{2}{3} \\ x = \frac{15}{19}y. \end{cases} \qquad (4.94)$$

Just a quick look at system (4.94) says that there are many numbers between 1666 and 2216, and also x and y must satisfy the third equation of the system.

Let us try to find some nonstandard way of solving system (4.93). We will use the properties of integers.

If x and y exist and satisfy system (4.93), x must be divisible by 15 and y must be divisible by 19.

In order that $\frac{4}{7}x$ and $\frac{3}{5}y$ be integers x also must be a multiple of 7 and y multiple of 5. So we can represent

$$\begin{aligned} x &= 15 \cdot 7 \cdot x_1 \\ y &= 19 \cdot 5 \cdot y_1, \end{aligned} \qquad (4.95)$$

where x_1 and y_1 are some unknown integers.

Replacing formulas (4.95) for x and y into system (4.93) we obtain:

$$\begin{cases} 7 \cdot 15 \cdot x_1 = \frac{15}{19} \cdot 5 \cdot 19 \cdot y_1 \\ \frac{4}{7} \cdot 7 \cdot 15 \cdot x_1 < 1000 \\ \frac{3}{5} \cdot 5 \cdot 19 \cdot y_1 > 1000, \end{cases}$$

where x_1 and y_1 are integer, or

$$\begin{cases} 7x_1 = 5y_1 \\ 3x_1 < 50 \\ 3 \cdot 19 \cdot y_1 > 1000. \end{cases} \qquad (4.96)$$

In order that the first equation of (4.96) has some integer solutions, x_1 must be a multiple of 5 and y_1 a multiple of 7.

Let us introduce new variables (do a new substitution),

$$x_1 = 5 \cdot x_2$$
$$y_1 = 7 \cdot y_2.$$

Now system (4.96) can be solved:

$$\begin{cases} 7 \cdot 5 \cdot x_2 = 5 \cdot 7 \cdot y_2 \\ 3 \cdot 5 \cdot x_2 < 50 \\ 3 \cdot 19 \cdot 7 \cdot y_2 > 1000, \end{cases}$$

where x_2 and y_2 are integers or

$$\begin{cases} x_2 = y_2 \\ x_2 < \frac{10}{3} = 3\frac{1}{3} \\ y_2 > \frac{1000}{399} = 2\frac{202}{399}. \end{cases} \tag{4.97}$$

From system (4.97) we notice that

$$2\frac{202}{399} < x_2 < 3\frac{1}{3}$$

and

$$2\frac{202}{399} < y_2 < 3\frac{1}{3}.$$

Since there is only one integer $x_2 = y_2 = 3$, then

$$x = 7 \cdot 15 \cdot x_1 = 7 \cdot 15 \cdot 5 \cdot x_2 = 7 \cdot 15 \cdot 5 \cdot 3 = 1575$$
$$y = 5 \cdot 19 \cdot y_1 = 5 \cdot 19 \cdot 7 \cdot y_2 = 5 \cdot 19 \cdot 7 \cdot 3 = 1995.$$

Answer. There are 1575 red balls and 1995 blue balls.

Problem 262 (Moscow State University Entrance Exam)

There are 600 more applicants to the University from high school students than from people who worked full time. There are 5 times as many ladies among the high school students than ladies among the nonstudents. And there are n times as many men among the high school graduates than there are in the group of nonstudents, such that $6 \leq n \leq 12$, where n is an integer. Find the total number of college applicants if there are 20 more men than women among the nonstudents.

Solution. In my number theory class my students always start by using the finite value of the parameter n from 6 to 12, and then they try to decide which one fits the condition of the problem. However, this way is not as good as it seems to be. There are 7 integers for n between 6 and 12, and you would solve 7 similar problems and it could take a while. Let us find a general approach to this problem.

Analysis of the condition shows that in order to translate the problem into a mathematical language, it is sufficient, besides n, to introduce just one other variable, say x, as a number of some subset of the applicants. Which one?

Recalling the previous problem maybe we will have to verify divisibility of the value x by some other number, and then replace x by some smaller integer, we conclude that the better choice is to introduce x as the number of female nonstudents. Then the number of any other type of applicants is going to be a positive integer as well. Of course, we assume that x is a natural number. If x is the number of female nonstudent applicants, then $(x+20)$ is the number of male nonstudent applicants, and the total of nonstudent applicants will be

$$(x+x+20), \tag{4.98}$$

$5x$ will be the number of female high school students, and the total number of applicants from high school will be represented by

$$5x+n(x+20). \tag{4.99}$$

By the condition of the problem expression (4.98) is 600 less than expression (4.99).

Thus, we have the equation,

$$2x+20+600 = 5x+n(x+20). \tag{4.100}$$

Solving (4.100) for x we obtain:

$$x = \frac{620-20n}{n+3}. \tag{4.101}$$

Recalling that x must be a positive integer, we rewrite the right part of (4.101) as:

$$x = \frac{620-20(n+3)+20\cdot3}{n+3} = \frac{680}{n+3}-20.$$

In order that x be a natural number, the number 680 must be divisible by $(n+3)$. A prime factorization of 680 gives: $680 = 2\cdot2\cdot2\cdot5\cdot17$. If $6 \le n \le 12$, then $9 \le (n+3) \le 15$, then n can be only 9, 10, 11, 12, 13, 14, or 15.

Notice that only $n+3 = 10$ is a factor of 680. This means that $n = 7$. Then

$$x = \frac{680}{10}-20 = 48,$$

and the total number of applicants is $2x+20+5x+7(x+20) = 832$.
Answer. 832 applicants.

Problem 263

Three ranchers came to the Fort Worth Stock Show to sell their yearling heifers. The first rancher brought 10 heifers, the second 16, and third 26. On the first day every rancher sold some of his heifers. Moreover, all ranchers sold their heifers at the same price, one that had not changed during the entire first day. On the second day the price for heifers went down and all three ranchers in fear of further reductions in price sold all their remaining heifers at a reduced price per heifer. What was the price per heifer on the first day and on the second day if each rancher took home $3500?

Solution. Let us introduce three variables x, y, and z as a number of heifers sold by the first, second, and third ranchers respectively on the first day. Because originally the first rancher had 10 heifers, the second 16, and the third 26, on the second day they would sell $(10-x)$, $(16-y)$, and $(26-z)$ heifers, respectively.

Let us create a table.

	Heifers brought	Heifers sold on the first day	Sold on the second day
Rancher 1	10	x	$(10-x)$
Rancher 2	16	y	$(16-y)$
Rancher 3	26	z	$(26-z)$

Table 4.8 Ranchers Stockyard Problem

Introducing two additional variables: t as the price for heifers on the first day of the Stock Show and p as the price for heifers on the second day, and using the conditions of the problem, we obtain the following system of three equations in five variables, x, y, z, t, and p.

$$\begin{cases} xt + (10-x)p = 3500, \ 1 \le x \le 9 \\ yt + (16-y)p = 3500, \ 1 \le y \le 15 \\ zt + (26-z)p = 3500, \ 1 \le z \le 26. \end{cases} \quad (4.102)$$

Combining like terms in each equation we obtain,

$$\begin{cases} x(t-p) + 10p = 3500 \\ y(t-p) + 16p = 3500 \\ z(t-p) + 26p = 3500 \end{cases}$$

Subtracting the last equation from the first and from the second, we have

$$(x-z)(t-p) - 16p = 0$$
$$(y-z)(t-p) - 10p = 0,$$

or

$$\begin{cases} (x-z)(t-p) = 16p \;\;(a) \\ (y-z)(t-p) = 10p \;\;(b) \end{cases} \tag{4.103}$$

Dividing (a) by (b) of (4.103) gives us the equality,

$$\frac{x-z}{y-z} = \frac{16}{10},$$

or

$$(x-z)\cdot 5 = (y-z)\cdot 8. \tag{4.104}$$

Using our previous experience of solving equations like (4.104) for integers x, y, and z we conclude that $(x-z)$ should be divisible by 8 and $(y-z)$ by 5. These can be written as

$$\begin{cases} x-z = 8n, \; n \in \mathbb{N}, \; 1 \le x \le 9 \\ y-z = 5m, \; m \in \mathbb{N}, \; 1 \le y \le 15 \\ \qquad\qquad\qquad 1 \le z \le 25. \end{cases}$$

Because $x > z$, then $x - z = 8$ only. On the other hand, from equation (4.104) we can see that $y - z = 5$ only. If $x - z = 8$ and $y - z = 5$, then $x = 9$, $z = 1$, and $y = 6$. It means that on the first day the first rancher sold 9 heifers, second 6 heifers, and third rancher sold only one heifer.

Knowing x, y, and z we can easily find t and p from equations (a) and (b) of (4.103). We have: $8(t - p) = 16p$ and $5(t - p) = 10p$, then $t - p = 2p$ or

$$t = 3p. \tag{4.105}$$

Using the first equation of system (4.102) and replacing t by $3p$ from (4.105) we obtain:

$$9 \cdot 2p + 10p = 3500$$
$$28p = 3500 \Rightarrow p = 125$$
$$t = 3 \cdot 125 = 375.$$

Answer. On the first day the price per heifer was \$375 and on the second day \$125.

Problem 264 (ASHME 1999)

At a classroom costume party, the average age of the b boys is g, and the average age of the g girls is b. If the average age of everyone at the party (all these boys and girls, plus their 42-year-old teacher) is $(b+g)$, what is the value of $(b+g)$?

Solution. Method 1. Let us consider the average age in the group of g girls. If we add up the ages of g girls (let us say A) and divide it by the number of girls (g) we obtain b. On the other hand, if we add up the ages of all b boys (B) and divide it by the number of boys (b) we obtain g. This can be written as

$$\frac{A}{g} = b, \quad \frac{B}{b} = g. \tag{4.106}$$

Then the average age of everyone at the party including the teacher will be

$$\frac{A+B+42}{b+g+1} = b+g. \tag{4.107}$$

Replacing A and B from (4.106), we can rewrite (4.107) as

$$\frac{2gb+42}{b+g+1} = b+g,$$

or multiplying both sides by $(b+g+1)$ we obtain:

$$2gb+42 = (b+g)(b+g+1). \tag{4.108}$$

Let $gb = S$. We are also given that the average of the boys' ages, the girls' ages and 42 (the teachers age) is $b+g$

$$2S+42 = (b+g)(b+g+1)$$
$$2S+42 = (b+g)((b+g)+1)$$
$$2S+42 = (b+g)^2 + (b+g)$$
$$2S = (b+g)^2 + (b+g) - 42$$
$$2S = ((b+g)+7)((b+g)-6).$$

Now look at all possible cases:

Case 1.

$$((b+g)+7) = 2S$$
$$((b+g)-6) = 1$$
$$2S-7 = 7$$
$$S = 7.$$

Case 2.

$$((b+g)+7) = 1$$
$$((b+g)-6) = 2S$$
$$-6 = 2S+6$$
$$S = -6.$$

Case 3.

$$((b+g)+7) = 2$$
$$((b+g)-6) = S$$
$$-5 = S+6$$
$$S = -11.$$

Case 4.

$$((b+g)+7) = S$$
$$((b+g)-6) = 2$$
$$S-7 = 8$$
$$S = 15.$$

Case 1 and Case 4 are the only ones to be considered, because the sum of the children's ages cannot be negative! Case 1 can also be eliminated, because if $S = 7$, then $b+g = 7$. The sum of the ages, S has to be divisible by both b and g. As 7 is prime, this will not work for any pair of numbers adding to 7.

Case 4 must then give the solution. If $S = 15$ then $b+g = 8$. The pair of numbers for b and g will be 3 and 5. There are either 3 boys and 5 girls, or 3 girls and 5 boys. The sum of ages, $S = 15$ is divisible by both 3 and 5.

Method 2. We have from (4.108) that $42 = b^2 + g^2 + b + g$. Multiply both sides by 4 we have $168 = 4b^2 + 4g^2 + 4b + 4g$. Completing the square we have:

$$170 = (2b+1)^2 + (2g+1)^2,$$

or

$$121 + 49 = (2b+1)^2 + (2g+1)^2,$$

then

$$(2b+1)^2 = 121$$
$$(2g+1)^2 = 49.$$

Therefore, $2b+1 = 11$ and $2g+1 = 7$, or $b = 5$ and $g = 3$. We make a conclusion:

- total age of 5 boys is $5 \cdot 3 = 15$,
- total age of 3 girls is $3 \cdot 5 = 15$,
- age of the teacher is 42,
- total age is 72,
- total number of people is $5 + 3 + 1 = 9$,
- average age is $72/9 = 8$, which is $(b+g)$.

Answer. $b+g = 8$.

Problem 265 (Moscow State University Entrance Exam)

An American tourist came to Moscow and was surprised at how difficult it was to pay for goods in Russian currency. Russians have kopeks and rubles. 100 kopeks is 1 ruble. It sounds similar to a dollar and cents but in Russia there are 1 kopek, 2 kopeks, 3 kopeks, 5 kopeks, 15 kopeks, 20 kopeks, and 50 kopeks. (Compare with cents: 1, 5, 10, and 25). The tourist put together on a plate some number (less than 15) of 3 kopeks and 5 kopeks coins and obtained 53 kopeks. He noticed that if in this set he replaces all 3-kopek coins by 5-kopek coins and all 5-kopek coins by 3-kopek coins, then the amount of money will decrease but not more than 1.5 times. How many 3-kopek coins did our tourist originally have?

Solution. Let n be the initial number of 3-kopek coins, and m be the initial number of 5-kopek coins. Then, we have:

$$3 \cdot n + 5 \cdot m = 53.$$

Solving this equation for n we obtain:

$$n = \frac{53 - 5m}{3}. \tag{4.109}$$

In order for n to be a positive integer $(53 - 5m)$ must be a positive multiple of 3. There are four such opportunities for m:

$$\begin{aligned} m_1 &= 1 \\ m_2 &= 4 \\ m_3 &= 7 \\ m_4 &= 10. \end{aligned}$$

Solving (4.109) for each m we obtain four possible ordered pairs for (m, n):

$$\{(1, 16),\ (4, 11),\ (7, 6), (10, 1)\}. \tag{4.110}$$

Only one of (4.110) will satisfy the problem condition:

$$\begin{cases} n + m < 15 \\ 1.5(5n + 3m) \geq 53. \end{cases} \tag{4.111}$$

Checking all possible ordered pairs (4.110) we notice that only $(7, 6)$ with $m = 7$ and $n = 6$ satisfies (4.111).

Answer. Our tourist had six 3-kopek coins and seven 5-kopek coins.

Chapter 5
Homework

5.1 Problems to Solve

1) Prove that $(n^5 - n)$ is divisible by 30.
   Hint: Use math induction or factor the given number.
2) Prove that $(n^3 + 3n^2 - n - 3)$ is divisible by 48 for any odd number n.
   Hint: Factor the number, and let $n = 2k + 1$.
3) Prove that $(k^3 + 5k)$ is divisible by 3 for any $k \in \mathbb{Z}$.
   Hint: Factor.
4) Find such natural numbers n, for which $N = \underbrace{1313\ldots13}_{2n\,\text{digits}}$ is divisible by 63.
5) Prove that a number $n(n+1)(n+2)(n+3) + 1$ is a perfect square for any $n \in \mathbb{N}$.
6) Find all pairs of integers p and q satisfying the equation: $4p^2 = q^2 - 9$.
   Hint: Factor as $(q - 2p)(q + 2p) = 9$.
7) Find all integer solutions of $x^2 = y^2 + 2y + 13$.
8) Find integers x and y such that $3x^2 + 11xy + 10y^2 = 7$.
9) Find all integer solutions of $3 \cdot 2^x + 1 = y^2$.
10) What are the last two digits of:

    a) $99^{100} - 51^{100}$,
    b) $99^{1999} - 51^{1999}$,
    c) $99^{2000} - 51^{1991}$.

11) Find the last digit of the number 2^{1000}.
12) Find the last two digits of 3^{2012}.
13) Find the last three digits of $3^{123456789}$.
14) Solve the equation $x^2 = 9y^2 + 7$ over the set of integers.
15) If N is an odd number, find a number x, such that $N^2 + x^2$ is a perfect square. Obviously, if N is a perfect square to start with, $x = 0$ would always work, but there may be others for $N > 1$. Find several examples and try to determine a general rule for x, if one exists.
16) Is it possible to rewrite 1974 as a difference of squares of two natural numbers?

© Springer International Publishing AG, part of Springer Nature 2018
E. Grigorieva, *Methods of Solving Number Theory Problems*,
https://doi.org/10.1007/978-3-319-90915-8_5

17) Find all integers n such that $n^5 - n$ is divisible by 120.
18) Numbers n and m are odd numbers. Can a number $n^2 + m^2$ be a perfect square?
19) Can a sum of the digits of a perfect square be equal to 1970?
20) Show that if n is an even integer and n is not divisible by 3 and 4, then $n^5 - 5n^3 + 4n$ is divisible by 1440.
21) Show that the number $3^{2n+2} - 8n - 9$ is divisible by 64 for any natural number n.
22) Prove that $7 \cdot 5^{2n} + 12 \cdot 6^n$ is divisible by 19 for all $n \in \mathbb{N}$.
23) Prove that $2^9 + 2^{99}$ is divisible by 100.
24) Find all integers x and y such that $6x^2 + 5y^2 = 74$.
25) Find all integers x and y satisfying $19x^2 + 28y^2 = 729$.
26) Find all integers x and y satisfying the equation $1 + x + x^2 + x^3 = 2^y$.
27) Find integers x, y, and z satisfying $1! + 2! + 3! + \cdots + x! = y^z$, where $n! = 1 \cdot 2 \cdot 3 \cdot \cdots \cdot (n-1) \cdot n$.
28) Find all integers n such that $\frac{n+1}{2n-1}$ is an integer.
29) Find all natural x, y, and z satisfying the equation $x + y + z = xyz$.
30) Find integers n such that the fraction $\frac{22n+3}{26n+4}$ is reducible.
31) What is greater:

   (a) 5^{300} or 3^{500},
   (b) 2^{700} or 5^{300}?

32) Find all natural numbers such that when divided by 17 give a remainder of 2, and when divided by 5 give a remainder of 3.
33) Prove that $43^{43} - 17^{17}$ is divisible by 10.
34) Solve the equation $33^{22} \equiv x \pmod{14}$.
35) Prove that $n^2 \equiv 1 \pmod{8}$ if n is odd.
36) If $k \equiv 1 \pmod{4}$ then what is $(6k+1)$ congruent to $\pmod{4}$?
37) Find any prime number k greater than 2 such that $(17k+1)$ is a perfect square. Solve this problem using congruence theorems.
38) Solve for integer x the equation $x^2 - 4x - 7 \equiv 0 \pmod{19}$.
39) Show that every prime odd number except 5 divides some number consisting of k digits of 1 ($\underbrace{111\ldots11}_{k\,\text{digits}}$).

   Hint: $\underbrace{111\ldots11}_{k\,\text{digits}} \equiv 0 \pmod{p}$.

40) Find the remainder of 15! when divided by 17.
41) Show that $18! \equiv 1 \pmod{437}$.
42) Given a prime number p, establish the congruence $(p-1)! \equiv (p-1) \pmod{(1+2+3+\cdots+(p-1))}$.
43) Find an integer solution to the following equation: $3xy + 16x + 13y + 61 = 0$.
44) Verify that $4 \cdot 29! + 5!$ is divisible by 31.
45) Find such integer that divided by 2 and 3 leaves a remainder of 1.
46) A farmer is on the way to market to sell eggs when a meteorite hits his truck and destroys all his produce. In order to file an insurance claim, he needs to know how many eggs were broken. He knows that when he counted the eggs by 2's there was 1 left over, when he counted them by 3's, there was 1 left over, when

he counted them by 4's there was 1 left over, when he counted them by 5's, there was 1 left over, and when he counted them by 6's, there was 1 left over, but when he counted them by 7's, there were none left over. What is the smallest number of eggs that were in the truck?

47) Find all integer solutions of $3x + 7y = 22$.
48) 2018! ends in how many zeros?
49) Solve $9x \equiv 8(\mod 34)$ using continued fractions and other methods.
50) Rewrite $\sqrt{13}$ as a continued fraction.
51) Find a number such that when divided by 7 leaves a remainder of 5, when divided by 12 leaves a remainder of 2 and when divided by 13 gives a remainder of 8.
52) Find the least positive residue of 10! $(\mod 143)$.
53) A man bought a dozen pieces of fruit apples and oranges for 99 cents. If an apple costs 3 cents more than an orange, and he bought more apples than oranges, how many of each did he buy?
54) The enrollment in a number theory class consists of sophomores, juniors, and seniors. If each sophomore contributes $1.25, each junior $0.90 and each senior $0.50, then the instructor has a fund of $25. There are 26 students, how many of each?
55) Solve in integers $x^2 - 3y^2 = 17$.
56) A number is ending in 1969. A number obtained by deleting the last four digits is an integer number of times less than the original number. Find all such numbers.
57) Find the remainders of $N = \underbrace{111\ldots1}_{2019\,\text{times}}$ when divided by $3, 4, 5, 7, 11$.
58) Prove that the sum of ten consecutive squares is not a square.
59) Rewrite 65 by all possible sums of two squares.
60) Given $n = x^2 + y^2$ prove that $17n$ can be written as sum of two squares. Find its representation.
61) Prove that the sum of the digits of all numbers between 1 and 10^n cannot be divisible by 9 .
62) Prove that $N = \underbrace{111\ldots1}_{2020\,\text{digits}}$ is not prime.
63) Prove that a square of an odd number divided by 8 leaves a remainder of 1.
64) Prove that $x^2 + y^2 + z^2 = 8n - 1$ has no solutions in integers. Hint. Consider divisibility by 8.
65) Given $(a, b) = 1$. Evaluate greatest common divisor, $(13a + 2b, 20a + 3b)$.
66) Solve the equation $x^2 - 2x - yx + 3y = 11$ in natural numbers.
67) Find all natural n for which the expression $n^2 + 17n - 2$ is divisible by 11.
68) Prove that the equation $x^3 + y^3 = 4(x^2y + xy^2 + 1)$ has no integer solutions.
69) Prove that any natural number $N = (1 + a^2)(1 + b^2)(1 + c^2)$ can be written as sum of two squares.
70) A quadratic equation with integer coefficients $x^2 + bx + c = 0$ has two real roots, x_1, x_2. Evaluate $(1 + x_1^2)(1 + x_2^2)$.
71) Find the largest power of 3 that divides 600!.
72) Can a number n be written by five squares? Justify your answer.
73) Find all integers, a, for which the number $\frac{a^2+1}{a+2}$ is also integer.

74) Find the maximum value of n for which 10^n divides 2019!
75) Find the smallest prime that divides $N = 3^{11} + 5^{13}$.
76) Find natural $x, y, z,$ and w satisfying the equation

$$x + \cfrac{1}{y + \cfrac{1}{z + \frac{1}{w}}} = \frac{34}{9}$$

77) Solve the equation $3x^2 + 6x - 7y + 5 = 0$ for all $x \in \mathbb{N}$, and $y \in \mathbb{N}$.
78) Rewrite $105/38$ as a continued finite fraction.
79) Solve the equation $x^2 - 3y^2 = 1$ in natural numbers.
80) Prove that $n^4 + 64$ is not prime.
81) Is $N = 2017^{1917} + 1917^{2017}$ divisible by 7?
82) Find all integer solutions to $x^2 - 4y^2 = 45$
83) Solve in integers: $x^2 - y^2 = 6$.
84) Find all solutions to $x^2 - 9y^2 = 7$.
85) Solve the equation $x^2 - 63y^2 = 1$ in integers.
86) Prove that there are infinitely many right triangles with integer sides, such that the hypotenuse is one unit longer than one leg.
87) Find exact formulas for the nth solution of $x^2 - 8y^2 = 1$.
88) Solve in integers: $-3xy + 10x - 16y + 45 = 0$.
89) Find all integer solutions to the equation: $n! + 6^n + 11 = k^2$.
90) Solve in integers: $n! + 5n = k^2 - 13$.
91) Find the last two digits of 2^{2010} and 3^{2010}.
92) In how many ways can 48 be written as a difference of squares?
93) Can 900 be represented by a difference of squares? If yes, justify your answer and find all possible pairs of n and m.
94) Rewrite 945 as a difference of two squares.
95) Factor 13561 using Fermat's factoring method and with the help of

$$233^2 \equiv 3^2 \cdot 5 \pmod{13561} \quad 1281^2 \equiv 2^4 \cdot 5 \pmod{13561}.$$

96) A box has red, white, and blue balls. The number of blue balls is at least the number of white balls and at most $1/3$ of the number of the red balls. The total number of white and blue balls is at least 55. What is the minimal possible number of red balls?
97) A frustrated bond investor tears a bond into eight pieces. Then she continues and cuts one of the pieces again into eight pieces. If she continues, can she get 2016 pieces?
98) Today in one hospital the average age of doctors and patients together is 40 years. The average age of the doctors is 35 and the average age of the patients is 50. Are there more doctors or patients? How many times more?
99) Find the number of ways to cut a convex polygon with n sides into triangles by cutting along non intersecting diagonals. How many triangles can be obtained if the figure is heptagon? What if it is n-gon?

100) A fruit farmer wants to plant trees. He has fewer than 1000. If he plants them in rows, 37 trees per row, then there will be 8 trees remaining. If he plants 43 per row, there will be 11 remaining. How many trees does he have? Solve the problem using Chinese Remainder Theorem.

101) Prove that

$$e = [2; 1, 2, 1, 1, 4, 1, 1, 6, \ldots],\qquad(5.1)$$

i.e., the elements a_n of the decomposition of an irrational number e into continued fraction have the following type:

$a_0 = 2, a_{3n} = a_{3n+1} = 1$ and $a_{3n-1} = 2n$.

102) The two cog wheels of which one has 30 teeth and the other 40 teeth are touching before beginning rotation. How many revolutions of each wheel are required before they will be in the same position as when they started (the rotation)?

103) Find a number divisible by 24 that has 33 total divisors.

104) Find the maximum number of zeros that the number x, represented as $x = 1^n + 2^n + 3^n + 4^n$ can end.

105) Given $a = 3^{2017} + 2$. Can $a^2 + 2$ be prime?

106) Solve the equation $n^3 - n = n!$ in natural numbers.

107) Solve the equation $n^2 - 1 = 8m$ in natural numbers.

108) Prove that $y^2 = 5x^2 + 6$ does not have solution in integers.

109) Find natural solutions to the equation $4z + 2xy = 4y^2z + zx^2$.

110) Prove that a number $N = \underbrace{11\ldots1}_{2n \text{ digits}} - \underbrace{22\ldots2}_{n \text{ digits}}$ is a perfect square.

5.2 Answers and Solutions to the Homework

1) **Method 1.** Using Math Induction.
   The prime factorization of 30 is $2 \cdot 3 \cdot 5$. If N is divisible by those factors, it is divisible by 30.

   $$N = n^5 - n = n(n^4 - 1) = n(n^2 + 1)(n^2 - 1) = (n^2 + 1)(n - 1)(n)(n + 1).$$

   $(n - 1)n(n + 1)$ is the product of 3 consecutive integers so it is divisible by both 2 and 3. Now we need to determine if N is divisible by 5.
   By $n = 1$ and $n = 2$ we see that $N = 0$ and $N = 30$ respectively, which are divisible by 5. Suppose by $n = k$, $N = k^5 - k$ is divisible by 5. We need to demonstrate that when $n = k + 1$, N is divisible by 5. We have

   $$\begin{aligned} N &= (k^2 + 2k + 2)k(k + 1)(k + 2) \\ &= (k^2 + 2k + 2)(k^3 + 3k^2 + 2k) = k^5 + 3k^4 + 2k^3 + 2k^4 + 6k^3 + 4k^2 + 2k^3 + 6k^2 + 4k \\ &= k^5 + 5k^4 + 10k^3 + 10k^2 + 4k \ = k^5 - k + 5k^4 + 10k^3 + 10k^2 + 5k. \end{aligned}$$

   Here $k^5 - k$ is divisible by 5 per our assumption. Each of the remaining terms is also divisible by 5, so N is divisible by 5. Therefore N is divisible by 5.
   Since N is divisible by 2, 3, and 5, N is divisible by 30.
   Method 2. Using factoring.

   $$\begin{aligned} n^5 - n &= n(n^4 - 1) = n(n^2 - 1)(n^2 + 1) \\ &= (n - 1)n(n + 1)(n^2 - 4 + 5) = (n - 1)n(n + 1)(n^2 - 2^2) + 5(n - 1)n(n + 1) \\ &= (n - 2)(n - 1)n(n + 1)(n + 2) + 5(n - 1)n(n + 1) = 30k. \end{aligned}$$

   Above, by rewriting $n^2 + 1 = [n^2 - 4] + 5$, after applying the difference of squares formula to the term inside brackets and then using distributive law, we represented $n^5 - n$ as sum of two terms. The first term is the product of five consecutive integers that is divisible by 5! and the second one is divisible by 30 as a product of 5 and three consecutive integers.

2) Factor the number as

   $$n^3 + 3n^2 - n - 3 = n^3 - n + 3(n^2 - 1) = n(n^2 - 1) + 3(n^2 - 1) = (n^2 - 1)(n + 3).$$

   Let $n = 2k + 1$ and then

   $$\begin{aligned} (n^2 - 1)(n + 3) \quad &= ((2k + 1)^2 - 1)(2k + 4) \quad = (2k + 1 - 1)(2k + 1 + 1)(2k + 4) \\ &= 2k(2k + 2)(2k + 4) = 2 \cdot 2 \cdot 2 \cdot k \cdot (k + 1) \cdot (k + 2) = 8 \cdot 6k_1 = 48 \cdot k_1. \end{aligned}$$

3)
   $$k^3 + 5k = k(k^2 - 1 + 6) = (k - 1)k(k + 1) + 6k.$$

4) **Answer.** $n = 9k, k \in \mathbb{N}$.

Solution. If a number N has $2n$ digits, then it consists of n repeated numbers 13, then the sum of all its digits equals $(1+3)n = 4n$. Since $63 = 7 \cdot 9$, then N must be simultaneously divisible by 9 and 7. If N is divisible by 9, then the sum of its digits must be divisible by 9, then $4n$ must be divisible by 9, then $n = 9k$, $k \in \mathbb{N}$. Let us see how many repeated 13 form a multiple of 7. Notice that 131313 is the smallest such number $131313 = 7 \cdot 18759$. This means that a number N is divisible by 7 if and only if it contains a whole number m of "blocks" [131313]. Then $2n = 6m$, then $n = 3m$. We find that n must be a multiple of 3. But $n = 9k$ is a multiple of 3. If a number N contains $9k$ repeated numbers 13 as $\underbrace{1313\ldots13}_{9k\,\text{digits}}$,

then N is divisible by 63. The smallest possible $N = \underbrace{131313131313131313}_{9\,\text{times}}$ has

a number 13 repeated 9 times.

To be divisible by 63, the number must be divisible by both 9 and 7. First look at divisibility by 9. If $N = 1313\ldots13$ with $2n$ digits, 13 is repeated n times. This makes the sum of the digits for N equal to $4n$. For N to be divisible by 9, then the sum of its digits must be divisible by 9. For $4n$ to be divisible by 9, it must be at least 36, therefore n has to be at least 9.

Now look at N when we let $n = 9$. We have $N = 131313131313131313$. We also need to recognize that 131313 is divisible by 7. Then 131313131313131313 can be rewritten as

$$131313 \cdot 10^{12} + 131313 \cdot 10^{6} + 131313$$
$$= 131313(10^{12} + 10^{6} + 1) = 7 \cdot 18759(10^{12} + 10^{6} + 1) = 7k,$$

therefore it is divisible by 7.

We have found that if $n = 9$, N is divisible by 9, and also divisible by 7, therefore N is divisible by 63. So, $n = 9k$ is the answer.

5) **Proof.** Changing the order of factors and combining the first two and the last two:

$$n(n+1)(n+2)(n+3)+1 = (n^2+3n)(n^2+3n+2)+1 = m(m+2)+1 = (m+1)^2$$

Therefore, it is a perfect square for any $m = n^2 + 3n$.

6) **Answer.** $\{(q,p) : (5,2,),(5,-2),(-5,-2),(-5,2),(3,0),(-3,0)\}$.

Solution. Begin with

$$4p^2 = q^2 - 9 \Leftrightarrow 4p^2 - q^2 = -9 \Leftrightarrow (2p-q)(2p+q) = -9.$$

Now look at all possible cases:
Case 1:
$$\begin{cases} 2p - q = +3, \\ 2p + q = -3 \end{cases} \Leftrightarrow 4p = 0 \Leftrightarrow p = 0, \ q = -3.$$

Case 2:

$$\begin{cases} 2p-q=-3, \\ 2p+q=+3 \end{cases} \Leftrightarrow 4p=0 \Leftrightarrow p=0,\ q=+3.$$

Case 3:

$$\begin{cases} 2p-q=+9, \\ 2p+q=-1 \end{cases} \Leftrightarrow 4p=8 \Leftrightarrow p=2,\ q=-5.$$

Case 4:

$$\{\,2p-q=-9, 2p+q=+1 \ \Leftrightarrow 4p=-8 \Leftrightarrow p=-2,\ q=+5.$$

Case 5:

$$\begin{cases} 2p-q=+1, \\ 2p+q=-9 \end{cases} \Leftrightarrow 4p=-8 \Leftrightarrow p=-2,\ q=-5.$$

Case 6:

$$\begin{cases} 2p-q=-1 \\ 2p+q=+9 \end{cases} \Leftrightarrow 4p=8 \Leftrightarrow p=2,\ q=5.$$

Therefore, we find all possible solutions of the equation.

7) **Answer.** $\{(x,y): (4,1),(4,-3),(-4,-3),(-4,1)\}.$

Solution. Let us factor the given equation as

$$x^2=y^2+2y+13 \Leftrightarrow x^2=(y+1)^2+12 \Leftrightarrow x^2-(y+1)^2=12 \Leftrightarrow (x+y+1)(x-y-1)=12.$$

The last equation has integer solutions only if x and y satisfy the following systems:

$$\begin{cases} x+y+1=6, \\ x-y-1=2 \end{cases} \Leftrightarrow x=4,\ y=1.$$

$$\begin{cases} x+y+1=2, \\ x-y-1=6 \end{cases} \Leftrightarrow x=4,\ y=-3.$$

$$\begin{cases} x+y+1=-6, \\ x-y-1=-2 \end{cases} \Leftrightarrow x=-4,\ y=-3.$$

$$\begin{cases} x+y+1=-2, \\ x-y-1=-6 \end{cases} \Leftrightarrow x=-4,\ y=1.$$

8) **Answer.** $\{(x,y): (9,-4),(-33,20),(-9,4),(33,-20)\}.$

9) **Answer.** $(x,y)=(3,5).$

Solution. Let us factor the equation:

$$3\cdot 2^x=(y-1)(y+1).$$

The left side is even, but the right side is a product of two numbers that differ by 2. In order to have solutions, we need to denote $y=2k+1$. Then

$$3\cdot 2^x=2k(2k+2)=2^2k(k+1) \Leftrightarrow 3\cdot 2^{x-2}=k(k+1).$$

Since 2 and 3 are two consecutive primes as $k(k+1)$ then $x-2=1$ and $k=2$, or $x=3$ and $y=5$.

10) **Answer.** The last two digits are: a) 00, b) 48, c) 50.
11) **Answer.** 6.
12) **Answer.** The last two digits are 96.

Solution. The last digit is 6 because $2012 = 4 \cdot 503$. The last two digits can be found under division by 25:

$$2^{2000} \equiv 1 \quad (\text{mod } 25), \qquad 2012 = 20 \cdot 100 + 12,$$
$$2^{12} = 4096 \equiv 21 \quad (\text{mod } 25), \ 2^{2012} \equiv 21 \quad (\text{mod } 25).$$

Then, the last two digits can be either 46 or 96. Since it is multiple of 4 then there is the unique variant 96 ($96 = 21 + 3 \cdot 25$).

13) **Answer.** $2^{123456789}$ ends in 12.

Solution. We have $123456789 = 4 \cdot 30864197 + 1$ then $2^{123456789} \equiv 2 (\mod 5)$. On the other hand, $123456789 = 6172839 \cdot 20 + 9$. So,

$$2^9 = 512 \equiv 12 (\mod 25) \Leftrightarrow 2^{123456789} \equiv 12 (\mod 25).$$

Out of two possible even numbers 12 and 62 ($62 = 2 \cdot 25 + 12$), we select 12 as a multiple of 4. Then, the last two digits must be 12.

14) **Answer.** $\{(x,y) : (4,1), (-4,1), (-4,-1), (4,-1)\}$.
    Hint: Factor as $(x - 3y)(x + 3y) = 7$.
15) **Answer.** $x = \pm 4(2t - 1)$, $m = -10t + 5$, $N = -6t + 3$, $t \leq 0$, $t \in \mathbf{Z}$.

Solution. Since N is an odd number, let $N = 2n + 1$.

$$N^2 + x^2 = m^2 \Leftrightarrow (2n+1)^2 + x^2 = m^2 \Leftrightarrow x^2 = m^2 - (2n+1)^2.$$

This equation will have solutions if and only if m is an odd number.
Let $m = 2k + 1$. Then

$$x^2 = (2k+1)^2 - (2n+1)^2 \Leftrightarrow x^2 = (2k+1-2n-1)(2k+1+2n+1)$$
$$\Leftrightarrow x^2 = 2^2(k-n)(k+n+1) \tag{5.2}$$

In order for (5.2) to be a perfect square, two cases must be considered.
Case 1. Let $k - n = k + n + 1$. No integer solutions.
Case 2. Let $4(k - n) = k + n + 1$. We have the equation

$$3k - 5n = 1. \tag{5.3}$$

One of the solutions of (5.3), $k = 2$, $n = 1$ can be found by inspection. This equation has infinitely many solutions. Let

$$k = 2 - 5t, \ n = 1 - 3t, \ t \in \mathbf{Z}$$

then $m = 2k + 1 = -10t + 5$ and $N = 2n + 1 = -6t + 3$. Therefore,

$$x^2 = 4(-10t + 5 + 6t - 3)(-10t + 5 - 6t + 3) \Leftrightarrow x^2 = 4(-2)(-8)(2t-1)(2t-1)$$
$$\Leftrightarrow x^2 = 16(2t-1)^2.$$

Hence, we obtain $x = \pm 4(2t-1)$.
We can find several particular solutions:

$$t = 0, \quad m = 5, \quad N = 3, \quad x = 4 \text{ or } x = -4,$$
$$t = -1, \, m = 15, \, N = 9, \quad x = 12 \text{ or } x = -12,$$
$$t = -2, \, m = 25, \, N = 15, \, x = 20 \text{ or } x = -20,$$
$$t = -3, \, m = 35, \, N = 21, \, x = 28 \text{ or } x = -28,$$
$$t = -4, \, m = 45, \, N = 27, \, x = 36 \text{ or } x = -36.$$

Note. You probably noticed that positive triples (m, N, x) are terms of arithmetic sequences with common difference of 10, 6, and 8, respectively: $(55, 33, 44)$, $(65, 39, 52)$, $(75, 45, 60)$, etc.

16) **Answer.** No.

Solution. Suppose that it is possible and $1974 = a^2 - b^2$. 1974 is divisible by 2 but not divisible by 4. If $a^2 - b^2 = 2n$ (even), then $(a - b)$ is even and $(a + b)$ is even, then $(a^2 - b^2)$ must be divisible by 4. We obtained a contradiction.

17) **Answer.** $n = 8k$ or $n = 2k - 1$, $k \in \mathbb{N}$.

18) **Answer.** No.

Solution. Because n and m are odd numbers, we can assume that $n = 2k + 1$ and $m = 2l + 1$. Then

$$n^2 + m^2 = 4(k^2 + l^2 + k + l) + 2.$$

From this formula we can notice that a number $(n^2 + m^2)$ is divisible by 2 but not divisible by 4. Or we can say that $(n^2 + m^2)$ is an even number, which in turn is not divisible by 4.
Notice that if a number $(n^2 + m^2)$ was a square of some integer, then it would be a square of an even number, because it is an even itself, and it would be divisible by 4, that is not true for our situation. Therefore we conclude that the sum of squares of two odd numbers cannot be a perfect square.

19) **Answer.** No.

Solution. Method 1. Assume that such a number exists. Let N be the number. It is known that a number divided by 3 leaves the same remainder as the sum of its digits. On one hand, the sum of its digits, 1970, can be written as $1970 = 3 \cdot 656 + 2$. Therefore, $N = 3k + 2$.
On the other hand, it was proven in Chapter 1 that any square divided by 3 leaves a remainder of 0 or 1. We can conclude that the sum of the digits of a perfect square cannot equal 1970. **Method 2.** Let N be a number n^2, where n can either

be written as $3k$, $3k+1$, or $3k+2$. Then, N is either $(3k)^2$, $(3k+1)^2$, or $(3k+2)^2$. It is easy to see that

$$
\begin{aligned}
(3k)^2 &= 9k^2 &&\equiv 0 \quad (\mathrm{mod}\ 3), \\
(3k+1)^2 &= 9k^2 + 6k + 1 &&\equiv 1 \quad (\mathrm{mod}\ 3), \\
(3k+2)^2 &= 9k^2 + 12k + 4 &&\equiv 1 \quad (\mathrm{mod}\ 3).
\end{aligned}
$$

Hence, $1970 \equiv 2 \ (\mathrm{mod}\ 3)$ which means the number whose digit sum is equal to 1970 is also $\equiv 2 \ (\mathrm{mod}\ 3)$, therefore this number cannot be a perfect square.

20) **Solution.** We have

$$
\begin{aligned}
n(n^4 - 5n^2 + 4) &= n(n-1)(n+1)(n-2)(n+2) \\
&= (n-2)(n-1)n(n+1)(n+2).
\end{aligned}
\tag{5.4}
$$

From (5.4) we can see that

a. (5.4) is divisible by 5 as a product of 5 consecutive numbers,
b. it is divisible by 3 as a product of 3 consecutive numbers,
c. either $(n-2)$ or $(n+2)$ is divisible by 3 because n is an even number,
d. one of $(n-2)$ and $(n+2)$ is divisible by 4,
e. n is divisible by 2 as an even number.

Then, $5 \cdot 9 \cdot 16 \cdot 2 = 1440$.

21) **Solution.** Let us prove this using mathematical induction.
Show that this is true for $n = 1$:

$$
3^{2 \cdot 1 + 2} - 8 \cdot 1 - 9 = 3^4 - 8 - 9 = 64
$$

is divisible by 64, so it is true for $n = 1$.
Assume it is true for any $n = k$ so that $3^{2k+2} - 8k - 9 = 64p$.
Show it is true for any $n = k+1$. We have

$$
\begin{aligned}
3^{2(k+1)+2} - 8(k+1) - 9 \quad &= 3^{2k+2+2} - 8k - 8 - 9 \\
= 3^{2k+2} \cdot 3^2 - 8k - 8 - 9 \quad &= 3^{2k+2} \cdot 9 - 9 \cdot 8k - 9 \cdot 9 + 8 \cdot 8k + 8 \cdot 9 - 8 \\
= 9(3^{2k+2} - 8k - 9) + 8(8k + 9 - 1).
\end{aligned}
$$

From the induction's assumption above, we can replace what is in the first parentheses by 64p:

$$
9(64p) + 8(8k + 8) = 64m + 64q = 64r.
$$

Since we have shown that the last expression is divisible by 64, it is true for all n, by the principle of mathematical induction.

22) **Solution.** We have $7 \cdot 5^{2n} + 12 \cdot 6^n = 19 \cdot 6^n + 7(25^n - 6^n)$. Every term of this expression is divisible by 19. For the second term we apply a property that

$$
a^n - b^n = (a - b)(a^{n-1} + a^{n-2}b + \cdots + ab^{n-2} + b^{n-1}).
$$

23) **Solution.** We have the expression:

$$2^9 + 2^{99} = 2^9(1 + 2^{90}) = 2^9(1 + (2^{10})^9) = 2^9(1 + 1024^9)$$
$$= 4 \cdot 2^7 \cdot 1025 \cdot C = 4100 \cdot 2^7 \cdot C \text{ divisible by } 100,$$

where C is a constant. Again we use $(1^9 + 1024^9) = (1 + 1024) \cdot C = 1025C$

24) **Answer.** $\{(3,2), (-3,-2), (3,-2), (-3,2)\}$.

Solution. Let us rewrite the equation in a different form $6x^2 - 24 = 50 - 5y^2$. After factoring it is

$$6(x^2 - 4) = 5(10 - y^2) \tag{5.5}$$

or $x^2 - 4 = 5u$ and $10 - y^2 = 6v$. After substitution of these expressions into (5.5) we obtain $6 \cdot 5u = 56 \cdot v$ or $u = v$.

Because $x^2 = 4 + 5u$ then $(4 + 5u) \geq 0$ and $u \geq -\frac{4}{5}$.

By analogy, we can write that $10 - y^2 = 6u$, then $(10 - 6u) \geq 0$ and $u \leq \frac{5}{3}$.

Therefore, integer number u must satisfy the double inequality $-\frac{4}{5} \leq u \leq \frac{5}{3}$, which gives us two possible values of u: $u = 0$ and $u = 1$.

If $u = v = 0$, we obtain that $y^2 = 10$, where y is integer, that is false.

If $u = v = 1$, then $x^2 = 9$ and $y^2 = 4$, which gives us four possible ordered pairs.

25) **Answer.** No solutions.

Solution. Since $(18x^2 + 27y^2) + (x^2 + y^2) = 729$, then $(x^2 + y^2)$ is divisible by 3. Hence, $x = 3u$ and $y = 3v$. Then, we rewrite the equation as $19u^2 + 28v^2 = 81$ that yields $u = 3t$ and $v = 3s$. These substitutions imply that

$$19t^2 + 28s^2 = 9. \tag{5.6}$$

However, equation (5.6) has no solution for integers s and t, therefore, the given equation does not have solution.

26) **Answer.** $\{(0,0), (1,2)\}$.

Solution. Factoring the left- and the right-hand sides of the equation we obtain

$$(x+1)(x^2+1) = 2^y = 2^m \cdot 2^{y-m}.$$

Let $x = 2^m - 1$ and $x^2 = 2^{y-m} - 1$ then

$$x^2 = 2^{2m} - 2^{m+1} + 1 = 2^{y-m} - 1,$$

or $2^{y-m} + 2^{m+1} - 2^m = 2$.

If $m = 0$, then $2^y + 2 - 1 = 2$ and $x = 0$, $y = 0$.

If $m > 0$, then $x = 1$ and $y = 2$.

27) **Answer.** $\{(x,y,z) : (1,1,z),\ \forall z \in \mathbb{N}; (3,3,2), (3,-3,2); (x, 1! + 2! + \cdots + x!, 1),$ $x \in \mathbb{N}\}$.

Solution. Let us consider two cases:

Case 1. Let $z = 2n$ (an even number), then the solution to our problem is similar to that of Problem 203 from Moscow Olympiad: $x = 1$, $y = 1$, z is any even number. Moreover, $x = 3$, $z = 2$, and $y = \pm 3$.

Case 2. Let $z = 2n + 1$ (an odd number). For $z = 1$ any value of x and the corresponding evaluated y, will satisfy.

Assume that $z \geq 3$ and $x > 8$. We have

$$9! + 10! + 11! + \cdots = 9!(1 + 10 + 11 + \ldots) = 27k$$

is divisible by 27.

On the other hand the first eight terms

$$1! + 2! + 3! + 4! + 5! + 6! + 7! + 8! = 46233 = 9m$$

are divisible by 9, not 27, therefore for any natural x we obtain that the number $1! + 2! + \cdots + x! = 9n$ is a multiple of 9.

In order for y^z to be divisible by 9 it is necessary that y be divisible by 3. Since $z \geq 3$, y^z must be divisible by 27. We obtained impossible situation and if $z \geq 3$ and $x \geq 8$ there are no solutions in integers.

Let us assume that $x < 8$. After checking all possible sums up to $x = 7$, we obtained that none of them is an integral power $z \geq 3$ of any natural number.

28) **Answer.** $n \in \{-1, 0, 1, 2\}$.

Solution. Let k be the values of the fraction

$$\frac{n+1}{2n-1} = k.$$

We rewrite this expression as

$$n + 1 = k(2n - 1), \tag{5.7}$$

or we have $n(2k - 1) = k + 1$. Then, we consider different variants:

a. if $n = 1$, then $2k - 1 = k + 1$, and it follows that $k = 2$;
b. if $n = 0$, then $k + 1 = 0$, and $k = -1$;
c. if $n = -1$, then $2k + 1 = -k + 1$, and $k = 0$;
d. if $n = 2$, then $2(2k - 1) = k + 1$, and $k = 1$.

Let us show that there are no other solutions. In order to have solutions for all $k > 2$ or $k < -1$ equation (5.7) can be rewritten as

$$|n + 1| = |k| \cdot |2n - 1| \geq |2n - 1|,$$

and at least the following inequality must be valid:

$$(n - 1)^2 \geq (2n - 1)^2 \Leftrightarrow 3n^2 - 6n \leq 0 \Leftrightarrow 0 \leq n \leq 2.$$

29) **Answer.** $\{(1,2,3),(1,3,2),(2,1,3),(2,3,1),(3,1,2),(3,2,1)\}$.

Solution. Let (a,b,c) be a solution for the equation. Then all triples as (a,c,b), (b,a,c), (b,c,a), (c,b,a), and (c,a,b) will be also the answer. How can we find the answer? It is clear that if $x = y = z = a$, will give us a solution $(0,0,0)$.
Since we are looking for natural solution only, then we will not consider this as an answer. Let us consider the following case: $x > 0$, $y = x + m$, $z = x + n$, $(m,n) = 1$.
The equation can be rewritten as

$$x(x+m)(x+n) = 3x+n+m \Leftrightarrow x^3 + (m+n)x^2 + (mn-3)x - (n+m) = 0.$$

This equation will have a solution in integers if and only if they are factors of the constant term, $-(m+n)$. Since $(m,n) = 1$, we will consider $x = 1$ as a solution. Then

$$1 + (m+n) + (mn-3) - (n+m) = 0 \Leftrightarrow km = 2 \Leftrightarrow k = 1, \ m = 2.$$

Therefore, $x = 1$, $y = 2$, $z = 3$ is one of the natural solutions. Using symmetry of the equation finally we have the desired solutions.

30) **Answer.** $n = 5k+1$, $k \geq 0$, $k \in \mathbb{Z}$.

Solution. If $k = 0$, then $n = 1$, and

$$\frac{22+3}{26+4} = \frac{25}{30} = \frac{5}{6},$$

then the fraction is reducible, $25/30 = 5/6$.
Let us rewrite the fraction so that one of the terms on the top will be 25 and one on the bottom 30:

$$\frac{22n+3}{26n+4} = \frac{22(x+1)+3}{26(x+1)+3} = \frac{25+22x}{30+26x}.$$

Since 22 and 26 are not multiples of 5 but 25 and 30 are, then in order for the fraction to be reducible, x must be a multiple of 5. Let $x = 5k$. Then $n = x+1 = 5k+1$ is the answer.

$$\frac{22n+3}{26n+4} = \frac{22(5k+1)+3}{26(5k+1)+3} = \frac{25+22\cdot 5k}{30+26\cdot 5k}$$
$$= \frac{5(5+22k)}{5(6+26k)} = \frac{22k+5}{26k+6}.$$

31) **Solution.** We have $(5^3)^{100} = 125^{100}$ and $(3^5)^{100} = 243^{100}$. Because $125 < 243$, we can conclude that $5^{300} < 3^{500}$.

Remark. This problem is trivial. Some of you, misleading by huge powers, tried to apply Euler Theorem. However, you have to compare the numbers, not their remainders. It would be a mistake because sometimes smaller numbers leave bigger remainders and vice versa. For example, $11 = 1*6 + 5$, but $31 = 5*6 + 1$, $5 > 1$, but $31 > 11$.

32) **Answer.** $n = 85k + 53$.

Solution. Method 1. Without using congruence.

We have $n = 17m + 2$ and $n = 5l + 3$. Then $5n = 85 \cdot m + 10$ and $17n = 85l + 51$. Because $51n - 50n = n$, then $10 \cdot 5n = 85 \cdot 10m + 100$ and $3 \cdot 17n = 85 \cdot 3l + 153$. Subtracting the last equation from the first we obtain $n = 85 \cdot (3l - 10m) + 53 = 85k + 53$.

Of course, if we could find at least one number satisfying the condition, say 53, we could get the answer right away: $N = 85k + 53$, where 85 is the *LCM* of 17 and 5.

Method 2. Using congruence.

We have $n = 17m + 2$ and $n = 5l + 3$. Then,

$$17m + 2 = 5l + 3 \Leftrightarrow 15m + 2m = 5l + 1 \Rightarrow 2m \equiv 1 \pmod 5, \ 2m \equiv 6 \pmod 5$$
$$\Rightarrow m \equiv 3 \pmod 5 \Rightarrow m = 5k + 3 \Rightarrow n = 17(5k + 3) + 2 \Leftrightarrow n = 85k + 53.$$

33) **Solution.** Number $43^{43} - 17^{17}$ is divisible by 10 if its last digit is zero. Thus, by Euler's Theorem we have:

$$43^4 \equiv 1 \pmod{10} \Rightarrow 43^{40} \equiv 1 \pmod{10} \ \ 43 \equiv 3 \pmod{10}$$
$$43^3 \equiv 27 \pmod{10} = 7 \pmod{10} \qquad 43^{43} \equiv 7 \pmod{10}$$

On the other hand,

$$17 \equiv 7 \pmod{10} \Rightarrow 17^4 \equiv 1 \pmod{10}$$
$$17^{16} \equiv 1 \pmod{10} \Rightarrow 17^{17} \equiv 7 \pmod{10}.$$

Subtracting both final expressions we obtain $43^{43} - 17^{17} \equiv 0 \pmod{10}$, which means that the difference is divisible by ten.

34) **Answer.** $x = 9$.

Solution. Since $(33, 14) = 1$ and 14 is not prime we will use Euler ϕ function:

$$\phi(14) = \phi(2 \cdot 7) = (2 - 1)(7 - 1) = 6.$$

Then

$$33^6 \equiv 1 \pmod{14} \Rightarrow (33^6)^3 = 33^{18} \equiv 1 \pmod{14} \Rightarrow 33^4 = 1185921 \equiv 9 \pmod{14}.$$

Therefore, $33^{22} \equiv 9 \pmod{14}$.

35) **Solution.** If n is odd the $n = 2k + 1$. We have

$$n^2 = (2k + 1)^2 = 4k^2 + 4k + 1 = 4k(k + 1) + 1.$$

Since k and $(k + 1)$ are two consecutive integers, then their product is divisible by two. Moreover,

$$n^2 = 4 \cdot 2m + 1 = 8m + 1.$$

Therefore, $n^2 \equiv 1 \pmod 8$.

36) **Solution.** If $k \equiv 1 \pmod 4$, then $6k \equiv 6 \pmod 4$. Moreover, $(6k + 1) \equiv 7 \pmod 4$ and $(6k + 1) \equiv 3 \pmod 4$.

37) **Answer.** $k = 19$.

   Solution. We have $n^2 \equiv 1 \pmod{17}$. Hence,

$$n^2 \equiv (1 + 19 \cdot 17) \pmod{17} = 324 \pmod{17}$$
$$n \equiv 18 \pmod{17}, \; n \equiv -18 \pmod{17}$$
$$n \equiv 1 \pmod{17} \Rightarrow n = 18, \; 324 = 17 \cdot 19 + 1, \; k = 19.$$

38) **Answer.** $x = 9; x = 14 \pmod{19}$.

   Solution. Method 1. Completing square on the left we obtain

$$(x^2 - 4x + 4 - 4 - 7) \equiv 0 \pmod{19} \; (x - 2)^2 \equiv 11 \pmod{19}$$
$$(x - 2)^2 \equiv (11 + 2 \cdot 19) \pmod{19} \quad (x - 2)^2 \equiv 49 \pmod{19}$$
$$(x - 2) \equiv 7 \pmod{19} \text{ or } (x - 2) \equiv -7 \pmod{19}$$
$$x \equiv 9 \pmod{19} \qquad \text{or } x \equiv -5 \pmod{19} = 14 \pmod{19}.$$

Method 2. Rephrasing the problem as follows: Find integer solutions of the equation $x^2 - 4x - 19y - 7 = 0$.

This equation can be considered as a quadratic in x where y is a parameter.

$$x = 2 \pm \sqrt{2^2 + (19y + 7)} = 2 \pm \sqrt{19y + 11}.$$

Then, x will be an integer if and only if discriminant is a perfect square. Hence, by $x = 2 \pm k$

$$k^2 = 19y + 11 \Rightarrow k^2 - 49 = 19y + 11 - 49 \Rightarrow (k - 7)(k + 7) = 19(y - 2).$$

Since 19 is prime, and $(k - 7)$ and $(k + 7)$ differ by 14, we must consider two possible cases:

Case 1. Let $k - 7 = 17m$ and $k + 7 = 19m + 14$. Then, $19m(19m + 14) = 19(y - 2)$ or $m(19m + 14) = y - 2$. Hence, $y = 19m^2 + 14m + 2$, where $m \in Z$.

We can make the following table.

m	$k = 19m + 7$	$y = 19m^2 + 14m + 2$	$x_1 = 2 + k$	$x_2 = 2 - k$
0	7	2	9	-5
1	26	35	28	-24
2	45	106	47	-43

Table 5.1 Case 1

Case 2. Let $k+7 = 19t$ and $k-7 = 19t - 14$. Then, $19t(19t - 14) = 19(y - 2)$ or $t(19t - 14) = y - 2$. Hence, $y = 19t^2 - 14t + 2$, where $t \in \mathbb{Z}$.
We again can make a new table.

t	$k = 19t - 7$	$y = 19t^2 - 14t + 2$	$x_1 = 2 + k$	$x_2 = 2 - k$
0	-7	2	-5	9
1	12	7	-10	14
2	31	50	-29	33

Table 5.2 Case 2

For example, two positive pairs of (x, y) are $(9, 2)$ and $(14, 7)$. There are infinitely many other solutions obtained by formulas above.

40) **Answer.** 1.

Solution. Using result of Problem 206, $2(p - 3)! + 1 \equiv 0 \pmod{p}$, we can rewrite it as $2(14)! + 1 \equiv 0 \pmod{17}$. Multiplying both sides by 15 we have:

$$2 \cdot 15(14)! \equiv -15 \pmod{17} \equiv 2 \pmod{17}.$$

Simplifying the left and right sides, this can be represented as

$$2 \cdot 15! \equiv 2 \pmod{17}.$$

Then, dividing both sides by 2 we obtain $15! \equiv 1 \pmod{17}$. Therefore, $15!$ divided by 17 leaves a remainder of 1.

41) **Solution.** First, we note that $437 = 23 \cdot 19$. Second, using the Wilson Theorem $(p - 1)! \equiv -1 \pmod{p}$ we can obtain several other congruence relationships. For example,

$$(p - 1)(p - 2)(p - 3)(p - 4)(p - 5)! \equiv -1 \pmod{p}$$
$$24(p - 5)! \equiv -1 \pmod{p}.$$

If $p = 23$, we have

$$24 \cdot 18! \equiv -1 \pmod{23}$$
$$24 \cdot 18! \equiv -24 \pmod{23} \tag{5.8}$$
$$18! \equiv -1 \pmod{23}$$

If $p = 19$, we can apply the Wilson Theorem directly as

$$18! \equiv -1 \pmod{19} \tag{5.9}$$

Since $18!$ leaves the same remainder divided by 19 and 23, then it will leave the same remainder when it is divided by the product of 19 and 23, or 437.
Therefore, combining (5.8) and (5.9) we state the desired result.

42) **Solution.** We want for a given prime number p to establish the congruence:

$$(p-1)! \equiv (p-1) \quad (\text{mod } (1+2+3+\cdots+(p-1))). \qquad (5.10)$$

First, we will add all integers from 1 to $(p-1)$:

$$1+2+3+\cdots+(p-1) = \frac{p}{2} \cdot (p-1).$$

Since p is prime then $p = 2k+1$ as an odd number, then

$$\frac{p}{2} \cdot (p-1) = \frac{p-1}{2} \cdot p = k(2k+1) = p \cdot k$$

and (5.10) can be rewritten as

$$(p-1)! \equiv (p-1) \quad (\text{mod } (p \cdot k)) \qquad (5.11)$$

We note that (5.11) is similar to

$$(p-1)! = p \cdot k \cdot l + (p-1) = p(kl+1) - 1 \text{ or}$$
$$(p-1)! \equiv -1 \quad (\text{mod } p).$$

The last statement is true since it is the Wilson Theorem. So as the original statement.

43) **Answer.** $\{(x,y) : (-4,3), (-6,7), (4,-5)\}$.

Solution. Multiply both sides by 3, then factor:

$$9xy + 3 \cdot 16x + 3 \cdot 13y + 3 \cdot 61 = 0$$
$$3y(3x+13) + 16(3x+13) + 3 \cdot 61 - 16 \cdot 13 = 0$$
$$(3x+13)(3y+16) = 25$$

Let $3x + 13 = a$ and $3y + 16 = b$ then we obtain the equation $ab = 25$, or we find that $a, b = \pm 25; \pm 5; \pm 1$. From this we have the desired result.

44) **Solution.** We have

$$5! = 120 \equiv -4 \quad (\text{mod } 31). \qquad (5.12)$$

It is easy to see that

$$2(p-2)! \equiv -1 \quad (\text{mod } p), \ 2(31-2)! \equiv -1 \quad (\text{mod } 31).$$

Then, we can find

$$(2 \cdot 28! \cdot 29) \equiv -29 \quad (\text{mod } 31),$$
$$(2 \cdot 29!) \equiv 2 \quad (\text{mod } 31),$$
$$29! \equiv 1 \quad (\text{mod } 31).$$

Hence,

$$(4 \cdot 29!) \equiv 4 \quad (\text{mod } 31). \qquad (5.13)$$

Adding (5.12) and (5.13) gives us the required result.

45) **Answer.** $n = 6k + 1$.

46) **Answer.** 301. (See Theorem 20)

47) **Answer.** $(x, y) = (5 + 7t, 1 - 3t)$, $t \in \mathbb{Z}$.

Solution. Let us rewrite the equation as $3(x - 5) = 7(1 - y)$. Since 3 and 7 are the mutually prime numbers, then we have $x - 5 = 7t$ and $1 - y = 3t$, where $t \in \mathbb{Z}$. From this we find the answer.

48) **Answer.** 501 zeros.

Solution. 402 factors of 5, 80 numbers divisible by 25, 16 numbers divisible by 125, and three numbers divisible by 625. Total of 501 factors of 5. Therefore it ends in 501 zeros. Please see similar problems in the book.

49) **Answer.** $x \equiv 16 \pmod{34}$.

Solution. Method 1. Continued Fractions.
For $9x \equiv 8 \pmod{34}$ we have $a = 9$, $b = 8$, $m = 34$. Since $(9, 34) = 1$, our congruence has exactly one solution!

$$\frac{34}{9} = 3 + \frac{7}{9} = 3 + \frac{1}{\frac{9}{7}} = 3 + \frac{1}{1 + \frac{2}{7}}$$

$$= 3 + \frac{1}{1 + \frac{1}{\frac{7}{2}}} = 3 + \frac{1}{1 + \frac{1}{3 + \frac{1}{2}}}.$$

$$\frac{P_0}{q_0} = \frac{3}{1}, \frac{P_1}{q_1} = \frac{3+1}{1} = \frac{4}{1},$$

$$\frac{P_2}{q_2} = 3 + \frac{1}{1 + \frac{1}{3}} = \frac{15}{4}, \frac{P_3}{q_3} = \frac{34}{9},$$

$$s = 3, \ s - 1 = 2, \ P_2 = 15 \tag{5.14}$$

$$x \equiv (-1)^s b P_{s-1} \pmod{m}$$
$$x \equiv (-1)^3 \cdot 8 \cdot 15 \pmod{34}$$
$$x \equiv -120 \pmod{34}$$
$$x \equiv (-120 + 34 \cdot 4) \pmod{34}$$
$$x \equiv 16 \pmod{34}.$$

Method 2. Standard.
We have:

$$9x \equiv 8 \pmod{34}$$
$$9x \equiv (8 + 34 \cdot 4) \pmod{34}$$
$$9x \equiv 144 \pmod{34}$$
$$x \equiv 16 \pmod{34}. \tag{5.15}$$

50) **Solution.** We have:

$$\sqrt{13} = 3 + \tfrac{1}{\alpha_1},$$

$$\alpha_1 = \frac{1}{\sqrt{13}-3} = \frac{\sqrt{13}+3}{4} = 1 + \tfrac{1}{\alpha_2},$$

$$\alpha_2 = \frac{1}{\alpha_1 - 1} = \frac{1}{\frac{\sqrt{13}+3}{4} - 1} = \frac{4}{\sqrt{13}-1}$$

$$= \frac{\sqrt{13}+1}{3} = 1 + \tfrac{1}{\alpha_3},$$

$$\alpha_3 = \frac{1}{\alpha_2 - 1} = \frac{1}{\frac{\sqrt{13}+1}{3} - 1} = \frac{3}{\sqrt{13}-2}$$

$$= \frac{\sqrt{13}+2}{3} = 1 + \tfrac{1}{\alpha_4},$$

$$\alpha_4 = \frac{1}{\alpha_3 - 1} = \frac{1}{\frac{\sqrt{13}+2}{3} - 1} = \frac{3}{\sqrt{13}-1}$$

$$= \frac{\sqrt{13}+1}{4} = 1 + \tfrac{1}{\alpha_5},$$

$$\alpha_5 = \frac{1}{\alpha_4 - 1} = \frac{1}{\frac{\sqrt{13}+1}{4} - 1} = \frac{4}{\sqrt{13}-3}$$

$$= \sqrt{13} + 3 = 6 + \tfrac{1}{\alpha_6},$$

$$\alpha_6 = \frac{1}{\alpha_5 - 6} = \frac{1}{\sqrt{13}+3-6} = \frac{1}{\sqrt{13}-3} = \alpha_1.$$

Then, we finally obtain:

$$\sqrt{13} = [3; \overline{1\,1\,1\,1\,6}]$$
$$= 3 + \cfrac{1}{1 + \cfrac{1}{1 + \cfrac{1}{1 + \cfrac{1}{1 + \cfrac{1}{6+\ldots}}}}}.$$

51) **Answer.** $x \equiv 866 \pmod{1092}$.

Solution. The condition of the problem can be written as

$$x \equiv 5 \pmod{7}, \tag{5.16}$$

$$x \equiv 2 \pmod{12}, \tag{5.17}$$

$$x \equiv 8 \pmod{13}. \tag{5.18}$$

Solving (5.16) and (5.17) we obtain

$$(7y + 5) \equiv 2 \pmod{12} \Rightarrow 7y \equiv -3 \pmod{12}$$
$$\Rightarrow 7y \equiv 21 \pmod{12} \Rightarrow y \equiv 3 \pmod{12},$$
$$x \equiv 7(3 \pmod{12}) + 5 \Rightarrow x \equiv 26 \pmod{84}.$$

Solving the last congruence with (5.18) we have

$$(84k + 26) \equiv 8 \quad (\text{mod } 13)$$
$$84k \equiv 8 \quad (\text{mod } 13) \Rightarrow 21x \equiv 2 \quad (\text{mod } 13),$$
$$21k \equiv 28 \quad (\text{mod } 13) \Rightarrow 3k \equiv 4 \quad (\text{mod } 13), \tag{5.19}$$
$$3k \equiv -9 \quad (\text{mod } 13) \Rightarrow k \equiv -3 \quad (\text{mod } 13),$$
$$x = 84k + 26 = 84(13m - 3) + 26 \Rightarrow x = 1092m - 226.$$

The smallest positive value of x happens if $m = 1$. It is $x = 866$. Now we check this:

$$866 = 123 \cdot 7 + 5$$
$$866 = 72 \cdot 12 + 2$$
$$866 = 66 \cdot 13 + 8$$

52) **Answer.** 32.

Solution. We have:

$$10! \equiv -1 (\quad \text{mod } 11). \tag{5.20}$$

Moreover, we obtain (see Problem 206):

$$2 \cdot 10! \equiv -1 \quad (\text{mod } 13), \ 2 \cdot 10! \equiv 12 \quad (\text{mod } 13),$$

and

$$10! \equiv 6 (\quad \text{mod } 13) \tag{5.21}$$

Substituting (5.20) into (5.21):

$$10! = (11n - 1) \equiv 6 \quad (\text{mod } 13).$$

Then, we find:

$$11n - 1 \equiv 6 \quad (\text{mod } 13) \Rightarrow 11n \equiv 7 \quad (\text{mod } 13)$$
$$\Rightarrow -2n \equiv 7 \quad (\text{mod } 13) \quad \Rightarrow 2n \equiv -7 \quad (\text{mod } 13)$$
$$\Rightarrow 2n \equiv 6 \quad (\text{mod } 13) \quad \Rightarrow n \equiv 3 \quad (\text{mod } 13) \Rightarrow n = 13k + 3.$$

Hence, we have $10! = 11(13k + 3) - 1 = 143k + 32$, or $10! \equiv 32 (\quad \text{mod } 143)$.
Note: This can be easily checked. Thus, $10! = 3628800 = 143 * 25376 + 32 = 3628768 + 32$.

53) **Answer.** He bought $x = 9$ apples at 9 cents each and $y = 3$ oranges at 6 cents each.

Solution. Let x be the number of apples and y the number of oranges, and p is the price of oranges. Then we have the following system:

$$\begin{cases} x + y = 12, & x > y \\ (p + 3)x + py = 99 \end{cases}$$

Solving the first equation for y, and considering that $x > y$, we obtain that $y = 12 - x$ and $x \in \{7; 8; 9; 10; 11; 12\}$. Substituting obtained expression into

$(p+3)x + py = 99$ we find $(p+3)x + p(12 - x) = 99$, or $3x + 12p = 99$. Hence, $x + 4p = 33$. Then, we have:

$$x \equiv 33 \pmod 4 \Rightarrow x \equiv 1 \pmod 4,$$

Moreover, from the inequality $7 \le x \le 12$ we obtain that $x = 9$. Then, $p = 6$ and $p + 3 = 9$. (Note that fruit price was very cheap that time.)

54) **Answer.** $\{(x, y, z) : (8, 15, 3), (16, 0, 10)\}$.

Solution. Let x be the number of sophomores and y the number of juniors, and z is the number of seniors. Then, we have the following system:

$$\begin{cases} 1.25x + 0.9y + 0.5z = 25 \\ x + y + z = 26 \end{cases}$$

Multiplying the first equation by 100 and the second by 125 we obtain:

$$\begin{cases} 125x + 90y + 50z = 2500 \\ 125x + 125y + 125z = 3250 \end{cases}$$

Subtracting equations we have $35y + 75z = 750$, or dividing it by 5 we find $7y + 15z = 150$. We can rewrite this equation as $7y = 15(10 - z)$.

Since 7 is prime then $y = 15k$, $k = 0$, or $k = 1$. If $k > 1$, then $15k > 30$ which contradicts the condition of the problem because the total number of the students is only 26. If $y = 0$ then $z = 10$ and $x = 16$, and if $y = 15$ then $z = 3$ and $x = 8$.

55) **Answer.** No solutions in integers.

Solution. We have:

$$x^2 - 3y^2 = 17 \Rightarrow x^2 - 17 = 3y^2 \Rightarrow x^2 \equiv 2 \pmod 3. \qquad (5.22)$$

Considering all possible $x = 3k$, $x = 3k + 1$ or $x = 3k + 2$, we understand that

If $x = 3k$,   then $x^2 \equiv (9k^2)(\mod 3) \equiv 0 \pmod 3$,
if $x = 3k + 1$, then $x^2 \equiv (9k^2 + 6k + 1)(\mod 3) \equiv 1 \pmod 3$,
if $x = 3k + 2$, then $x^2 \equiv (9k^2 + 12k + 4)(\mod 3) \equiv 1 \pmod 3$.

No square can leave a remainder of two when divided by 3.
Therefore equation (5.22) can never have a solution in integers, so as the original equation.

56) **Answer.** $N = 11969$; $N = 111969$; $N = 1791969$; $N = 19691969$.

Solution. Without loss of generality, assuming the a number written by deleting four last digits is x, the condition of the problems can be written as

$$10^4 \cdot x + 1969 = k \cdot x.$$

or in a factorized form:

$$1969 = x \cdot (k - 10000)$$

Since 1969 is a multiple of 11, $1969 = 11 \cdot 179$, where 179 is prime, then solution to the equation can be obtained by solving the four systems:

a.

$$\begin{cases} x = 1969 \\ k - 10000 = 1 \end{cases} \Leftrightarrow k = 10001, \ x = 1969, \ N = 10001 \cdot 1969 = 19,691,969.$$

b.

$$\begin{cases} x = 1 \\ k - 10000 = 1969 \end{cases} \Leftrightarrow k = 11969, \ x = 1, \ N = 11,969.$$

c.

$$\begin{cases} x = 11 \\ k - 10000 = 179 \end{cases} \Leftrightarrow k = 10179, \ x = 11, \ N = 10179 \cdot 11 = 111,969.$$

d.

$$\begin{cases} x = 179 \\ k - 10000 = 11 \end{cases} \Leftrightarrow k = 10011, \ x = 179, \ N = 10011 \cdot 179 = 1,791,969.$$

57) **Answer.** $r = 0$; $r = 3$; $r = 1$; $r = 6$; $r = 1$.

Solution. a. Divisibility by 3. Since $2019 = 3 \cdot 673 = 3n$, then $N = \underbrace{111 \ldots 1}_{2019 \text{ times}}$ is evenly divisible by 3.

b. Divisibility by 4. The last two digits make 11, $11 = 4 \cdot 2 + 3$, then the remainder of N divided by 4 is 3.

c. Divisibility by 5. The last digit is 1 so as the remainder of N divided by 5.

d. Divisibility by 7. Because the number 111111 is divisible by 7, we will look for the whole blocks of such six-digit numbers within N. Since $2019 = 7 \cdot 288 + 3$, then the remainder of N divided by 7 equals the remainder of last three ones number at the end, 111 divided by 7. Denote $M = 111111 = 7k$

$$N = \underbrace{111 \ldots 1}_{2019 \text{ times}} = \underbrace{111 \ldots 1}_{2016 \text{ times}} 111$$
$$= 10^{2018} + 10^{2017} + 10^{2016} + 10^{2015} + 10^{2014} + \ldots$$
$$+ 10^8 + 10^7 + 10^6 + 10^5 + 10^4 + 10^3 + 111$$
$$= 10^{2013} \cdot (10^5 + 10^4 + \ldots + 10 + 1)$$
$$+ 10^{2007} \cdot (10^5 + 10^4 + \ldots + 10 + 1)$$
$$+ \ldots + 10^3 \cdot (10^5 + 10^4 + \ldots + 10 + 1) + 111$$
$$= [10^{2013} \cdot M + 10^{2007} \cdot M + \ldots + 10^3 \cdot M + 7 \cdot 15] + 6$$
$$= 7k + 6.$$

Therefore the remainder of N divided by 7 is $r = 6$.

e. Divisibility by 11. Because 2019 is odd, then by Property 1 the remainder is $r = 1$.

58) **Proof.** Let

$$N = (n-4)^2 + (n-3)^2 + (n-2)^2 + (n-1)^2 + n^2$$
$$+ (n+1)^2 + (n+2)^2 + (n+3)^2 + (n+4)^2 + (n+5)^2$$

be the sum of ten consecutive squares.

After simplification and collecting like terms, N can be written as

$$N = 10n^2 + 10n + 85 = 5 \cdot (2n^2 + 2n + 17).$$

Obviously, if N is a perfect square then at least the following must be true:

$$2n^2 + 2n + 17 = 5m$$

Let us check if it is possible by substituting $n = 5k$, $n = 5k \pm 1$, $n = 5k \pm 2$ into the left side. We obtain:

$$
\begin{aligned}
n = 5k &\Rightarrow 2n^2 + 2n + 17 = \quad\;\; 50k^2 + 10k + 17 \neq 5l \\
n = 5k - 1 &\Rightarrow 2n^2 + 2n + 17 = 50k^2 - 10k + 17 \neq 5l \\
n = 5k + 1 &\Rightarrow 2n^2 + 2n + 17 = 50k^2 + 30k + 21 \neq 5l \\
n = 5k - 2 &\Rightarrow 2n^2 + 2n + 17 = 50k^2 - 30k + 21 \neq 5l \\
n = 5k + 2 &\Rightarrow 2n^2 + 2n + 17 = 50k^2 + 50k + 29 \neq 5l.
\end{aligned}
$$

We can see that $2n^2 + 2n + 17$ is never a multiple of 5. Therefore the sum of the squares of ten consecutive natural numbers is never divisible by 5 and hence it cannot be a perfect square.

59) **Answer.** $65 = 8^2 + 1^2 = 4^2 + 7^2$.

Solution. Using prime factorization, 65 can be written as

$$
\begin{aligned}
65 &= 5 \cdot 13 = (2+i)(2-i)(3+2i)(3-2i) \\
65 &= [(2+i) \cdot (3-2i)] \cdot [(2-i) \cdot (3+2i)]] = [(2+i) \cdot (3+2i)] \cdot [(2-i) \cdot (3-2i)]] \\
65 &= (8-i) \cdot (8+i) = (4+7i) \cdot (4-7i) \\
65 &= 8^2 + 1^2 = 4^2 + 7^2.
\end{aligned}
$$

60) **Proof.** Let $x^2 + y^2 = n$. It is easy to see that

$$(4x+y)^2 + (x-4y)^2 = 17(x^2 + y^2) = 17n.$$

61) **Proof.** All numbers between 1 and 10^n, except the last one, can be written as n digit number, $\overline{a_n a_{n-1} \ldots a_1 a_0}$ and paired with another number of the form

$$\overline{(9-a_n)(9-a_{n-1})...(9-a_1)(9-a_0)}$$

There are precisely $5 \cdot 10^{n-1}$ such pairs, and the sum of the digits in each pair is the same and equals $9 \cdot n \cdot 5 \cdot 10^{n-1}$. The last number in the sequence, 10^n, is written by n zeros and one 1, and we need to add this 1 to the previous sum of the paired numbers. Finally, the sum is $9 \cdot n \cdot 5 \cdot 10^{n-1} + 1 = 9 \cdot k + 1$. The statement is proven. Additionally such a sum is not divisible by 3, 5, and 2. Moreover, it will not be divisible by any power of 2 from 1 to $(n-1)$, and any power of 5 from 1 to n. For many n, the sum is prime.

62) **Proof.** Since $2020 = 4 \cdot 505$, and $1111 = 11 \cdot 101$, then the given number consists of 505 blocks, each divisible by 11. Then the entire number is divisible by 11. Therefore it is not prime.

63) **Proof.** Any odd number n can be written in one of the following forms:

$$8k \pm 1, \ 8k \pm 3, \ 8k \pm 5, \ 8k \pm 7,$$

then its square will satisfy the congruence:

$$n^2 \equiv 1 \pmod 8.$$

Therefore $n^2 = 8m + 1$.

64) **Proof.** As we established the square of any odd number divided by 8 leaves a remainder of 1. Any even number can be written as

$$8k, \ 8k \pm 2, \ 8k \pm 4, \ 8k \pm 6,$$

So its square would be either divisible by 8 or leave a remainder of 4, i.e., as $n = 18 = 8 \cdot 2 + 2$, $n^2 = 324 = 8 \cdot 40 + 4$. If we add three squares, the sum of any three squares in any combination (odd,odd, odd), (odd,odd,even), (odd,even,even), etc. divided by 8 would leave any remainder from 0 to 6 but never 7. A number that divided by 8 leaves a remainder of 7 also can be written as $8m - 1$. Therefore this case is not possible.

65) **Answer:** 1.

Solution. We will find the greatest common divisor of two expressions by the procedure of consecutive subtraction:

$$(13a + 2b, 20a + 3b) = (7a + b, 13a + 2b) = (6a + b, 7a + b)$$
$$= (a, 6a + b) = (a, 5a + b) = (a, 4a + b) = ... = (a, b) = 1.$$

66) **Answer:** $(x, y) \in \{(1,6), (2,11), (5,2), (7,6), (11,11)\}$

Solution. We can transform the given equation and factor it as follows:

$$x^2 - 3x + 3y - xy + x - 3 = 11 - 3$$
$$x(x-3) - y(x-3) + (x-3) = 8$$
$$(x-3) \cdot (x - y + 1) = 8$$
$$x - y + 1 = \frac{8}{x-3}$$

Because we need to find only natural solutions, (x,y), we can check all divisors of 8 by solving

$$x - 3 = -2, -1, 1, 2, 4, 8$$
$$y = (x-3) - \frac{8}{x-3} + 4.$$

$(x,y) \in \{(1,6), (2,11), (5,2), (7,6), (11,11)\}$.

67) **Answer.** $n = 11k - 3$.

Solution. We can rewrite this number as

$$n^2 + 17n - 2 = n^2 + 6n + 9 + 11n - 11$$
$$= (n+3)^2 + 11 \cdot (n-1),$$

from we can see that $n + 3 = 11k$ or $n = 11k - 3$, $k \in \mathbb{N}$.

68) **Proof.** Suppose a contradiction and that a solution exists. Because the right-hand side is a multiple of 4, then either both variables are even or both odd. Assume that both variables are even, and that $x = 2n$ and $y = 2m$, then the left side is divisible by 8, but the right-hand side will be written as

$$4 \cdot (8n + 8m + 1) = 8t + 4$$

will leave a remainder of 4 when divided by 8. Hence both variables cannot be even. Let us factor the equation as

$$(x+y) \cdot (x^2 - xy + y^2) - 4xy \cdot (x+y) = 4$$

or in a simplified form:

$$(x+y) \cdot ((x+y)^2 - 7xy) = 4.$$

Under assumption that both variables can be only odd, and using the fact that the right-hand side equals 4, then we need to consider only $x + y = 2$ or $x + y = 4$. Both are impossible. Therefore this equation has no solutions.

69) **Proof.** The number can be written using complex numbers as

$$N = (1 + a^2)(1 + b^2)(1 + c^2) = [(a+i)(b-i)] \cdot [(b+i)(a-i)] \cdot (c+i)(c-i)$$

After manipulation we will have the result as $A^2 + B^2$.

70) **Answer.** $(x_1^2 + 1)(x_2^2 + 1) = (c - 1)^2 + b^2$.

Solution. Consider a quadratic polynomial $p(x) = x^2 + bx + c$ and its roots x_1, x_2. On one hand,

$$p(i) \cdot p(-i) = (i^2 + b \cdot i + c)(((-i)^2 - b \cdot i + c)$$
$$= (c - 1 + ib)(c - 1 + ib) = (c - 1)^2 + b^2.$$

On the other hand, quadratic polynomial can be written in a factorized form as

$$p(x) = (x - x_1)(x - x_2)$$

Evaluate again $p(i) \cdot p(-1)$:

$$p(i) \cdot p(-i) = (i - x_1)(i - x_2)(-i - x_1)(-i - x_2)$$
$$= (x_1 - i)(x_1 + 1)(x_2 - i)(x_2 + i)$$
$$= (x_1^2 + 1)(x_2^2 + 1).$$

Equating both results, we finally obtain the answer:

$$(x_1^2 + 1)(x_2^2 + 1) = (c - 1)^2 + b^2.$$

71) **Answer:** 297.

Solution. The largest power of 3 that divided 600! Is

$$n = \left[\frac{600}{3}\right] + \left[\frac{600}{3^2}\right] + \left[\frac{600}{3^3}\right] + \left[\frac{600}{3^4}\right] + \left[\frac{600}{3^5}\right]$$
$$= 200 + 66 + 22 + 7 + 2 = 297.$$

72) **Solution.** Any number can be written as the sum of five squares. By Lagrange Theorem 28, any number can be written as sum of four squares and zero square can be always added.

73) **Answer.** $a = 3$, $a = -7$, $a = -1$, $a = -3$.

Solution. Let us rewrite it as $\frac{a^2 + 1}{a + 2} = a - 2 + \frac{5}{a + 2}$, then $a + 2 = 5n = \pm 1; \pm 5$, hence,

$$a = 3, \ a = -7, \ a = -1, \ a = -3.$$

74) **Answer:** $n = 502$.

Solution. Because $10 = 2 \cdot 5$, then we need to find the power of 5 in prime factorization of 2019!

$$n = \left[\frac{2019}{5}\right] + \left[\frac{2019}{5^2}\right] + \left[\frac{2019}{5^3}\right] + \left[\frac{2019}{5^4}\right]$$
$$= 403 + 80 + 16 + 3 = 502.$$

$$2019! = 10^{502} \cdot k.$$

75) **Answer:** 2.

Solution. Because the number is the sum of two odd numbers, it is an even number, $N = 2k$. Its smallest prime factor is 2.

76) **Answer:** $x = 3, y = 1, z = 3, w = 2.$

Solution. Rewriting a fraction on the right as a finite continued fraction (unique representation) we have the following:

$$\frac{34}{9} = 3 + \frac{7}{9} = 3 + \frac{1}{\frac{9}{7}} = 3 + \frac{1}{1+\frac{2}{7}} = 3 + \frac{1}{1+\frac{1}{\frac{7}{2}}} = 3 + \frac{1}{1+\frac{1}{3+\frac{1}{2}}}$$

$$x = 3, y = 1, z = 3, w = 2.$$

77) **Answer.** $(x, y) = \{(7n + 4, 21n^2 + 30n + 11), (7k + 1, 21k^2 + 12k + 2)\}$, n, $k = 0, 1, 2, 3, \ldots$

Solution. Method 1. Let us consider this equation as quadratic in x, where y is a parameter. Thus,

$$3x^2 + 6x + (5 - 7y) = 0. \tag{5.23}$$

Since the coefficient of x is even, we can use the formula for discriminant divided by 4:

$$\frac{D}{4} = \left(\frac{b}{2}\right)^2 - ac,$$
$$x = \left(-\frac{b}{2} \pm \sqrt{\frac{D}{4}}\right)/a.$$

For our problem we find:

$$\frac{D}{4} = \left(\frac{b}{2}\right)^2 - ac = 3^2 - 3(5 - 7y) = 21y - 6.$$

In order (5.23) to have integer solutions $\frac{D}{4}$ must be a perfect square! We obtain that $21y - 6 = k^2$. Subtracting 1 from both sides we find

$$21y - 7 = k^2 - 1$$
$$7(3y - 1) = (k - 1)(k + 1).$$

Let $m = k - 1$, then for the equation below we obtain

$$7(3y - 1) = m(m + 2).$$

We will consider two cases:

a. $y = 2n$ (even number), then $7(6n - 1) = m(m + 2)$. Since 7 is prime number, we have two subcases:
   - $m = 7$ and $m + 2 = 6n - 1$, which has no solutions,
   - $m + 2 = 7$, and $m = 6n - 1$, then $m = 5$, $n = 1$, and $y = 2$.
b. $y = 2n + 1$ (odd number), then $7(3(2n + 1) - 1) = m(m + 2)$, or $14(3n + 1) = m(m + 2)$. The last equation has solution in integers only if $3n + 1 = m + 2$ and $14 = m$, which give us $n = 5$, and $y = 11$.

Next, we find corresponding x using a quadratic formula:

$$x_1 = \frac{-3 \pm \sqrt{21 \cdot 2 - 6}}{3} = \frac{-3 \pm 6}{3} = 1, -2$$
$$x_2 = \frac{-3 \pm \sqrt{21 \cdot 11 - 6}}{3} = \frac{-3 \pm 15}{3} = 4, -6.$$

Finally, we will select only positive integers.
$\{(x,y) : (1,2), (4,11)\}$.

Method 2. Using congruence notation we can rewrite it as

$$3x^2 + 6x + 5 \equiv 0 \ (mod\,7)$$

Multiplying which by 3 we obtain a new congruence:

$$3 \cdot 3x^2 + 3 \cdot 6x + 15 \equiv 0 \ (mod\,7)$$
$$(3x+3)^2 + 6 \equiv 0 \ (mod\,7)$$

The last congruence can be rewritten and solved as follows:

$$(3x+3)^2 \equiv 1 \ (mod\,7)$$

$3x+3 \equiv 1 (mod\,7)$	or	$3x+3 \equiv -1 (mod\,7)$
$3x \equiv 5 (mod\,7)$		$3x \equiv -4 (mod\,7)$
$3x \equiv 12 (mod\,7)$		$3x \equiv 3 (mod\,7)$
$x \equiv 4 (mod\,7)$	or	$x \equiv 1 (mod\,7)$

Finally, we will substitute each x into the given equation and find the corresponding y:
1) $x = 4 + 7n$, then

$$3 \cdot (7n+4)^2 + 6 \cdot (7n+4) + 5 = 7y$$
$$y = 21n^2 + 30n + 11.$$

2) $x = 1 + 7k$, then

$$3 \cdot (7k+1)^2 + 6 \cdot (7k+1) + 5 = 7y$$
$$y = 21k^2 + 12k + 2.$$

78) **Solution.** Let us use Euclidean algorithm:

$$105 = 38 \cdot \underline{2} + 29$$
$$38 = 29 \cdot \underline{1} + 9$$
$$29 = 9 \cdot \underline{3} + 2$$
$$9 = 2 \cdot \underline{4} + 1$$
$$2 = 1 \cdot \underline{2} + 0$$

Recording all underlined quotients (on the right), we can represent 105/38 as a finite continued fraction (unique representation):

$$\frac{105}{38} = 2 + \cfrac{1}{1 + \cfrac{1}{3 + \cfrac{1}{4 + \frac{1}{2}}}}$$

79) **Answer:** $(1,0) \rightarrow (2,1) \rightarrow (7,4) \rightarrow (26,15) \rightarrow \dots$

    Solution. $x^2 - 3y^2 = 1$ Nontrivial solution is $(2,1)$. Then we will get matrix $B = \begin{pmatrix} 2 & 3 \\ 1 & 2 \end{pmatrix}$ and obtain the iterative formula $(x,y) \rightarrow (2x + 3y, x + 2y)$

80) **Proof.** Let us add and subtract $16n^2$ and then factor the obtained expression:

$$n^4 + 64 = n^4 + 16n^2 + 64 - 16n^2 = \left(n^2 + 8\right)^2 - (4n)^2$$
$$= (n^2 - 4n + 8) \cdot (n^2 + 4n + 8).$$

Therefore the given number cannot be prime.

81) **Proof.**

$$2017 \equiv 1 \,(mod\,7) \quad \Rightarrow \quad 2017^{1917} \equiv 1 \,(mod\,7)$$
$$1917 \equiv -1 \,(mod\,7) \quad \Rightarrow \quad 1917^{2017} \equiv (-1)^{2017} \equiv -1 \,(mod\,7)$$
$$2017^{1917} + 1917^{2017} \equiv 0 \,(mod\,7)$$

Yes, the number is divisible by 7.

82) **Answer:**

$$(23, 11), \ (-23, 11), \ (23, -11), \ (-23, -11),$$
$$(9, 3), \ (-9, 3), \ (9, -3), \ (-9, -3),$$
$$(7, 1), \ (-7, 1), \ (7, -1), \ (-7, -1).$$

Solution. This can be factored as

$$(x - 2y)(x + 2y) = 45$$

Factoring 45 we have to solve the following systems:

$$\begin{cases} x - 2y = 1 \\ x + 2y = 45 \end{cases} \Rightarrow x = 23, \ y = 11$$
$$\begin{cases} x - 2y = 3 \\ x + 2y = 15 \end{cases} \Rightarrow x = 9, \ y = 3$$
$$\begin{cases} x - 2y = 5 \\ x + 2y = 9 \end{cases} \Rightarrow x = 7, \ y = 1$$

Because of the symmetry, we know that if (a,b) is a solution, then $(-a,b)$, $(a,-b)$, $(-a,-b)$ are also solutions to the equation.

83) **Answer:** No solutions.

Solution. Consider the equation $x^2 - y^2 = 6$. If the difference of squares divisible by 2, then it must be divisible by 4, however, 6 is not a multiple of 4. We have no solutions.

84) **Answer:** $(x,y) = \{(4,1), (-4,1), (4,-1), (-4,-1)\}$.

Solution. Factor $(x-3y)(x+3y) = 7$. Using symmetry we obtain
$(x,y) = \{(4,1), (-4,1), (4,-1), (-4,-1)\}$.

85) **Answer:** $(1,0) \rightarrow (8,1) \rightarrow (127,16) \rightarrow (2024,255) \rightarrow \ldots$

Solution. We can see that $x = 8, y = 1$ satisfies the given equation $x^2 - 63y^2 = 1$. All other solutions can be generated by matrix : $B = \begin{pmatrix} 8 & 63 \\ 1 & 8 \end{pmatrix}$ which also can be found by iteration as follows $(x,y) \rightarrow (8x+63y, x+8y)$.

86) **Proof.**

Let the hypothenuse be y, then one leg will be $(y-1)$ and let the other leg be x. By Pythagorean theorem the following is true:

$$x^2 + (y-1)^2 = y^2.$$

After simplification this can be written as

$$x^2 = 2y - 1.$$

We need to find such x and y that the square of x is an odd number. There are infinitely many such cases. We can simply start from $x = 3, 5, 7, \ldots 2k+1, k \in N$. Substituting

$$x = 2k + 1$$

into equation, we have

$$(2k+1)^2 = 2y - 1$$
$$4k^2 + 4k + 1 = 2y - 1 \quad y = 2k^2 + 2k + 1, \ k \in N$$

Since such a triangle is defined by the triple $(x, y-1, y)$, then the answer is

$$(2k+1; 2k^2+2k, 2k^2+2k+1), \ k \in N.$$

Substituting different values for k, we obtain the first three such triangles

$$(3,4,5), \ (5,12,13); \ (7,24,25)$$

There are infinitely many of such triangles.

87) **Solution.** Because $(3,1)$ is a solution then all other solutions can be obtained from the trivial $(1,0)$ by formula:

$$(x,y) \longrightarrow (3x+8y, x+3y)$$

which means that we have the recurrent formulas:

$$x_{n+1} == 3x_n + 8y_n$$
$$y_{n+1} = x_n + 3y_n.$$

In order to find exact formulas, consider irrational number

$$\alpha = 3 + 1 \cdot \sqrt{8}$$

and its conjugate

$$\overline{\alpha} = 3 - 1 \cdot \sqrt{8}$$

Consider their nth powers:

$$(3+\sqrt{8})^n = x_n + y_n\sqrt{8}$$
$$(3-\sqrt{8})^n = x_n - y_n\sqrt{8}$$

Adding and subtracting the left and right sides, we finally get the exact formulas:

$$x_n = \frac{(3+\sqrt{8})^n + (3-\sqrt{8})^n}{2}$$
$$y_n = \frac{(3+\sqrt{8})^n - (3-\sqrt{8})^n}{\sqrt{8}}.$$

88) **Answer.** $x = 3, y = 3; x = -5, y = -5; x = -7, y = 5.$

Solution. The equation can be written as

$$x(10 - 3y) = 16y - 45$$

Solve it for x and extract the integer part in the numerator of the fraction:

$$x = \frac{16y-45}{10-3y} = -5 + \frac{y+5}{10-3y}$$
$$3x = -15 - \frac{3y+15}{3y-10} = -15 - \frac{(3y-10)+25}{3y-10}$$
$$3x = -16 + \frac{25}{10-3y}$$

In order for x to be an integer number, the denominator of the fraction must be a divisor of 25 ($\pm 1, \pm 5, \pm 25$) and additionally, the entire expression on the right-hand side must be a multiple of 3. Additionally, for any value of $10 - 3y = a$, we need to check that $y = \frac{10-a}{3}$ is an integer. Hence, the following values of the denominator are possible:

$$10 - 3y = -5, \ y = 5, \ x = -7$$
$$10 - 3y = 25, \ y = -5, \ x = -5$$
$$10 - 3y = 1, \ y = 3, \ x = 3.$$

89) **Answer.** $n = 2, k = \pm 7$.

Solution. This problem is similar to one from the book. Consider the last digits of the left and right sides: $6^n + 11$ will always end in 7. Since for $n \geq 5$, $n!$ ends in zero, then if $n \geq 5$, $n! + 6^n + 11$ will end in 7. We all know that no perfect square ends in 7, then there will be no solutions for this case.
We need to check solutions for all $n < 5$.
The answer is $n = 2, k = \pm 7$.

90) **Answer.** $n = 2, k = \pm 5$.

Solution. Move 13 to the left side and compare again the last digits of the left and right sides of the equation:

$$n! + 5n + 13 = k^2$$

If $n \geq 5$, $n!$ ends in zero, and since $5n$ ends in 5, then $n! + 5n + 13$ will end in 8. No solutions. If $n < 5$, then there is one solution at $n = 2$ we obtain $k = \pm 5$

91) **Answer.** 24 and 49.

Solution.

$$2^{2010} = (5-1)^{1005} = 5^{1005} - 1005 \cdot 5^{1004}$$
$$+ \frac{1005 \cdot 1004}{2} \cdot 5^{1003} + ... + 1005 \cdot 5 - 1 = 25k - 1$$

This number divided by 5 or 25 leaves a remainder -1, then the last digit is 4 and the last two are **24**.

$$3^{2010} = (10-1)^{1005} = 10^{1005} - 1005 \cdot 10^{1004}$$
$$+ \frac{1005 \cdot 1004}{2} \cdot 10^{1003} + ... + 1005 \cdot 10 - 1 = 25k - 1 = 50m - 1$$

the last digit is 9, the last two digits are **49**.

92) **Answer.** It can be done in three ways.

Solution. First, we will factor 48 into two even numbers, find s and p, then find n and m:

$$2 \cdot 24$$
$$4 \cdot 12$$
$$6 \cdot 8$$

It can be done in three ways. (See Table 5.3)

93) **Answer.** $900 = 226^2 - 224^2 = 78^2 - 72^2 = 50^2 - 40^2 = 34^2 - 16^2$

Solution. Factoring 900 into two even numbers $s > p$, we will fill out the Table 5.4:
$900 = 226^2 - 224^2 = 78^2 - 72^2 = 50^2 - 40^2 = 34^2 - 16^2$

2s	2p	s	p	n=s+p	m=s−p
24	2	12	1	13	11
12	4	6	2	8	4
8	6	4	3	7	1

Table 5.3 Difference of squares for 48

2s	2p	s	p	n=s+p	m=s−p
450	2	225	1	226	224
150	6	75	3	78	72
90	10	45	5	50	40
50	18	25	9	34	16

Table 5.4 Difference of squares for 900

94) **Answer.** There are exactly eight possible representations of 945 as a difference of two squares.

Solution. Let us factor 945 and first represent it as a product of two numbers t and s, $s > t$. We will make the Table 5.5.

t	s	$n = \frac{s+t}{2}$	$m = \frac{s-t}{2}$
1	945	473	472
3	315	159	156
5	189	97	92
7	135	71	64
9	105	57	48
15	63	39	24
21	45	33	12
27	35	31	4

Table 5.5 Difference of squares for 945

For any odd number of similar table, there will be always one pair of consecutive integers $(473, 472)$. The number of the rows in the table will depend on possible factoring pairs and the table will stop when one of the numbers, $s > \sqrt{2k+1}$. In this table the last (the 8th) row has a pair $(27, 35)$ and we can see that $35 > \sqrt{945} \approx 30.74$

$$945 = 473^2 - 472^2 = 159^2 - 156^2 = 97^2 - 92^2 = 71^2 - 64^2$$

95) **Answer.** $13561 = 71 \cdot 191$.

Solution. Multiplying the left and right sides of two equations,

$$233^2 \equiv 3^2 \cdot 5 \pmod{13561}$$
$$1281^2 \equiv 2^4 \cdot 5 \pmod{13561},$$

we obtain

$$(233 \cdot 1281)^2 \equiv 60^2 \pmod{13561}$$

that can be written as the difference of two squares:

$$298473^2 - 60^2 = 298413 \cdot 298533 \equiv \quad \pmod{13561}$$

One of the factors on the left-hand side contains factor of 13561. We will find it using Euclidean algorithm:

$$298413 = 22 \cdot 13561 + \underline{71}$$

Thus 71 is a factor of 13561. Dividing it by 71 we obtain

$$13561 = 71 \cdot 191.$$

96) **Answer**. The box has 28 white, 28 blue, and 84 red balls.

Solution. Let r, w, and b be the number of red, white, and blue balls, respectively. By the condition of the problem the following is valid:

$$\begin{cases} b \geq w \\ b \leq \frac{r}{3} \end{cases}$$

This also can be written as a double inequality:

$$w \leq b \leq \frac{r}{3}.$$

Adding b to all sides of the inequality, we obtain

$$w + b \leq 2b \leq \frac{r}{3} + b.$$

Using the information about the total number of blue and white balls and the second inequality of the system, we have the following:

$$55 \leq b + w \leq \frac{r}{3} + w \leq \frac{r}{3} + \frac{r}{3} = \frac{2r}{3}$$

This can be written as

$$55 \leq b + w \leq \frac{2r}{3}$$
$$2r \geq 165$$
$$r \geq 82.5$$

Because the number of the balls can be only a natural number, then $r=83$ will be the first possible choice to check. If we substitute it into the first double inequality, we will obtain $w \leq b \leq 27$ but this contradicts the condition $b + w \geq 55$. Let us check the next natural number for r, $r = 84$, we obtain that

$$w \leq b \leq \frac{84}{3} = 28$$

And that $w = b = 28$ would give us the minimal number of balls in the box.

97) **Answer**. It is impossible.

Solution. If one piece is broken into 8 pieces then the number of pieces is increased by 7 pieces. At the beginning she had one piece, one bond, and after the first cut she had $1 + 7 = 8$ pieces. If she continued, she would get $(1+7k)$ pieces. Unfortunately, 2016 is divisible by 7 and cannot give us a remainder of 1. Therefore, the bond investor cannot tear the bond into 2016 pieces, but could cut it into 2017 pieces. $(2017 = 7 \cdot 288 + 1)$

98) **Answer**. There are twice as many doctors as patients.

Solution. Assume that d is the number of doctors in the hospital and p is the number of patients. Also assume that $p = k \cdot d$. Then the total age of the doctors is $35d$ and the patients is $50kd$. Then the quantity $35d + 50kd$ will represent the doctors and patients together and the following is true:

$$35d + 50kd = 40(d + kd)$$
$$5d(7 + 10k) = 5d(8 + 8k)$$
$$2k = 1$$
$$k = \tfrac{1}{2}$$

There are twice as many doctors as patients.

99) **Solution.** If we draw several convex polygons, we will notice that this problem is similar to one where we had to place parentheses between numbers a, b, c, d. The difference is that since each polygon is convex then all vertices are connected with their neighbors. This is a problem on application of Catalan numbers. Denote the nth number by $c(n)$. We have the following formulas:

$$c(4) = k_3 = 1 \cdot 1 + 1 \cdot 1 = 2$$
$$c(5) = k_4 = 1 \cdot 2 + 1 \cdot 1 + 2 \cdot 1 = 5$$
$$c(6) = k_5 = 1 \cdot 5 + 1 \cdot 2 + 2 \cdot 1 + 5 \cdot 1 = 14$$
$$c(7) = k_6 = 1 \cdot 14 + 1 \cdot 5 + 2 \cdot 2 + 5 \cdot 1 + 14 \cdot 1 = 42.$$

In general, the nth -gon can be cut into triangles in $c(n)$ ways:

$$c(n) = k_{n-1} = \frac{2^{n-2} \cdot (3 \cdot 5 \cdot 7 \cdot (2n - 5))}{(n - 1)!}$$

100) **Answer**. 785 trees.

Solution. Let x be the number of trees. Using congruence and Chinese Remainder Theorem (Theorem 20), we also can solve this problem. We have the system of congruences:

$$\begin{cases} x \equiv 8 \,(mod\,37) \\ x \equiv 11 \,(mod\,43) \end{cases} \Rightarrow a_1 = 8, m_1 = 37, \ a_2 = 11, m_2 = 43.$$

Next, we need to evaluate corresponding M_1, M_2 and solving the following congruences to find corresponding N_1, N_2 :

$$M = 37 \cdot 43 = 1591$$
$$M_1 = \frac{M}{m_1} = 43 \qquad 43 \cdot N_1 \equiv 1 \,(mod\,37)$$
$$M_2 = \frac{M}{m_2} = 37 \qquad 37 \cdot N_2 \equiv 1 \,(mod\,43)$$

As soon as we know N_1, N_2, the answer will be found from the formula:

$$x \equiv (a_1 N_1 M_1 + a_2 N_2 M_2) \,(mod\,M).$$

Solving each congruence, we obtain

$$43 \cdot N_1 \equiv 1 \,(mod\,37) \qquad 37 \cdot N_2 \equiv 1 \,(mod\,43)$$
$$37 \cdot 36 + 1 = 43 \cdot 31 \qquad 43 \cdot 6 + 1 = 37 \cdot 7$$
$$43 \cdot N_1 \equiv 43 \cdot 31 \,(mod\,37) \qquad 37 \cdot N_2 \equiv 37 \cdot 7 \,(mod\,43)$$
$$N_1 \equiv 31 \,(mod\,37) \qquad N_2 \equiv 7 \,(mod\,43)$$

Now, we find the answer as

$$x \equiv (8 \cdot 31 \cdot 43 + 11 \cdot 7 \cdot 37) \,(mod\,1591),$$
$$x \equiv 13513 \,(mod\,1591)$$
$$(13513 = 8 \cdot 1591 + 785)$$
$$x \equiv 785 \,(mod\,1591)$$
$$x = 785.$$

01) Proof. Denote fractions close to the right-hand side of (5.1) by $\frac{P_n}{Q_n}$, and those approaching the left-hand side of (2.14) by $\frac{R_n}{S_n}$ $(n = 0, 1, 2, 3, ...)$. Let us prove that

$$\frac{R_n}{S_n} = \frac{P_{3n+1} + Q_{3n+1}}{P_{3n+1} - Q_{3n+1}}.$$

Considering the value of the elements of the continued fraction (5.1), we have:

$$P_{3n+1} = P_{3n} + P_{3n-1}, P_{3n} = P_{3n-1} + P_{3n-2},$$
$$P_{3n-1} = 2nP_{3n-2} + P_{3n-3}, P_{3n-2} = P_{3n-3} + P_{3n-4},$$
$$P_{3n-3} = P_{3n-4} + P_{3n-5},$$

from which we find:

$$P_{3n+1} = 2P_{3n-1} + P_{3n-2} = (4n+1)P_{3n-2} + 2P_{3n-3}$$
$$= (4n+2)P_{3n-2} + P_{3n-3} - P_{3n-4} = (4n+2)P_{3n-2} + P_{3n-5}.$$

We also have a similar relationship for Q_{3n+1}, so that

$$\begin{cases} P_{3n+1} = (4n+2)P_{3n-2} + P_{3n-5}, \\ Q_{3n+1} = (4n+2)Q_{3n-2} + Q_{3n-5} \end{cases} \tag{5.24}$$

Let us prove using Mathematical Induction in n, that

$$R_n = \frac{1}{2}(P_{3n+1} + Q_{3n+1}). \tag{5.25}$$

It follows from (2.14) and (5.1) by evaluation that $R_0 = 2, R_1 = 13, P_1 = 3, P_4 = 19, Q_1 = 1, Q_4 = 7$, so that the relationship (5.25) is correct at $n = 0, n = 1$. Suppose that the relationship (5.25) is true for all R terms with numbers less than n, where $n \geq 2$, so for example,

$$R_{n-1} = \frac{1}{2}(P_{3n-2} + Q_{3n-2}), R_{n-2} = \frac{1}{2}(P_{3n-5} + Q_{3n-5}),$$

Then using the equalities (5.24), obtain:

$$R_n = (4n+2)R_{n-1} + R_{n-2} = \frac{1}{2}((4n+2)(P_{3n-2} + Q_{3n-2}) + P_{3n-5} + Q_{3n-5}) = \frac{1}{2}(P_{3n+1} + Q_{3n+1}\cdot).$$

Regarding mathematical Induction Principle, the equality (5.25) is true for all n. By analogy, we can prove that

$$S_n = \frac{1}{2}(P_{3n+1} - Q_{3n+1}).$$

Considering now the limit of the quotient of R_n and S_n, we find:

$$\frac{\lim \frac{P_{3n+1}}{Q_{3n+1}} + 1}{\lim \frac{P_{3n+1}}{Q_{3n+1}} - 1} = \lim \frac{P_{3n+1} + Q_{3n+1}}{P_{3n+1} - Q_{3n+1}} = \lim \frac{R_n}{S_n} = \frac{e+1}{e-1}, \text{ that is } \lim \frac{P_{3n+1}}{Q_{3n+1}} = e.$$

Because the continued fraction on the right-hand side of (5.1) converges, we will also have that $\lim \frac{P_n}{Q_n} = e$, which proves the statement.

102) **Answer.** 120 revolutions.

   Solution. This will happen after 120 revolutions, which is the Least Common Multiple of two given numbers, $LCM = 120$.

103) **Answer.** The number is 9216.

   Solution. Let x be the unknown number and $\tau(x)$ be the total number of its factors, then by the condition of the problem

$$x = 24n, \quad \tau(x) = 33 = 3 \cdot 11.$$

From the second formula we can see that the prime factorization of x must contain only two factors:

$$x = p^\alpha \cdot q^\beta.$$

Additionally, we can find that

$$\tau(x) = 3 \cdot 11 = (\alpha + 1)(\beta + 1), \quad \alpha = 2, \ \beta = 10.$$

Working with x we obtain that

$$x = 2^3 \cdot 3 \cdot n = p^2 \cdot q^{10},$$

then the only values of p, q, and n are

$$p = 3, \ q = 2 \ n = 3 \cdot 2^7$$

Finally $x = 3^2 \cdot 2^{10} = 9216$.

104) **Answer.** The given number ends in maximum of two zeros.

Solution. Consider the first three natural powers:

- $n = 1$, then $x = 1 + 2 + 3 = 4 = 10$ and it ends in one zero.
- $n = 2$, then $x = 1 + 4 + 9 + 16 = 30$ again ends in one zero.
- $n = 3$, then $x = 1^3 + 2^3 + 3^3 + 4^3 = (1 + 2 + 3 + 4)^2 = 100$ ends in two zeros.

Can x end in more than two zeros? For example, can it end in three zeros? We know that if a number ends in three or more zeros, then it must be divisible by 8. Let us look at the remainders of different powers of 3 when divided by 8:

$$3 \equiv 1 \pmod 8$$
$$3^2 = 9 \equiv 1 \pmod 8$$
$$3^3 = 27 \equiv 3 \pmod 8$$
$$3^4 = 81 \equiv 1 \pmod 8$$
$$3^5 \equiv 3 \pmod 8$$
$$3^6 \equiv 1 \pmod 8$$

$$- - - - - - - - - - - - - - - - - - - -$$

$$3^{2k-1} \equiv 3 \pmod 8$$
$$3^{2k} \equiv 1 \pmod 8.$$

If $n > 2$ and even, then both 2^n and 4^n are divisible by 8. 3^{2k} divided by 8 leaves a remainder of 1, if n is even, and a remainder of 3 if n is odd. 1 in any power is 1. Then in the case of an even power n, $x = 8m + 2$ is not divisible by 8. If the power $n = 2k - 1$ is odd, then $x = 8m + 3 + 1 = 8m + 4$
Therefore for any power $n > 2$, the given number x divided by 8 leaves either a remainder of 2 or 4 so it cannot end in three zeros.

105) **Solution.** The given number can be written as

$$3 \cdot 3^{2017} + 2 = 3n + 2, \ n = 3^{2017}.$$

If we raise it to the second power, we will obtain

$$a^2 + 2 = (3n + 2)^2 + 2 = 9n^2 + 12n + 4 + 2 = 3 \cdot (3n^2 + 4n + 2),$$

which is a multiple of 3 and cannot be prime.

106) **Answer.** $n = 5$.

   Solution. The given equation can be written as

$$n! = (n-1) \cdot n \cdot (n+1),$$

and because $n = 1$ is not a solution, we can simplify it as

$$n+1 = (n-2)!$$

or

$$n+1 = (n-2)(n-3)(n-4) \cdot \ldots \cdot 3 \cdot 2 \cdot 1.$$

This equation has a solution at $n = 5$ because $6 = 3!$ There are no other solutions for $n > 5$, because the following inequality is always true:

$$n+1 < 2 \cdot (n-2) < (n-2) \cdot (n-3) \cdot \ldots \cdot 2 \cdot 1.$$

107) **Answer.** $(n,m) = (4l+1, l \cdot (2l+1))$, $l \in \mathbb{N}$.

   Solution. Using the difference of squares applied to the left side, the equation can be rewritten as

$$(n-1) \cdot (n+1) = 8m.$$

Because the factors on the left-hand side differ by 2, and the right side is a multiple of 8, then each factor on the left side must be even. Denote

$$\begin{aligned} n-1 &= 2k, \quad \Rightarrow \quad \boxed{n = 2k+1} \\ n+1 &= 2k+2 = 2 \cdot (k+1) \end{aligned}$$

We have new equation:

$$4k \cdot (k+1) = 8m$$
$$\boxed{k = 2l}, \; k+1 = 2l+1$$
$$4 \cdot 2l \cdot (2l+1) = 8m$$
$$\boxed{m = l \cdot (2l+1)}$$

Finally, we have a solution

$$(n,m) = (4l+1, l \cdot (2l+1)), \; l \in \mathbb{N}.$$

Taking different values of l we can obtain the following solutions:

$$\begin{aligned} l &= 1, \quad n = 5, \; m = 3 \\ l &= 2, \quad n = 9, \; m = 10 \\ l &= 3, \quad n = 13, \; m = 21 \end{aligned}$$

$(n,m) = (4l+1, l \cdot (2l+1))$, $l \in \mathbb{N}$.

108) **Proof.** Assume that a solution exists. Let us rewrite the given equation in a different form as follows:

$$y^2 - x^2 = 4x^2 + 6$$
$$(y - x) \cdot (y + x) = 4x^2 + 6$$

We can see that the right side is divisible by 2, then the left side must be divisible by 2. If $(y - x) = 2n$, then the second factor on the left is also even and vice versa. Hence, the left side must be divisible by 4. However, the right side is not divisible by 4. Contradiction. No solutions.

109) **Answer.** $(x, y, z) = \{(1, 1, 2), (2, 1, 1)\}$

Solution. This equation can be factored as

$$z \left(x^2 + 4y^2 - 4\right) = 2xy$$

Next, we can solve it for $z \neq 0$ because the factor inside parentheses is not zero. Obviously, $2xy \neq 0$ because x, y, $z \in \mathbb{N}$ are natural numbers. We obtain

$$z = \frac{2xy}{x^2 + 4y^2 - 4}.$$

Since all variable are natural numbers, this expression must be greater or equal to one, which can be written as

$$z = \frac{2xy}{x^2 + 4y^2 - 4} \geq 1$$

Or in its equivalent form:

$$x^2 + 4y^2 - 4 - 2xy \leq 1.$$

Completing the square, this inequality will be rewritten as follows:

$$(x - y)^2 + 3y^2 \leq 4.$$

From this we obtain that $y \leq 1$, $(x - y)^2 \leq 1$.
Finally we obtain that $y = 1$, $(x - 1)^2 \leq 1 \Rightarrow x = 1$, or $x = 2$.
$(x, y, z) = \{(1, 1, 2), (2, 1, 1)\}$

110) **Proof.** It can be shown that $\underbrace{11\ldots1}_{n \text{ digits}} = \frac{10^n - 1}{9}$, then the given number can be represented as

$$M = \frac{10^{2n} - 1}{9} - 2 \cdot \frac{10^n - 1}{9}$$
$$= \frac{10^{2n} - 2 \cdot 10^n + 1}{9} = \left(\frac{10^n - 1}{3}\right)^2$$

Because $10^n - 1 = 3k$, $\forall n \in \mathbb{N}$, the proof is complete.

Erratum to: Pythagorean Triples, Additive Problems, and More

Erratum to:
Chapter 4 in: E. Grigorieva, *Methods of Solving Number Theory Problems*,
https://doi.org/10.1007/978-3-319-90915-8_4

In the original version of the book, the term "convergence" has to be updated as "congruence" under Sect. 4.3.2. in Chapter 4. The erratum chapter and the book have been updated with the change.

The updated online version of this chapter can be found at
https://doi.org/10.1007/978-3-319-90915-8_4

References

1. Sabbagh, K.: The Riemann Hypothesis. Farrar, Straus and Giroux. NY (2004).
2. Williams, K.S., Hardy, K.: The Red Book of Mathematics Problems. (Undergraduate William Lowell Putnam competition). Dover (1996)
3. Hardy, G.H.: On the Representation of a Number as the Sum of Any Number of Squares, and in Particular of Five. http://www.ams.org/journals/tran/1920-021-03/S0002-9947-1920-1501144-7/S0002-9947-1920-1501144-7.pdf
4. Hardy, G.H., Write, E.M.: An Introduction to the Theory of Numbers, 6th edn. Oxford University Press, Oxford (2008)
5. Arnold, V.I.: Lectures and Problems: A Gift to Young Mathematicians, AMS (2016)
6. Grigorieva, E.V.: Methods of Solving Complex Geometry Problems. Birkhauser (2013)
7. Grigorieva, E.V.: Methods of Solving Nonstandard Problems. Burkhauser (2015)
8. Grigorieva, E.V.: Methods of Solving Sequence and Series Problems. Burkhauser (2016)
9. Dudley, U.: Number Theory, 2nd edn. Dover Publications (2008).
10. Barton, D.: Elementary Number Theory, 6th edn. McGraw Hill (2007).
11. Grigoriev, E. (ed.): Problems of the Moscow State University Entrance Exams, pp. 1–132. MAX-Press (2002). (In Russian).
12. Grigoriev, E. (ed.): Problems of the Moscow State University Entrance Exams, pp. 1–92, MAX-Press (2000). (In Russian).
13. Grigoriev, E. (ed.): Olympiads and Problems of the Moscow State University Entrance Exams. MAX-Press (2008). (In Russian).
14. Savvateev, A.: Video Lecture Notes, 2014 (In Russian).
15. Lidsky, B., Ovsyannikov, L., Tulaikov, A., Shabunin, M.: Problems in Elementary Mathematics. MIR Publisher, Moscow (1973).
16. Vinogradova, Olehnik, Sadovnichii.: Mathematical Analysis, Factorial (1996). (In Russian).

© Springer International Publishing AG, part of Springer Nature 2018   377
E. Grigorieva, *Methods of Solving Number Theory Problems*,
https://doi.org/10.1007/978-3-319-90915-8

17. http://takayaiwamoto.com/Sums_and_Series/sumint_1.html.
18. Dunham, W.: Euler, The Master of Us All. The Mathematical Association of America. Washington D.C. (1999).
19. Eves, J.H.: An introduction to the History of Mathematics with Cultural Connections, pp. 261–263. Harcourt College Publishers, United States (1990).
20. W. Sierpinski, 250 Problems in Elementary Number Theory. Elsevier, New York, (1970).
21. Alfutova, N., Ustinov, A.: Algebra and Number Theory, MGU, Moscow (2009). (In Russian).
22. Agahanov, N.K., Kupzhov, L.P.: Russian Mathematical Olympiads 1967–1992. Moscow, Prosveshenie (1997). (In Russian).
23. Sadovnichii, V.A., Podkolzin, A.S.: Problems of Students Mathematics Olympiads. Moscow, Nauka (1978). (In Russian).
24. Linnik, Yu. V.: An Elementary Solution of the Problem of Waring by Schnirelman's Method. Mat. Sb. **12**, 225–230 (1943).
25. Vinogradov, I.M.: Foundation of Number Theory, Nauka, Moscow (1952). (In Russian).
26. Barbeau, E.: Pell's Equation, pp 16–31. Springer (2003).
27. Nesterenko, Y., Nikiforov, E.: Continued Fractions, Quantum, V.10, N3, pp. 21–27 (2000).
28. Falin, G., Falin, A.: Linear Diophantine Equations, Lomonosov Moscow State University (2008). (In Russian).
29. Zolotareva, N.D., Popov, Y., Sazonov, V., Semendyaeva, I., Fedotov, M.: Algebra: Olympiads and MGU Entrance Exams. Moscow State University, Moscow (2011). (In Russian).
30. Shestopal, G.: How to Detect a Counterfeit Coin? Quantum, N 10, pp. 21–25 (1979).
31. Shklarsky, D.O., Chentzov, N.N., Yaglom, I.M.: The USSR Olympiad Problem Book: Selected Problems and Theorems of Elementary Mathematics. Dover (1993).

Books and Contest Problems for Further Reading:

1. Gardiner, A.: The Mathematical Olympiad Handbook. Oxford University Press, New York (2011).
2. Galperin, G.: Moscow Olympiads. Moscow. Nauka (2005). (In Russian).
3. Andreescu, T., Andrica, D., Cucureseanu, I.: An Introduction to Diophantine Equations. Burkhauser (2010).
4. Andreescu T., Feng Z., Loh, P.-S., (eds.) USA and International Mathematical Olympiads. MAA (2003).
5. Andreescu T., Feng Z., Loh P.-S., (eds.) USA and International Mathematical Olympiads. MAA (2004).
6. Andreescu T., Feng Z., Loh P.-R., (eds.) Mathematical Olympiads 2001–2002: Problems and Solutions from Around the World. MAA (2004).

7. Andreescu T., Feng Z., Lee, G. Jr (eds.) Mathematical Olympiads 2000–2001: Problems and Solutions from Around the World. MAA (2003).
8. The William Lowell Putnam Mathematics Examination 1985–2000 by Kedlaya, Poonen, Vakil. (The Putnam is the primary mathematics competition for undergraduates.)
9. Hungarian Problem Book III by Liu. Intermediate and challenging problems from an old Hungarian competition.

Appendix 1. Historic Overview of Number Theory

Important properties of integers were established in ancient times. In Greece, the Pythagorean school (6th century BC) studied the divisibility of numbers and considered various categories of numbers. Some of the number categories considered were simple (prime), composite, perfect, and amicable (friendly). In his "Elements" Euclid (3rd century BC) gives an algorithm for determining the greatest common divisor of two numbers, outlines the main properties of divisibility of integers, and proves the theorem that primes form an infinite set. Eratosthenes (3rd century BC) gave a way to extract prime numbers from a series of natural numbers (Eratosthenes sieve) and took a further step in the theory of primes. Of great importance were the works of the Greek mathematician Diophantus of Alexandria (about 3rd century AD). He devoted much of his work to the solution of indefinite equations in rational numbers (in China, from the second century, indefinite equations were also a subject of great interest).

Further flourishing of the theory of numbers begins in modern times and is associated with the name of the great 17th-century mathematician Pierre Fermat. Fermat, influenced by the works of Diophantus, studied the solution of many such equations in integers. Fermat is most often associated with and known for Fermat's Little Theorem and Fermat's Last Theorem. Fermat's Last Theorem is so well known because it remained without proof until recently when Andrew Wiles proved it using elliptical curves. But, before we discuss that, let's consider some other great mathematicians who contributed to the theory of numbers over the past 300 years. The great Swiss mathematician Leonard Euler proved almost all of Fermat's theorems that remained without proof. Of Euler's many works (more than 800), more than 100 papers involve number theory. Further in his 1742 correspondence with Goldbach (at that time a member of St. Petersburg Academy of Sciences) included a very complicated additive problem. Goldbach conjectured that every odd integer greater than five is the sum of three primes. Euler strengthened the conjecture and made it a more difficult problem by adding that every even number greater than three is the sum of two primes. The Russian mathematician I.M. Vinogradov used his method of trigonometric sums to prove the last assertion for almost all number

© Springer International Publishing AG, part of Springer Nature 2018
E. Grigorieva, *Methods of Solving Number Theory Problems*,
https://doi.org/10.1007/978-3-319-90915-8

$N \geq N_0 = e^{e^{16038}}$)). This conjecture is also named as binary Goldbach's conjecture and is still unsolved.

In 1770, English mathematician Waring formulated without proof: Any positive integer $N > 1$ can be represented as a sum of n-th powers of positive integers, i.e., as $N = x_1^n + x_2^n + \ldots + x_k^n$; the number of terms k depends only on n. A particular case of the Waring problem is the Lagrange theorem, which concerns the fact that each N is the sum of four squares. The first general solution (for any n) of the Waring problem was given by D. Hilbert (1909) with a very rough estimate of the number of terms k in relation to n. More exact estimates of k were obtained in the 1920s by G. Hardy and J. Littlewood, and in 1934,1. M. Vinogradov, by the method of trigonometric sums that he created, obtained results that were close to definitive. An elementary solution of the Waring problem was given in 1942 by Russian mathematician L. V. Linnik. At present the Waring problem for $n = 2$, which concerns representing a natural number with two, three, four or more squares, is entirely solved. The special significance of the Waring problem consists in the fact that during its investigation powerful methods of analytic number theory have been created.

In the 1930s, Russian mathematician Shnirelman discovered a new method of adding numerical sequences that is very important in additive number theory. Using this method, he proved that every natural number, except 1, is a sum of not more than C prime numbers, where C does not depend on a given number ($C < 210$).

Fermat's Last Theorem, asserted that the equation $a^n + b^n = c^n$ does not have integer solutions if $abc \neq 0$ and $n > 2$. This statement has been verified for over three hundred years, for all $n \leq 150000$ but until recently, in the general case, it had not been proven. The history of the relationship between the Fermat's Last Theorem and elliptic curves begins in 1955, when the Japanese mathematician Yutaka Taniyama (1927–1958) formulated a problem that was a slightly weakened version of the following:

Hypothesis 1 (Taniyama). Any elliptic curve defined over the field of rational numbers is modular.

In this form, the hypothesis of Taniyama appeared in the early 60s in the works of Goro Shimura. In subsequent years, Shimura and the French mathematician A. Weil showed the fundamental connection of the Taniyama hypothesis with many sections of the arithmetic of elliptic curves. At the turn of the 60s and 70s, the French mathematician Yves Hellegouarch compared the elliptic curve

$$y^2 = x(x - a^n)(x - c^n)$$

to the Fermat equation and used the results concerning Fermat's theorem to examine points of finite order on elliptic curves. Further developments showed that the comparison of the Fermat equation with the elliptic curve was truly revolutionary...

In 1985, German mathematician Gerhard Frey suggested that the curve corresponding to the counterexample to Fermat theorem couldn't be modular (in contradiction with Taniyama hypothesis). Frey himself failed to prove this assertion, but soon the proof was obtained by American mathematician Kenneth Ribet. In other words, Ribet has shown that Fermat's theorem is a consequence of the Taniyama hypothesis. On June 23, 1993, a mathematician from Princeton, Andrew Wiles, speaking at a conference on number theory in Cambridge, announced the proof of the Taniyama conjecture for a wide class of elliptic curves (so-called semistable curves), in which all curves of the form above are included. Thus, he stated that he had proved Fermat's theorem.

Further events developed quite dramatically. In the beginning of December 1993, a few days before the manuscript of Wiles' work was supposed to go to press, errors were found in his proof. Their correction took more than a year. The text with the proof of the Taniyama hypothesis, written by Wiles in collaboration with Taylor was published in the summer of 1995. Let us formulate the statement, proven by Andrew Wiles:

Hypothesis 2 (Taniyama's conjecture for quasi-stable (semistable) curves). Every semistable elliptic curve defined over the field of rational numbers is modular.

Therefore, the proof of Fermat's Last Theorem follows from Taniyama's conjecture for semistable elliptic curves. Note, finally, that the significance of Taniyama's conjecture is by no means confined to its connection with Fermat's theorem. From the Taniyama hypothesis, the Hasse-Weil conjecture follows (for elliptic curves over the field of rational numbers, these hypotheses are equivalent) and together they open new horizons in the study of the arithmetic of elliptic curves.

Appendix 2. Main Directions in Modern Number Theory

To address challenges of number theory, a variety of research and numerical methods have evolved, which are also taken as the basis for classification of its directions. From the point of view of methods and applications, seven main directions in number theory can be distinguished:

1. **Elementary Methods of Number Theory** (elementary theory of numbers—theory of divisibility, theory of congruences, theory of forms, and indefinite equations). Elementary methods include those that use mainly information from elementary mathematics and, at most, elements of analysis of infinitesimal mathematics. Elementary number theory considers methods of the theory of comparisons, created by the great German mathematician Carl F. Gauss (1777–1855), the methods of continued fractions, developed by the French mathematician J. Lagrange (1736–1813), and many other methods. It must be borne in mind that the elementarily of the method does not yet speak of its simplicity. In number theory, the great merit belongs to the Russian mathematician P. Chebyshev (1821–1894), the French mathematicians J. Liouville (1809*–1882) and S. Hermit (1822–1901), the Norwegian mathematicians A. Thue and V. Brun, and Danish mathematician A. Selberg. The fundamental method in additive number theory was created by the Russian mathematician LG. Schnirelmann (1905–1938), who sought to prove Goldbach's conjecture. In 1930, using the Brun sieve, he proved that any natural number greater than 1 can be written as the sum of not more than C prime numbers, where C is an effectively computable constant.

2. **Analytic Number Theory.** Analytic number theory uses mathematical analysis, complex analysis, theory of numbers, probability theory, and other topics in mathematics. The founder of this direction is Leonard Euler (1707–1783). Gauss's work had a significant impact on the development of this theory as well. In the real world, the analytical methods were developed by German mathematician L. Dirichlet (1805–1859) and by Russian mathematician P.l. Chebyshev. The work of the German mathematician B. Riemann (1826–1866) played an important role in the development of analytical methods related to the theory of complex variables. In addition, the works of the German

© Springer International Publishing AG, part of Springer Nature 2018
E. Grigorieva, *Methods of Solving Number Theory Problems*,
https://doi.org/10.1007/978-3-319-90915-8

mathematician G. Weil (1885–1955) and the Russian mathematician G. Voronoi (1868–1908) were of great importance. Great successes were achieved by Indian mathematician S. Ramanujan (1887–1920), English mathematicians G. Hardy (1877–1947) and J. Littlewood (1885–1977), and German mathematician K. Siegel (1896–1981). The powerful methods in analytic number theory were also created by Russian mathematicians I.M. Vinogradov (1891–1983), A.O. Gelfond (1906–1968), e^π is Gelfond constant), and Yu.V. Linnik (1915–1972).

3. **Algebraic Number Theory.** This theory, starting from the concept of an algebraic number, was created in the works of J. Wallis (1616–1703), J. Lagrange and L. Euler; it uses the techniques of abstract algebra to study the integers, rational numbers, and their generalizations. Number-theoretic questions are expressed in terms of properties of algebraic objects such as algebraic number fields and their rings of integers, finite fields, and function fields. These properties, such as whether a ring admits unique factorization, the behavior of ideals, and the Galois groups of fields, can resolve questions of primary importance in number theory, like the existence of solutions to Diophantine equations. Especially important are the works of the German scientists C. Gauss, E. Kummer (1810–1893), R. Dedekind (1831–1916), and L. Kronecker (1823–1891) and outstanding Russian scientists E.I. Zolotarev (1847–1878) and G.F. Voronoi (1868–1908). It is necessary to mention the works of G. Hasse (1898–1979), K. Siegel. The periods of investigation in analytic number theory during the 1930 are closely related to the discoveries of Paul Erdős and Aurel Wintner. Among Russian mathematicians major successes in this field were achieved by N.G. Chebotarev (1894–1947), B.V. Venkov (1900–1962), and in particular I.R. Shafarevich (1923–2017).

4. **Geometric Number Theory.** In this theory, the so-called spatial lattices or systems of integer points that have integers as coordinates in a given Cartesian coordinate system are used. This theory, used in geometry and crystallography, is associated with the theory of quadratic forms in number theory. The founders of this theory are G. Minkowski (1864–1909), G.F. Voronoi (1868–1908) and F. Klein (1849–1925). These methods have been successfully applied by Russian mathematicians, especially B. N. Delone (1890–1980) and B.A. Venkov.

5. **Probabilistic Number Theory.** The founders of this theory, which explicitly uses probability to answer questions of number theory, are P. Erdos (1913–1996), A. Wintner (1903–1958), and J.P. Kubilius (1921–2011). One basic idea underlying it is that different prime numbers are, in some serious sense, like independent random variables. A systematic construction of probabilistic number theory, given at an application angle to the distribution of values of unstable additive functions, was carried out by J.P. Kubilius in his monograph "Probabilistic methods in number theory", and a systematic presentation of this theory is contained in the monographs by P.D.T.A. Elliott "Probabilistic number theory", "Theorems on mean values. Theorems about values", "Probabilistic number theory. The Central limit theorem", and "Arithmetic functions and product functions."

6. **Topological Number Theory** (polyadic analysis). A theory of polyadic number is constructed by considering the fundamental relationship between additive and

multiplicative properties of integers based on the theory of congruences and the axioms of a topological ring. This theory of numbers first appeared in 1924 in German mathematician Hans Prufer's article "New Foundations of Algebraic Number Theory". The constructions of polyadic numbers were also proposed by Herne Hensel, J. von Neumann, and in more detail and thoroughly by EV Novosyolov. A detailed account of the theory of polyadic numbers is contained in the book of E. Hewitt and K. Ross and in articles by E. Novoselov.

7. **Applied Number Theory** (semi-generated algorithms, algorithmic problems in number theory-cryptography). Fundamentals of the theory of information security—digital encryption, algorithms for fast multiplication (comparison of arithmetic operations by their complexity), discrete logarithms, etc. (see "Introduction to modern cryptography" by J. Katz and Y. Lindell and "Modern Cryptography Primer Theoretical Foundations and Practical Applications", by C. Koscielny and M. Kurkowski, 2013, Springer.)

Index

A

algebraic geometry, 245, 251, 253, 254, 257
Amicable, 1, 43, 49, 50, 381
Application of factoring, 165
Application of the Euler's Formula, 120
approximation of irrational numbers, 78
Archimedes, 53, 141, 142, 204
Arithmetic and geometric mean, 153
arithmetic progression of only prime numbers, 131
arithmetic sequence, 129, 130, 154, 344
Arithmetica, 141, 143

B

Babylonians, 43, 63, 70, 72, 78, 123, 124, 143, 177
Bezout's Identity, 145
binomial coefficients, 168
Brahmagupta, 245
Brouncker, 205

C

Catalan numbers, 107, 135, 137, 139
Chebyshev, 63, 385
Chinese Remainder Theorem, 55, 112, 163, 339, 353, 370
circuit construction, 83
complex numbers, 2
Congruence, 18, 53–57, 59, 110, 112, 113, 119, 122, 123, 156–158, 160, 162–164, 239–241, 243, 244, 336, 349, 351–354, 359, 363, 370, 371, 385, 387
convergent fractions, 80, 89, 159, 216, 236

D

Decimal Representation of a Natural Number, 13, 95

Determinant, 59–61, 209, 212, 213, 233
Difference of Cubes, 165, 182
difference of squares, 4, 11, 25, 31, 100, 165, 166, 171–177, 179, 183, 187, 188, 191, 205, 210, 335, 338, 340, 365, 368, 374
different bases, 63
Diophantine equations, 142, 144, 156, 157, 378, 386
Direct proof, 127, 133
discriminant, 88–90, 169, 170, 176, 177, 196–198, 212, 213, 350, 362
Divisibility by 2, 3, 4, 5, 8, 9, 11, 13

E

Egyptians, 70, 123
electrical circuit, 83
Eratosthene, 1, 142, 204, 381
Euclid, 1, 26, 43, 48, 49, 141, 381
Euclidean algorithm, 63–67, 69, 71, 73, 78, 79, 82, 146, 148, 172, 173, 243, 363, 369
Euler, 1, 26, 28, 30, 44, 48, 63, 110, 204, 212, 215, 349, 381, 385, 386
Euler zeta function, 28
Euler's formula, 63, 117–122, 162, 164, 165
Eulers Four Squares Identity, 280, 282
Eulers Theorem on Solution of Quadratic Congruence, 304

F

Factorials, 108, 238
Fermat, 1, 2, 63, 108, 141, 171–174, 204, 205, 212, 215, 382, 383
Fermat prime, 43
Fermat's factorization method, 25, 171, 172, 174
Fermat's Last theorem, 141, 382, 383

© Springer International Publishing AG, part of Springer Nature 2018
E. Grigorieva, *Methods of Solving Number Theory Problems*,
https://doi.org/10.1007/978-3-319-90915-8

Fermat's Little Theorem, 5, 25, 26, 63, 108, 110, 113, 115, 117, 119, 122
Fermat-Pell's equation, 141
Fibonacci, 229, 232
finding the last digits, 101, 120, 123
finite continued fractions, 79, 83
Frenicle de Bessy, 205

G
Gauss, 1, 53, 127, 237, 385, 386
Gaussian elimination, 155
Generating Functions, 135
geometric sequence, 45, 154
Germain, Sophie, 182
golden ratio, 78
greatest common divisor, 1, 36–41, 63–67, 69, 70, 145, 172, 337, 359, 381
greatest integer, 24, 32, 35, 82
Greeks, 50, 51, 63, 78, 123–125, 142, 201, 204

H
Hardy, 382
Hilbert, 382
homogeneous linear equation, 143
homogeneous polynomial, 179, 180, 182, 199
hyperbola, 209, 212, 213, 232, 233

I
infinite continued fraction, 63, 78, 84–87, 92, 215, 219, 220
irrational, 2, 63, 84–90, 92, 94, 124, 130, 215, 216, 223, 339, 366

J
Jacobi symbol, 247, 306, 314, 315

L
Lagrange, 1, 88, 89, 205, 215, 219, 234, 237, 239
Lagrange four squares theorem, 282
Lagrange Theorem, 63, 88, 89, 361, 382
last three digits, 14, 98, 122, 123, 192, 335
last two digits, 14, 97, 101, 104–106, 115, 117, 120–123, 192, 238, 335, 338, 343, 357, 367
least common multiple, 36–40, 42, 56, 66, 112, 372
Least nontrivial solution, 221, 223
Legendre formula, 35, 36
Legendre symbol, 306–313, 315
Legendres Three Squares Theorem, 277, 282
linear equation, 67, 80, 81, 146, 156, 158, 204
Linnik, 382, 386
Littlewood, 382, 386

Lucas, 44

M
Mathematical Induction, 132, 133, 135, 345, 372
Matrix Iterative Approach, 207
matrix of linear transformation, 207, 209, 210, 223, 226, 230, 231
Mersenne prime, 29, 43, 44
MGU Entrance Exam, 153, 170, 325, 326
Moscow State University Entrance Exam, 33

N
Natural, 1–3, 5–7, 9–13, 21, 24–28, 31, 42, 43, 45, 46, 51, 57, 61, 69, 71, 82, 91, 95, 99, 124, 125, 127–130, 132, 134, 141, 142, 144, 149, 150, 164, 167, 174, 176, 179, 184, 185, 188–190, 194, 195, 205, 210–215, 219, 221–223, 226–228, 234, 236, 237, 335–339, 347, 348, 360, 370, 373, 375, 382, 385
Newton Binomial Theorem, 5
nonhomogeneous linear equation, 144
nonlinear equations, 143, 152, 165, 170, 182, 208, 209
nonresidues, 242
nontrivial solution, 31, 202, 204, 206, 209–212, 220, 223, 224, 230, 231, 234, 364
Number of divisors, 36, 45, 47
Number Raised to a Power, 63, 75, 101, 108

O
orthogonal transformation, 209

P
Pell's equation, 31, 85, 142, 201, 202, 204–207, 211–213, 215–217, 225, 226, 229, 230, 233, 236
Pell's Type Equation, 211, 224, 225, 227, 230
perfect numbers, 43, 44, 48, 49, 174
periodic fraction, 223
periodic fraction with period k, 88
Physics, 83
prime factorization, 23, 24, 30, 33, 34, 36, 39, 40, 44–46, 49, 65, 74, 119, 171, 340, 358, 361, 372
prime number, 1, 18, 23–30, 43, 110, 113, 114, 130, 132, 184, 187, 199, 236, 239, 336, 352, 353, 362, 381, 382, 385, 386
Prime Representation as Sum of Two Squares, 269
Proof by contradiction, 130, 190
proper divisors, 43–45, 47, 48

Properties of Jacobi Symbol, 314
Properties of Legendre symbol, 307
Pythagoras, 43, 48, 49, 141
Pythagorean quadruples, 245
Pythagorean triples, 226

Q
quadratic congruence, 293, 294, 297, 301,
    303–307, 311
quadratic form, 207, 208, 212, 215, 231, 386
quadratic irrationalities, 63, 87, 88, 91, 215
quadratic residuals, 305
quadratic equation, 78, 87–89, 91, 93, 169,
    176–179, 197, 337

R
rational points on the unit circle, 245, 255
rational points on the unit sphere, 258, 260
rational number, 2, 79, 81, 84, 86, 130, 141,
    381–383, 386
real number, 2, 78, 86, 88, 210, 215
reciprocity relationship, 313
reducing a fraction, 63, 64
relatively prime, 34, 37, 66, 68, 88, 109, 114,
    117–119, 131, 144, 159, 163, 183
remainder, 4, 6, 9–12, 18, 42, 53–55, 57–59,
    61, 64–66, 68–70, 72, 82, 95–101,
    103–105, 107–112, 115–117, 119–121,
    146, 156, 161, 191, 192, 200, 239, 240,
    244, 336, 337, 344, 348, 351, 356, 357,
    359, 360, 367, 370, 373
representing an integer number as a sum of
    cubes, 245
representing an integer number as a sum of
    other numbers, 245
representing an integer number as a sum of
    squares, 245
Riemann, 28, 29
Riemann Hypothesis, 29

S
similar triangles, 261, 263
solution generator, 223
Solving Linear Congruence Using Continued
    Fractions, 159
Solving Linear Equations and Systems Using
    Congruence, 156
square number, 50–52, 201
square of an odd number divided by 8, 337
stereographic projection, 257, 259, 261
Sum of Cubes, 165
Sum of divisors, 45, 49, 50
sum of four squares, 361, 382

T
Thales Theorem, 248
triangular numbers, 50–52, 201–203, 207
Trigonometric Approach, 256
two consecutive natural numbers, 195, 202
two consecutive triangular numbers, 51, 52,
    201

U
unit resistors, 83
Using complex numbers, 360
Using Euclidean Algorithm to Solve Linear
    Equations, 146

V
Vieta's theorem, 41, 171, 177–179
Vinogradov, 381, 382, 386

W
Waring, 382
Waring's problem, 382
Wilson's Theorem, 25, 26, 237, 239, 241, 243
word problems, 42, 143, 144, 147

Printed in the United States
By Bookmasters